Noriega / Rauwendaal

Troubleshooting the Extrusion Process

Maria del Pilar Noriega E.
Chris Rauwendaal

Troubleshooting the Extrusion Process

A Systematic Approach to Solving Plastic Extrusion Problems

3rd Edition

HANSER

Hanser Publishers, Munich Hanser Publications, Cincinnati

The Authors:

Maria del Pilar Noriega E., Calle 16 A Sur No. 29 B - 40, Medellin, Colombia, South America

Chris Rauwendaal, 10556 Combie Rd PMB 6677 Auburn, CA 95602-8908, USA

FSC
www.fsc.org
MIX
Papier aus verantwortungsvollen Quellen
FSC® C016439

Distributed in the Americas by:
Hanser Publications
414 Walnut Street, Cincinnati, OH 45202 USA
Phone: (800) 950-8977
www.hanserpublications.com

Distributed in all other countries by:
Carl Hanser Verlag
Postfach 86 04 20, 81631 Munich, Germany
Fax: +49 (89) 98 48 09
www.hanser-fachbuch.de

Library of Congress Control Number: 2019946876

Editor: Dr. Julia Diaz Luque
Production Management: Jörg Strohbach
Coverconcept: Marc Müller-Bremer, www.rebranding.de, Munich
Coverdesign: Max Kostopoulos
Typesetting: Kösel Media GmbH, Krugzell
Printed and bound by Hubert & Co. GmbH und Co. KG BuchPartner, Göttingen
Printed in Germany

ISBN: 978-1-56990-775-7
E-Book ISBN: 978-1-56990-776-4

Acknowledgments

From author **Chris Rauwendaal:**

I want to thank my family for their continuing support and patience when I work on these time-consuming projects. I would also like to thank the hundreds of companies I have worked with. Each troubleshooting problem is not only a challenge but also an opportunity to learn. After working in the extrusion industry for over 45 years I have had the good fortune to learn more than I ever expected. Also, I would like to thank the thousands of people that have attended my extrusion seminars. I have found that the best way to learn a subject well is to teach it – this is one of many enjoyable and rewarding aspects of teaching. In most seminars there is very good interaction. As a result, not only the attendees learn from the instructor but the instructor also learns from the attendees – truly a win-win situation.

I would like to acknowledge Verena Resonnek at the department of Kunststofftechnik Paderborn, University of Paderborn, Germany for providing information on the development of a controller capable of automatic optimization of extruder barrel temperatures.

Author **Maria del Pilar Noriega Escobar** thanks her husband, family, and best friends for their valuable support and patience during all these years of applied research, industrial applications, and innovation in polymer processing. She is grateful to Mr. Hans Udo Steinhäuser, President of the Plastics and Rubber Research Institute (ICIPC) and her Institute's colleagues for their support during the preparation of this book.

Norberto Montoya is acknowledged for micrographs in the first edition and Juan Carlos Gallego for micrographs in the second edition.

Eberhard Grünschloss from IKT (Institut für Kunststofftechnik, Stuttgart) and Tim Andreas Osswald (UW-Madison) are acknowledged for the valuable discussions and, finally, Diana María Angel is gratefully acknowledged for the figures in the three editions.

Preface

One of the greatest challenges in actual extrusion operations is efficient and rapid problem-solving. Extrusion problems often result in downtime and/or out-of-spec product, and this can be very costly. However, because of the nature of the extrusion process, it is often quite difficult to determine the cause of the problem and find the proper solution, particularly if it must be done quickly. Despite the industrial importance of extrusion troubleshooting, no book currently deals exclusively with this topic. This book is an attempt to rectify this situation.

Both authors have worked in extrusion for many years and have been involved in numerous troubleshooting projects. Although it is impossible to discuss all possible extrusion problems, it is possible to discuss the main categories and to develop a systematic and methodical approach to solving extrusion problems. In this book, the authors frequently use flow charts and fishbone charts to allow systematic troubleshooting.

The authors added a substantial amount of new material to this third edition, including:

- Chapter 1: new section on collection and interpretation of extrusion process data
- Chapter 2: data acquisition systems section substantially expanded and updated with cloud-based DAS and systems that can automatically detect machine problems; new sections on rotational rheometry and the smartphone
- Chapter 3: new sections covering how screw design can affect extruder performance and melt temperature variation; additionally, barrel temperature profiles for many polymers from LDPE to PEEK
- Chapter 4: ten new case studies
- Appendix 3: new section with information on barrel temperature optimization for PP and HDPE for a 2.5-inch (63.5 mm) extruder and a description of recent research on automatic optimization of extruder barrel temperatures conducted at the department of Kunststofftechnik Paderborn, University of Paderborn, Germany by Verena Resonnek

- Appendix 4 (new): process signal analysis using Fast Fourier Transform

The authors welcome feedback from readers, along with additional material on extrusion troubleshooting. This will allow more information to be incorporated into future editions of this book.

List of Acronyms

ABS	Acrylic-butadiene-styrene copolymer
BTP	Barrel temperature profile
DAS	Data acquisition system
DOE	Design of experiments
FPVC	Flexible polyvinylchloride
HDPE	High density polyethylene
L/D	Length to diameter ratio
LDPE	Low density polyethylene
LLDPE	Linear low density polyethylene
OTE	One-at-a-time experiments
PC	Polycarbonate
PE	Polyethylene
PEEK	Polyether ether ketone
PET	Polyethylene terephthalate
PMMA	Polymethyl methacrylate (acrylic)
PP	Polypropylene
PS	Polystyrene
PVC	Polyvinylchloride
RPVC	Rigid polyvinylchloride
SPC	Statistical process control

Contents

Acknowledgments V

Preface VII

List of Acronyms IX

1 Requirements for Efficient Troubleshooting 1

1.1 Instrumentation 1

1.2 Understanding the Extrusion Process 2

1.3 Collection and Analysis of Historical Data (Time Line) 4

1.4 Team Building 5

1.5 Condition of the Equipment 6

1.6 Information about the Feed Stock 6

1.7 Problem-Solving Techniques 8

1.8 Collection and Interpretation of Extrusion Process Data 9

1.8.1 Introduction 9

1.8.2 Vital Signs of the Extrusion Process 9

1.8.2.1 Melt Pressure 10

1.8.2.2 Melt Temperature 16

1.8.2.3 Training 25

1.8.3 Conclusions 25

2 Tools for Troubleshooting 27

2.1 Temperature Measurement Devices 27

2.2 Data Acquisition Systems 28

2.2.1 Portable Data Collectors/Machine Analyzers 29

2.2.2 Fixed-Station Data Acquisition Systems 30

2.3 Light Microscopy 33

2.4 Thermochromic Materials 34
2.5 Thermal Analysis, IR Spectroscopy, and Rheometry 35
2.5.1 Differential Thermal Analysis and Differential Scanning Calorimetry 35
2.5.2 Thermogravimetric Analysis 37
2.5.3 Fourier Transform Infrared Spectroscopy 38
2.5.4 Thermomechanical Analysis 39
2.5.5 Torque Rheometry 40
2.5.6 High Pressure Capillary Rheometry 41
2.5.7 Rotational Rheometry 43
2.5.8 Other Thermal Characterization Techniques 46
2.6 Miscellaneous Tools 46
2.6.1 Infrared Thermography 47
2.6.2 The Smartphone 48
2.6.3 Power Measurements 49

3 Systematic Troubleshooting 51

3.1 Upsets versus Development Problems 51
3.2 Machine-Related Problems 51
3.2.1 The Drive System 52
3.2.2 The Feed System 53
3.2.3 The Heating and Cooling System 53
3.2.4 How Screw Design Can Affect Extruder Performance 54
3.2.5 Wear Problems 65
3.2.5.1 Wear Mechanisms 66
3.2.5.2 Test Methods for Wear 67
3.2.5.3 Causes of Wear 72
3.2.5.4 Solutions to Wear Problems 77
3.2.5.5 Rebuilding Worn Screws and Barrels 83
3.2.6 Screw Binding 87
3.2.6.1 Extrusion of Fluoropolymers 88
3.2.6.2 The Mechanics of Screw Binding 88
3.2.6.3 Changes in Clearance Due to Temperature Differences 88
3.2.6.4 Analysis of Temperature Distribution in Extruder Screws 91
3.2.6.5 Change in Clearance Due to Compressive Load 92
3.2.6.6 Results from Analysis 92
3.3 Polymer Degradation 93
3.3.1 Types of Degradation 93
3.3.1.1 Thermal Degradation 94

3.3.1.2 Mechanical Degradation ... 94
3.3.1.3 Chemical Degradation ... 96
3.3.2 Degradation in Extrusion ... 97
3.3.2.1 Residence Time Distribution ... 97
3.3.2.2 Temperature Distribution Simple Calculations ... 101
3.3.2.3 Temperature Distribution Numerical Calculations ... 107
3.3.2.4 Reducing Degradation ... 114
3.4 Extrusion Instabilities ... 115
3.4.1 Frequency of Instability ... 116
3.4.1.1 High-Frequency Instabilities ... 117
3.4.1.2 Screw Frequency Instabilities ... 120
3.4.1.3 Low-Frequency Instabilities ... 122
3.4.1.4 Very Slow Fluctuations ... 123
3.4.1.5 Random Fluctuations ... 123
3.4.2 Functional Instabilities ... 124
3.4.2.1 Solids-Conveying Instabilities ... 125
3.4.2.2 Plasticating Instabilities ... 125
3.4.2.3 Melt-Conveying Instabilities ... 126
3.4.2.4 Devolatilization Instabilities ... 126
3.4.2.5 Mixing Related Instabilities ... 127
3.4.2.6 Distributive Mixing Sections ... 127
3.4.2.7 Dispersive Mixing Sections ... 132
3.4.2.8 Solving Mixing Problems ... 139
3.4.2.9 Melt Temperature Variation ... 140
3.4.3 Solving Extrusion Instabilities ... 149
3.5 Air Entrapment ... 151
3.6 Gel Problems ... 153
3.6.1 Measuring Gels ... 153
3.6.2 Gels Created in the Extrusion Process ... 154
3.6.3 Removing Gels Produced in Polymerization ... 155
3.7 Die-Flow Problems ... 156
3.7.1 Melt Fracture ... 156
3.7.2 Die-Lip Buildup ("Die Drool") ... 158
3.7.3 V- or W-Patterns ... 159
3.7.4 Specks and Discoloration ... 160
3.7.5 Lines in Extruded Product ... 161
3.7.5.1 Weld Lines ... 162
3.7.6 Optical and Appearance Properties ... 163

4 Case Studies **165**

4.1 Film Coextrusion – Degradation in the Middle Layer 165
4.1.1 Description of the Problem 165
4.1.2 Analysis of the Problem 166
4.1.3 Solution 167

4.2 Film Coextrusion with Interfacial Problems 169
4.2.1 Description of the Problem 169
4.2.2 Analysis of the Problem 169
4.2.3 Solution 170

4.3 Lines in the Extruded Film 170
4.3.1 Description of the Problem 170
4.3.2 Analysis of the Problem 171
4.3.3 Solution 172

4.4 Color Variation in Polypropylene Carpet Fiber 172
4.4.1 Description of the Problem 172
4.4.2 Analysis of the Problem 173
4.4.3 Solution 174

4.5 Plastic Film with Poor Transparency 175
4.5.1 Description of the Problem 175
4.5.2 Analysis of the Problem 176
4.5.3 Solution 177

4.6 Wear Problem in Film Extrusion 178
4.6.1 Description of the Problem 178
4.6.2 Solution 178

4.7 Multilayer Film – Several Appearance Problems 179
4.7.1 Description of the Problem 179
4.7.2 Analysis of the Problem 179
4.7.3 Solution 180

4.8 Dispersion Problem in a High-Density Polyethylene Bottle 181
4.8.1 Description of the Problem 181
4.8.2 Analysis of the Problem 181
4.8.3 Solution 182

4.9 Polymer Degradation 184
4.9.1 Description of the Problem 184
4.9.2 Analysis of the Problem 184
4.9.3 Solution 185

4.10 Heat-Sealing Problems in a Coextruded Film 187
4.10.1 Description of the Problem 187
4.10.2 Analysis of the Problem 188
4.10.3 Solution 189

4.11 Output Problem in a Blown Film Line 190
4.11.1 Description of the Problem 190
4.11.2 Analysis of the Problem 190
4.11.3 Solution 192
4.12 Masterbatch Selection 193
4.12.1 Description of the Problem 193
4.12.2 Analysis of the Problem 194
4.12.3 Solution 195
4.13 Pipe Extrusion Problem 196
4.13.1 Description of the Problem 196
4.13.2 Analysis of the Problem 196
4.13.3 Solution 197
4.14 Gel Formation in a Coextruded Film 199
4.14.1 Description of the Problem 199
4.14.2 Analysis of the Problem 199
4.14.3 Solution 200
4.15 Agglomerates and Grammage Variation in a PP Sheet 200
4.15.1 Description of the Problem 200
4.15.2 Analysis of the Problem 201
4.15.3 Solution 203
4.16 Insufficient Melting and Mixing in a Plasticating Unit 204
4.16.1 Description of the Problem 204
4.16.2 Analysis of the Problem 205
4.16.3 Solution 207
4.17 High Melt Temperature and Insufficient Output in Coextrusion 208
4.17.1 Description of the Problem 208
4.17.2 Analysis of the Problem 209
4.17.3 Solution 210
4.18 Deficient Solids Conveying and Dispersion 213
4.18.1 Description of the Problem 213
4.18.2 Analysis of the Problem 213
4.18.3 Solution 215
4.19 Instability of Formation at the Die 216
4.19.1 Description of the Problem 216
4.19.2 Analysis of the Problem 217
4.19.3 Solution 219
4.20 Intermittent Pumping in a Vented Extruder 222
4.20.1 Description of the Problem 222
4.20.2 Analysis of the Problem 223
4.20.3 Solution 225

4.21 Melt Fracture or Sharkskin in m-PE 227
4.21.1 Description of the Problem 227
4.21.2 Analysis of the Problem 227
4.21.3 Solution 228
4.22 Scale-up of LLDPE Single Screw Extruder 229
4.22.1 Description of the Problem 229
4.22.2 Analysis of the Problem 229
4.22.3 Solution 230
4.23 Non-homogeneous Melt in Blow Molding 232
4.23.1 Description of the Problem 232
4.23.2 Analysis of the Problem 232
4.23.3 Solution 232
4.24 High Melt Temperature in Sheet Extrusion 234
4.25 Gear Pump Speed Variation in Sheet Extrusion 238
4.26 Melt Temperature Variation in Tubing Extrusion 244
4.27 Black Specks in Tubing Extrusion 247
4.28 Mechanical Degradation in TPE Extrusion 252
4.29 Degradation in a Long Adapter 255
4.30 Shrink Voids in Rod Extrusion 258
4.31 Improper Preheating of Extruders 268

Appendix 1: Systematic Problem Solving 285

Appendix 2: Machine Troubleshooting and Maintenance 287
A2.1 Check the Oil 287
A2.2 Unusual Noises 288
A2.3 Vibration Monitoring 288
A2.4 Drive Motors and Belts 289
A2.5 Spare Parts 290
A2.6 Screw and Barrel 290
A2.7 Extruder Maintenance Checklist 291

Appendix 3: Extruder Barrel Temperatures 295
A3.1 Setting Extruder Barrel Temperatures 295
A3.2 Extruder Barrel Temperature Profile Optimization 296
A3.2.1 Introduction 296
A3.2.2 Facts about Barrel Temperature Profile (BTP) 296

A3.2.3 Typical Process Temperatures for Different Plastics 298
A3.2.4 Guidelines and Considerations for Setting Barrel Temperatures 299
A3.2.5 BTP Optimization by Design of Experiments (DOE) 304
A3.2.6 BTP Optimization by One-at-a-Time Experiments (OTE) 305
A3.2.7 Dynamic BTP Optimization 305
A3.2.8 Other Studies on Optimization of Extruder Barrel Temperature Profiles 307
A3.2.9 Conclusions 321

Appendix 4: Process Signal Analysis Using Fast Fourier Transform 323

References 335

Index 347

1 Requirements for Efficient Troubleshooting

Before acting on a specific extrusion problem, the troubleshooter should address certain issues. It is important to diagnose a developing extruder problem quickly and accurately to minimize downtime or off-quality product. Good instrumentation and a solid understanding of the extrusion process are important requirements for efficient troubleshooting. Instrumentation is very important in process control, but absolutely essential in troubleshooting. Without good instrumentation, troubleshooting is a guessing game at best, even when the troubleshooter fully understands the entire process. Thus, lack of instrumentation can prove very costly when it causes a particular problem to remain unsolved for even a limited length of time.

Important prerequisites to an efficient problem solving process are:

- Good instrumentation
- Good understanding of the extrusion process
- Collection and analysis of historical data
- Team building
- Good information about the condition of the equipment
- Good information about the feed stock

1.1 Instrumentation

The extrusion process is, to a large extent, a black box process. It is not possible to visually observe what goes on inside the extruder. We can see material going into the extruder and material coming out of the extrusion die. However, what happens between the feed opening and the die exit cannot be seen in normal extruders because the process is obscured by the extruder barrel. Therefore, we are largely dependent on instrumentation to determine what happens inside the extruder. We can think of instrumentation as our "window to the process". Because it is very difficult to determine what is happening inside the extruder without good instrumentation, successful problem solving requires good instrumentation.

It is not sufficient to have ample instrumentation on the extruder; it is also important to confirm that the sensors and readouts are working correctly. For instance, if a temperature zone along the extruder is showing an excessively low or high temperature, we should verify that the temperature reading is correct. The measuring instruments must be correctly calibrated, and we should establish that the instrument is capable of measuring the variation of interest. In statistical process control (SPC), specific procedures have been developed to determine the capability of the measuring instrument [1].

1.2 Understanding the Extrusion Process

A good understanding of the extrusion process is necessary to solve extrusion problems efficiently. It is recommended for the reader new to extrusion to take classes covering the material characteristics of plastics, typical features of extrusion machinery, instrumentation and operating controls, the inner workings of extruders, and screw and die design. Classes and continuing education short courses on extrusion are available from a variety of colleges and organizations. There are also a number of training programs available [2] as videos, interactive computer-based instruction, and web-based instruction.

In many extrusion operations, the primary mode of training is on-the-job training. However, on-the-job training is often the least effective and most expensive method of training. Extruders are expensive machines that must be operated correctly to produce good parts. If an extruder is not operated correctly, out-of-spec parts may be produced, or the extruder may be damaged. It is also important to realize that extruders are potentially dangerous devices. Serious accidents can occur when extruders are not operating properly. Therefore, it is imperative that people who operate extrusion equipment receive comprehensive safety training.

A short review of the extrusion process implies the understanding of the functional zones. The functional zones of a conventional single-screw extruder are shown in Figure 1.1:

- *feed hopper*, this zone is designed to operate under gravity flow and it feeds the granules or the particulate solids into the extruder,
- *solids conveying zone*, this zone is designed to transport and to slightly compress the granules or the particulate solids,

- *plasticating zone*, polymer melting takes place in this zone due to viscous dissipation and heat transfer by conduction from the extruder barrel,
- *melt conveying zone* (or *metering zone*), this zone is intended to transport the molten polymeric material and to achieve pressure buildup,
- *mixing zone*, this zone is designed to improve melt homogeneity by means of shear mixing elements, such as distributive, dispersive, or elongational mixers,
- *die forming*, the forming or shaping of the extrudate takes place in this zone for subsequent post-extrusion processes, such as calibration, cooling, winding or cutting, among others, and
- *venting*, this zone is only present in two-stage screw extruders or vented extruders and is used to remove any residual moisture or volatiles from the polymer melt.

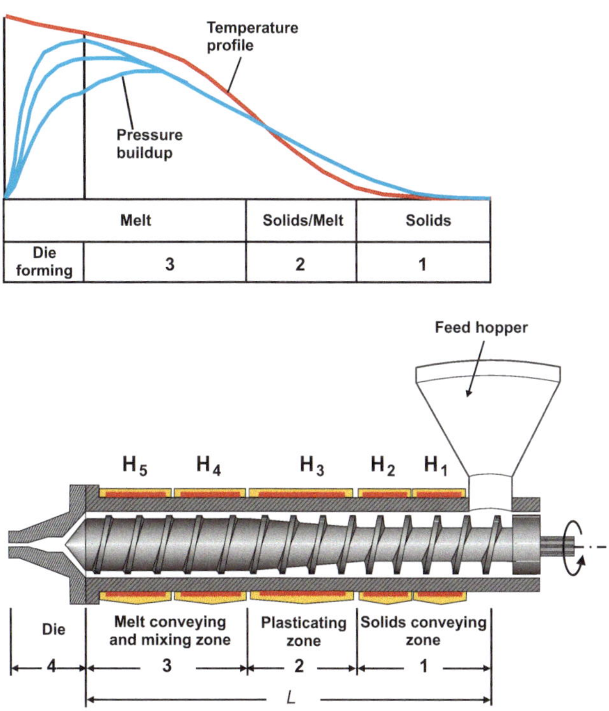

Figure 1.1 Functional zones in a single-screw extruder

1.3 Collection and Analysis of Historical Data (Time Line)

To understand why a process is not behaving correctly, we must compare the current process conditions with previous conditions when the problem did not exist. This is known as constructing a time line. The data to be collected include not only process information from the extruder, such as temperatures, pressures, motor load, line speeds, barrel dimensions, screw dimensions, and such, but also information about the material as well as any other variables that can affect the process. Changes in the process can occur not only as a result of deviations in machine parameters, but also as a result of material variations. For instance, a change in the stabilizer level of the plastic can cause degradation problems in the absence of changes in the machine conditions and settings (Section 1.6).

The time line is constructed on the basis of the fact that the process was running well for a certain period of time. Therefore, there must be an identifiable change or changes that precipitated the process upset. The task is to identify these changes and correct them, and thus get the process back under control. The time line creation process starts during a period of process stability and ends some time after the problems in the process were noticed. All events even remotely connected to the process are listed on the time line. Once the time line is complete, it becomes a helpful tool for identifying the event that precipitated the problem.

It should be noted that not all events result in an immediate problem. In some cases there can be a considerable incubation time before the effects of a change become noticeable. This, of course, complicates the troubleshooting process; it is important to keep this in mind and not jump to conclusions. The author (CR) experienced a case when a disastrous wear problem was the result of an event that took place four months earlier. The wear remained insignificant until approximately four months after a new feed housing was installed. However, once rapid wear commenced, a screw was destroyed within 48 hours.

Figure 1.2 shows an example of a time line leading up to a gel problem. In constructing the time line, the troubleshooter should be sure to list all events that could potentially affect the process. Events such as a power outage, the installation of a new or refurbished extruder screw, or a new resin lot should obviously be included. Some events (such as construction in the area, changes in material handling, maintenance on the plant water system, operator training, or power surges) are less obvious but may still affect the process.

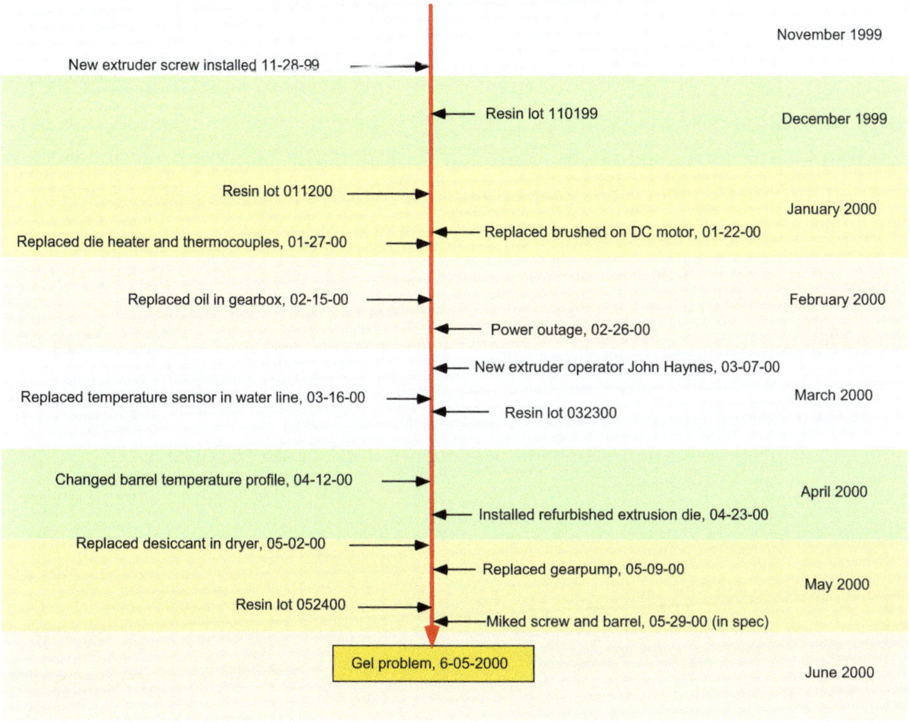

Figure 1.2 Example of a time line leading up to a gel problem

1.4 Team Building

If the scope of a problem is small, a single individual can perform the problem-solving process, and there is no need to organize a team. In many cases, however, problems involve several different departments and functions and require a wide range of skills for solution. In such cases, a team effort is needed. Extrusion problems often require input from materials quality control (QC), purchasing, maintenance, engineering, and possibly from other departments.

1.5 Condition of the Equipment

When a problem develops on an extruder, it is important to have good information on the condition of the equipment. Extruders should be well maintained, and good maintenance records should be available to assess the condition of the various components of the machine. Maintenance recommendations from the extruder manufacturer should be followed to ensure good performance.

Extruder screws and barrels will wear over time. The wear rate depends on many factors. Extruder screws can last for several years or only several weeks. It is important to measure the internal diameter (ID) of the barrel and the outer diameter (OD) of the screw on a regular basis, at least once a year, so that the life of the screw and barrel can be predicted. This allows the screw and the barrel to be replaced at predetermined intervals preventing unpleasant surprises.

1.6 Information about the Feed Stock

The performance of an extruder is determined as much by the characteristics of the feed stock as by the characteristics of the machine. Feed stock properties that affect the extrusion process include bulk flow properties, melt flow properties, and thermal properties. Important bulk flow properties are the bulk density, compressibility, particle size, particle shape, external and internal coefficients of friction, and agglomeration tendency. Important melt flow properties are the shear and elongational viscosities as a function of strain rate and temperature. The commonly used melt indexer provides only limited information on the melt viscosity. Important thermal properties include the specific heat, the glass transition temperature, the crystalline melting point, the latent heat of fusion, the thermal conductivity, the density, the degradation temperature, and the induction time as a function of temperature.

In general, polymer material data is dependent on temperature and pressure. In other cases they are also dependent on surface roughness, sliding velocity, granulometry, shear rate, strain rate, and time. It is important to point out that thermal material data of polymers differ strongly depending on whether the polymer is an amorphous or a semi-crystalline material.

A change in the material can cause a problem in extrusion when it affects one or more polymer properties that determine the extrusion behavior of the material. If a material problem is suspected, the troubleshooter should first examine the QC records on incoming material to look for any record of a change in feed stock properties. Unfortunately, often, the only QC test on incoming material is a melt index (MI) test. This test detects only a very limited number of material-related extrusion problems. Thus, in many cases, more extensive material testing than the regular QC testing may be required.

There are a number of problems associated with making measurements on the critical properties of the feed stock. Approximately ten properties should be measured, and some of the measurements are fairly time-consuming. Thus, when a quick solution is required, it may take too much time to fully characterize the extrusion properties of a material. Another problem is that some important properties are difficult to measure with a high degree of accuracy and reproducibility. The most notable property in this category is the external coefficient of friction. Instruments to measure all the pertinent properties may not be readily available. Not all companies can afford to maintain a fully instrumented laboratory. Finally, even after a material is fully characterized, and no significant changes in properties have been found, there is no guarantee that the extrusion problem is not material-related. Because most tests are done on samples of about 0.01 kg or less, and most extruders run at a throughput of several hundred to several thousand kg/h, there is a considerable chance that the test sample may not have been representative of the entire feed stock.

A practical test for a material-related extrusion problem is to extrude some material from a previous batch to see if the problem will disappear. If this is indeed the case, it strongly indicates that the problem is material-related. For this reason, it is helpful to retain some material from older batches; this also provides a reference for more detailed measurements. If the problem is material-related, there are two possible solutions. The easiest solution is to change the material back to the way it was before the problem developed. However, this may not always be possible. Thus, if the material-related change is permanent, then the extrusion process will have to be adjusted to accommodate the material change. At this point, the nature of the problem may change from an upset to a development problem. The chance of solving the problem will depend on the nature and the magnitude of change in the material.

1.7 Problem-Solving Techniques

When analyzing complex extrusion problems that appear to have many interconnected causes it is important to use a so-called *cause-effect* diagram. There are several possibilities to represent *cause-effect* diagrams, including fishbone diagrams or Ishikawa diagrams, problem tree diagrams, problem checklists, mind maps, concept maps, among others.

The fishbone was introduced by the Japanese professor Kaoru Ishikawa in 1943. The diagram is drawn with an arrow that will serve as the backbone from which further possible causes (bones) will be categorized and related. Occasionally, the detailed causes may come from other origins. The diagram allows the identification of potential causes for the problem and it helps to determine which ones are most likely resolve the problem.

A problem tree diagram is a way of visualizing the *cause-effect* relationships with regard to a specific technical problem. In such a representation the causes are shown at lower levels and the effects at upper levels. The problem tree diagram facilitates the organization of problems into a logical sequence which will lead to logical conclusions and the identification of technically cost-effective solutions.

The checklist is a very common tool and easy to apply to systems, projects, and problems. It is a list of activities to be noted, checked, or remembered. The objective is to ensure that a long list of tasks is not forgotten. In the plastics industry, checklists are used to check process compliance, to prevent failures or errors, to assure product quality, and to solve technical problems.

Mind mapping is also a useful technique for troubleshooting. This tool was developed by Tony Buzan in the 1970s. A mind map uses pictures and/or word phrases to organize and develop thoughts in a non-linear manner. The elements of a mind map are arranged intuitively according to the importance of the idea or task. The map allows establishment of connections that are not obvious. It helps to "observe" a problem and its solution.

A concept map is a diagram that allows visualizing of relationships among concepts. It is also a graphical tool and is normally used to represent knowledge. Concepts are represented as boxes and are connected with labeled arrows in a hierarchical structure. The relationship between concepts can be done with linking phrases. Figure 1.3 shows a concept map visualizing a whitening problem in a plastic product.

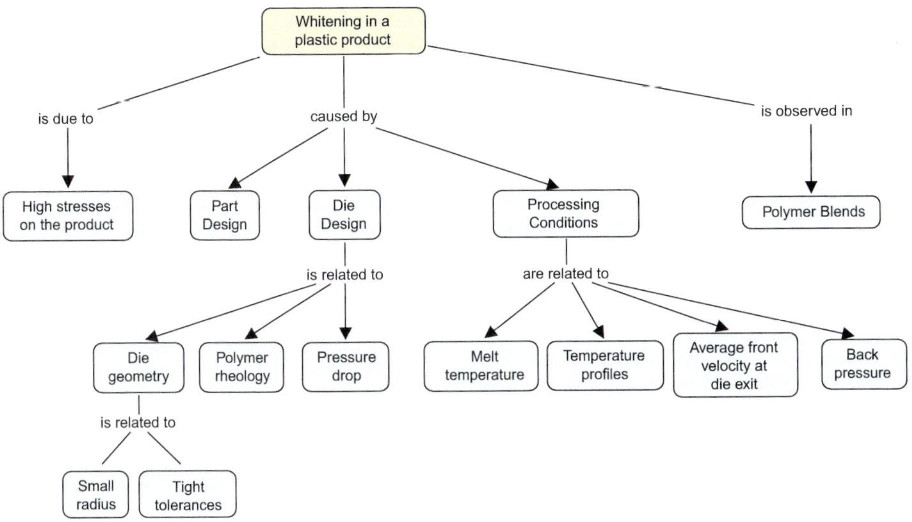

Figure 1.3 Example of a concept map visualizing a whitening problem

1.8 Collection and Interpretation of Extrusion Process Data

1.8.1 Introduction

Many modern extrusion lines are equipped with data acquisition systems (DAS). These systems can present extruder operators and process engineers with a large amount of data. There are several critical issues in the proper use of a DAS:

- Are the correct process variables measured and monitored?
- Are these process variables measured correctly?
- Is the data collection rate appropriate?
- Is the operating personnel capable of proper interpretation of the data?

These issues will be addressed in this section.

1.8.2 Vital Signs of the Extrusion Process

The three most important extrusion process variables are melt pressure (P), melt temperature (T), and motor load (I). These three process variables are the vital signs of the extruder. When the extruder is not functioning correctly, one or two or all three of these variables will show abnormal behavior. It is no coincidence that these three process variables are similar to the vital signs of the human body: blood pressure, body temperature, and pulse. The human body functions because the heart pumps blood through our vascular system. Blood pressure and pulse are measures of how well our blood is being pumped. In extrusion P, T, and I are measures of the pumping efficiency of the extruder.

One of the most important process measurements is the specific energy consumption (SEC). This is the motor power divided by the throughput. In the U.S. the SEC is typically expressed in units of hp·h/lb – more commonly, the SEC is expressed in kW·h/kg. The SEC is a measure of the amount of energy from the motor introduced into the polymer. As a result, the SEC is closely related to the increase in stock temperature in the extrusion process. Higher levels of SEC lead to higher melt temperatures.

The minimum SEC needed to extrude a polymer is determined by the thermal properties. It is the specific heat multiplied by the temperature rise in the extrusion process – this is the increase in enthalpy. For semi-crystalline polymers the minimum SEC value is about 0.10 hp·h/lb or about 0.16 kW·h/kg. It is good practice to monitor the SEC because it is a good measure of the energy efficiency of the extrusion process. If the SEC in a single-screw extrusion process is above 0.2 hp·h/lb (0.32 kW·h/kg), this indicates poor energy efficiency. This will likely result in high melt temperatures.

1.8.2.1 Melt Pressure

The most important process variable is melt pressure. Die inlet pressure is a direct measure of the flow rate through the die. Variation of die inlet pressure causes variation in flow rate; this, in turn, will cause dimensional variation of the extruded product, see Figure 1.4. A graphical display of die inlet pressure versus time provides a clear picture of the process stability – or lack thereof. A simple display of pressure (analog or digital) is much less useful.

When the extruder operator looks at a digital pressure readout with numbers typically changing ten times per second or more, it is impossible to get a clear picture of the pattern. The human mind has very limited ability to analyze large numbers of numerical data. On the other hand, when the operator can look at a graph of pressure versus time, the trend of the pressure pattern becomes immediately obvious. The human mind has remarkable ability to interpret graphical information. For this reason, trend plots are a very important element of a good DAS.

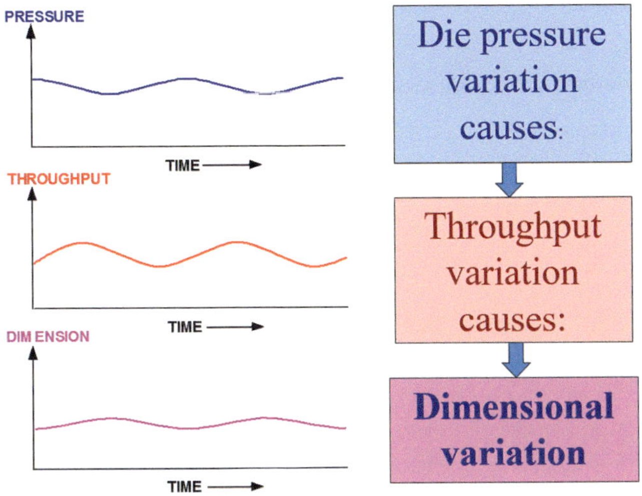

Figure 1.4 Effect of die inlet pressure variation on dimensional variation of extrudate

Pressure is measured with a pressure transducer. These pressure transducers have a thin diaphragm that is in contact with the molten plastic, see Figure 1.5.

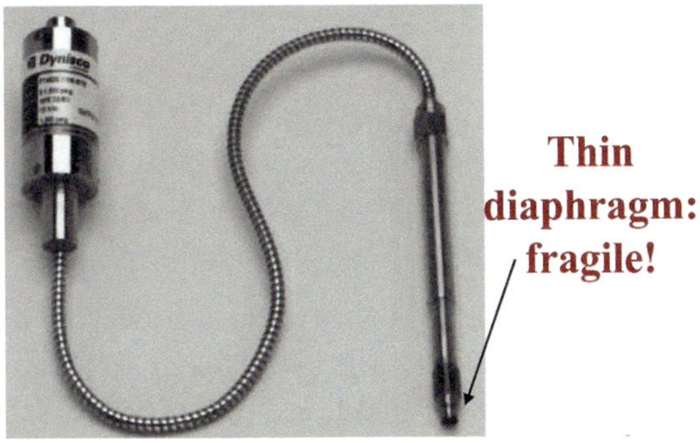

Figure 1.5 Typical pressure transducer

Increasing melt pressure will deflect this diaphragm and this deflection is a measure of the amount of pressure acting on the diaphragm. There are several different types of pressure transducers. Table 1.1 lists several types with a comparison in terms of robustness, maximum temperature, dynamic response, and error.

Table 1.1 Comparison of Several Types of Pressure Transducers

Type	Robustness	T max. [°C]	Dynamic response	Error %
Bourdon	Poor	200	Poor	3-5
Pneumatic	Good	400	Poor	1.5
Capillary*	Fair	400	Fair	0.5-3
Pushrod	Fair	400	Poor	3
Piezo-electric	Good	120	Poor	0.5-1
Piezo-resistive	Good	400	Good	0.2-0.5
Optical	Good	600	Good	1.0

* Safety concern with mercury

Many strain gage capillary pressure transducers are filled with mercury. Clearly, this is a safety concern – for this reason an asterisk is placed next to this transducer in Table 1.1. The diaphragm of a capillary pressure transducer is quite thin, usually from 0.100 to 0.125 mm – this is about the thickness of a typical sheet of paper. As a result, the diaphragm is easily damaged. When the diaphragm of a mercury filled pressure transducer ruptures the liquid will enter into the workplace and contaminate the product. For that reason, mercury-filled transducers should not be used for medical extrusion and food packaging extrusion.

It is worth noting that the capillary pressure transducer is the most commonly used transducer in the extrusion industry. Table 1.1 clearly shows that from a technical point of view the capillary pressure transducer is not the best – in fact, it is not even second best. Considering that pressure measurement is the most important in the extrusion process, it is surprising that the most commonly used pressure transducer is one that has some clear deficiencies.

At a very minimum, pressure should be measured at two locations: before and after the breaker plate. Usually a screen pack is placed against the breaker plate to capture contaminants and to make sure that these contaminants do not end up in the extruded product. As contamination builds up on the screen pack, the resistance to flow increases and the extruder output will drop. This, of course, will reduce the extrudate dimensions. To counteract this problem a pressure feedback control is typically used. In this control, the pressure after the screen pack and breaker plate is measured and controlled by the screw speed, see Figure 1.6.

When contamination builds up on the screen pack, the pressure after the breaker plate will reduce. With pressure feedback control the screw speed is adjusted such that the pressure after the breaker plate (die inlet pressure) remains constant. Pressure feedback control has been available in extruders for decades. It is also used when the extruder is equipped with a gear pump – in this case, the inlet pressure of the gear pump is controlled by the screw speed.

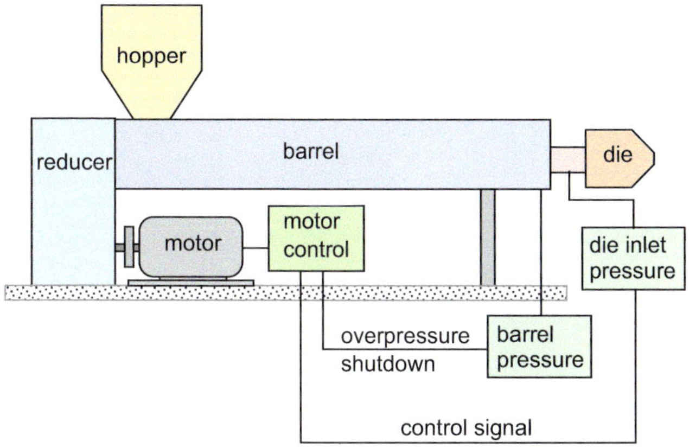

Figure 1.6 Pressure feedback control in extrusion

One limitation of pressure feedback control is that it can only act on very slow changes in pressure – changes occurring over ten minutes or longer. Pressure feedback control cannot reduce short-term pressure variations because the control loop is too slow. However, there are several other methods that can be used to reduce short-term pressure variation.

With pressure feedback control there is a gradual increase in the screw speed and the pressure just before the screen pack and breaker plate. This pressure is often referred to as the barrel pressure. When the screw speed increases, the viscous heating in the extruder increases and this will generally result in an increase in melt temperature. It is important to understand that pressure feedback control will result in changes in the extrusion process, in particular the melt temperature. Changes in melt temperature can affect the properties of the extruded product even if the product dimensions do not change.

When contamination on the screen pack builds up rapidly, this can affect the extrusion process significantly. This problem can be avoided to a large extent by using a continuous screen changer.

It is also important to have an overpressure shutdown as shown in Figure 1.6. This is a critical safety feature because under certain circumstances pressures can increase very rapidly – much faster than an extruder operator can act to shut down the extruder. Excessive pressures can cause the die to blow off the extruder in an explosive manner. Clearly, this is a problem one should avoid. For this reason, overpressure shutdown is an important safety feature. Another important safety feature is a rupture disk, see Figure 1.7. The rupture disk is typically placed in the extruder barrel just before the breaker plate and screen pack – this is where the highest pressure is likely to occur.

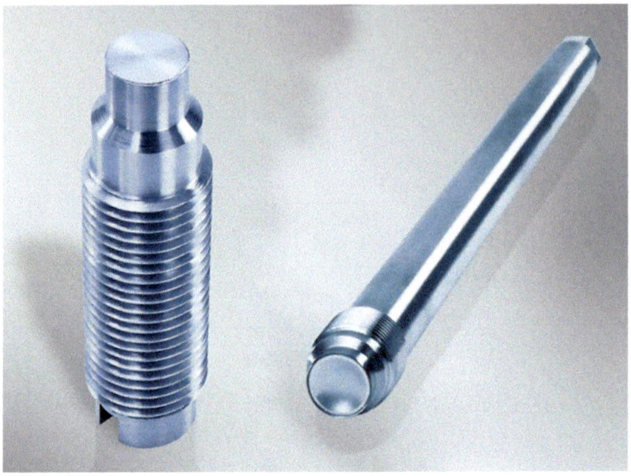

Figure 1.7 Rupture disk (courtesy of Striko)

For proper process analysis and troubleshooting it is very helpful to have pressure transducers along the length of the extruder. Unfortunately, this is rare in the extrusion industry – without pressure transducers along the barrel our ability to troubleshoot extrusion problems effectively is substantially diminished.

When the extruder is equipped with a gear pump, it is very helpful when two pressure transducers are located downstream of the gear pump some distance apart. With this setup the melt viscosity in the polymer line can be easily determined. The gear pump speed is a direct measure of the volumetric flow rate. The shear rate in the line is determined from the flow rate and the geometry of the flow channel. The shear stress can be determined from the difference in pressure between the two pressure transducers. The melt shear viscosity is simply the ratio of shear stress and shear rate. Such a setup provides real-time monitoring of the melt viscosity at minimal cost.

Pressures inside the Extruder

The pressures inside the extruder reflect the flow that occurs in the extruder. Figure 1.8 shows the flow in a cross section of the screw channel in the metering section of the extruder. At the top of the channel there is a drag flow toward the pushing flight flank and the pressure increases from P_4 at the trailing flight flank to P_1 at the pushing flight flank. At the pushing flight flank, material will move toward the root of the screw by pressure flow. As a result, pressure P_1 will be greater than pressure P_2 at the bottom left of the screw channel. At the lower portion of the screw channel material will move to the trailing flank of the flight by pressure flow because pressure P_2 is greater than pressure P_3. There is pressure flow along the trailing flank of the flight because pressure P_3 is greater than pressure P_4.

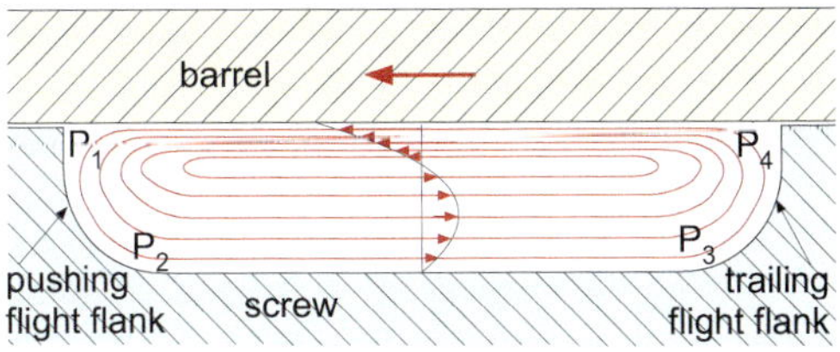

Figure 1.8 Cross channel flow in single-screw extruder

The highest pressure in the channel is pressure P_1 and the lowest pressure is pressure P_4. This is the pressure drop across the flight. This pressure drop ΔP_x ($P_1 - P_4$) can be expressed as follows for a Newtonian fluid:

$$\Delta P_x = \frac{6\mu v_b W_x \cos\phi}{H^2} \tag{1.1}$$

In this expression μ is the melt viscosity, v_b is the barrel velocity, W_x is the width of the channel, ϕ is the flight helix angle, and H is the depth of the channel. This pressure drop ΔP_x depends on the viscosity, screw geometry, and screw speed. When the screw geometry and screw speed are known, the viscosity can be determined from the pressure drop ΔP_x.

When pressure is measured in this part of the extruder, the pressure will have a saw tooth pattern as shown in Figure 1.9.

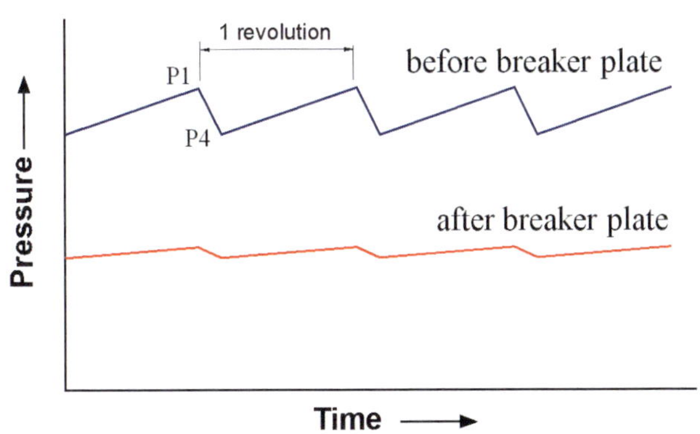

Figure 1.9 Pressure versus time in the metering section

The time between peaks equals the time for one screw revolution for a single flight screw. This means that the screw speed can be determined from the pressure trend. For a double-flighted screw there will be two pressure peaks for each revolution of the screw.

The pressure pattern in the melting zone of the extruder will be different because the screw channel is not only filled with molten polymer but also unmelted polymer. In a single-screw extruder the melting usually occurs by contiguous solids melting or CSM.

In contiguous solids melting the solid particles are compressed together in a solid bed, see Figure 1.10. The solid bed forms a helical solid ribbon that wraps around the screw. As melting progresses the size of the solid bed reduces while, at the same time, the size of the melt pool increases. There is a thin melt film between the solid bed and the barrel. Most melting takes place at the interface between the solid bed and the melt film.

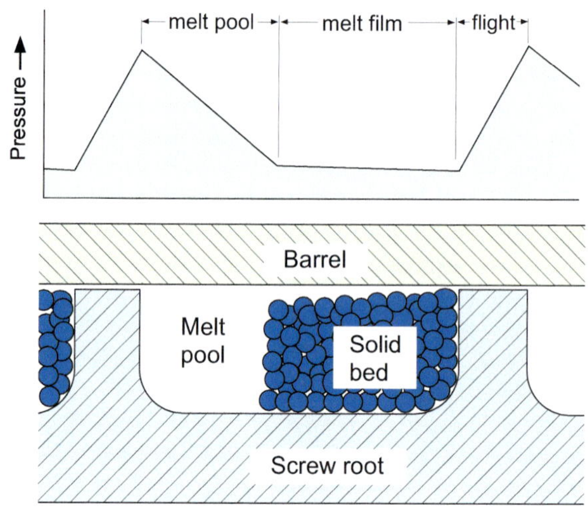

Figure 1.10 Contiguous solids melting and pressures

If pressures are measured in the melting zone, the pattern will be different from that of the melt conveying zone. The pressure gradient in the melt film will be different from the pressure gradient in the melt pool as shown in Figure 1.10. Therefore, when pressure is measured midway along the barrel, the pressure pattern reflects the location of the solid bed, the melt pool, and the flight. By locating several pressure transducers along the length of the barrel, the progress of melting along the length of the extruder can be analyzed.

1.8.2.2 Melt Temperature

Melt temperature is usually measured with an immersion melt temperature probe. The sensing stem of the probe should be immersed at least 5 mm to provide a good measure of the melt temperature. Some melt temperature probes have a stem with adjustable depth, see Figure 1.11.

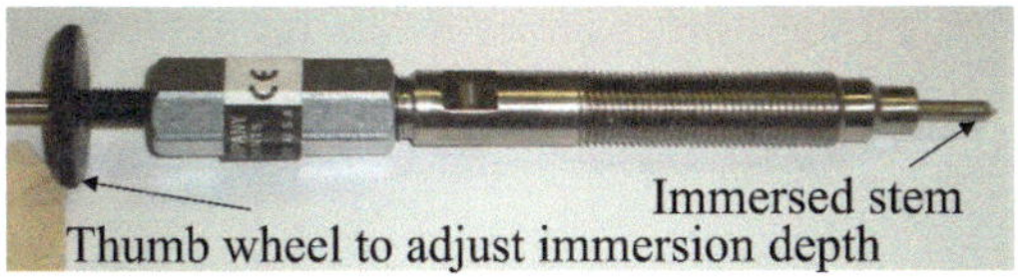

Figure 1.11
Melt temperature probe with adjustable depth

These adjustable melt temperature probes can be used to measure melt temperature at different locations across the depth of the channel. It is good practice to completely retract the sensing stem before shutdown to minimize the chance of damaging the melt temperature probe.

Immersion melt temperature probes have some drawbacks. Since they protrude into the melt stream, they alter the velocities in the melt stream. This, in turn, will change the viscous heating and the melt temperatures of the melt stream. Therefore, the immersion probe will measure a melt temperature that is different from the melt temperature that would occur without a probe immersed in the melt stream.

Another drawback is the fact that an immersion probe will create some degree of stagnation downstream of the probe. This can create problems in thermally sensitive materials like rigid PVC and EVOH. A further drawback is the slow thermal response of an immersion probe. For this reason, an immersion probe cannot detect rapid changes in melt temperature – these are changes occurring over a few seconds or less. This is a serious limitation of immersion probes because rapid melt temperature changes occur in every extrusion process. This will be discussed in more detail in the next section.

Another method of measuring melt temperature is by measuring infrared radiation (IR). Infrared thermometers are readily available and relatively inexpensive, see Figure 1.12.

Figure 1.12
IR thermometer (courtesy of Raytek)

Inside the extruder or the die, the melt temperature can be measured through a transparent window as shown in Figure 1.13.

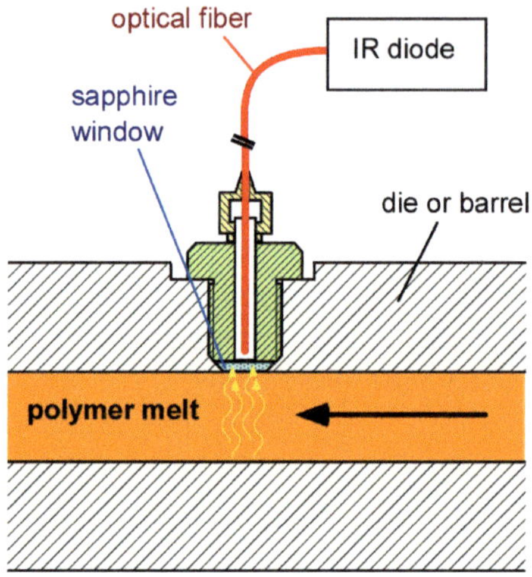

Figure 1.13 IR melt temperature measurement

Another way to measure the melt temperature with IR is to use a probe with a sapphire window at the end in contact with the molten polymer, see Figure 1.14. The transparent window is mounted in a threaded stem. The window is made out of sapphire so that it can handle high temperatures (up to 600 °C) and pressures (up to 1500 bar or about 22,000 psi). The probe fits in a standard pressure transducer hole, making installation quite easy. These probes are commercially available.

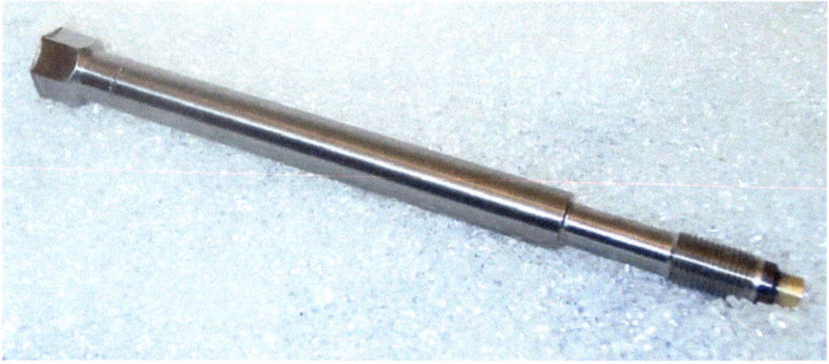

Figure 1.14 Sapphire window mounted in threaded stem (photo courtesy of FOS Messtechnik)

The threaded probe with the sapphire window can be used in combination with a simple IR thermometer. This allows IR melt temperature measurement in the extruder barrel or in the die.

IR temperature measurement provides two important benefits. One, the response time is fast, about 10 milliseconds. Two, the measurement does not disturb the flow as is the case with an immersion probe. Because of the fast response, the IR measurement can detect rapid changes in melt temperature that cannot be detected by an immersion melt temperature probe. Melt temperature variation will be discussed next.

Melt Temperature Variation in Extrusion

The melt temperatures in extrusion result from viscous dissipation and heat transfer. In drag flow the shear rates and viscous dissipation are relatively uniform. However, the heat transfer to the screw and barrel results in large temperature differences within the polymer melt. The largest temperature gradients tend to occur at the barrel surface.

The melt temperature distribution can be determined by performing a non-isothermal analysis of the flow in the screw channel. In the case of a non-Newtonian polymer melt this is typically done using finite element analysis (FEA). Figure 1.15 shows a color contour plot of melt temperatures in the screw channel of a 63-mm single-screw extruder running a 0.2 melt index high density polyethylene (HDPE) at 100 rpm. The vertical axis has been stretched about eight times to make it easier to examine the temperature distribution.

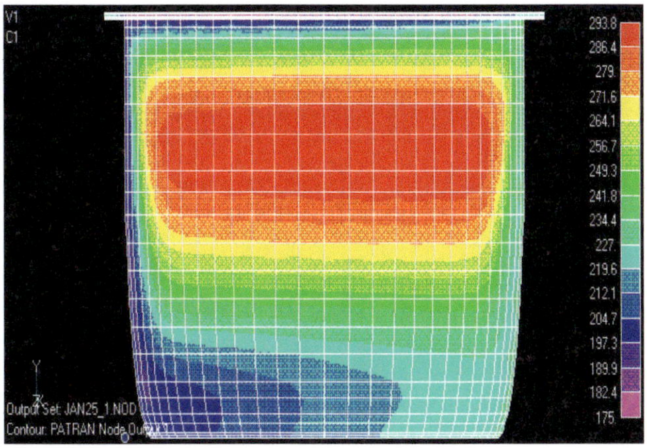

Figure 1.15 Melt temperature distribution in 63-mm extruder running at 100 rpm

The top surface is the barrel, the bottom surface is the root of the screw, the left vertical boundary is the pushing flank of the flight, and the right vertical boundary is the trailing flank of the flight. It is clear that the melt temperatures are highly non-uniform. The temperatures close to the barrel are relatively low because of the heat transfer to the barrel - the barrel cools down the melt that is in close proximity to the barrel. The temperatures at the pushing flight flank are low because the cool melt close to the barrel moves down along the flight when it reaches the pushing flight flank.

At the root of the screw, the cool melt from the pushing flight flank moves toward the trailing side of the flight. If there is no heat transfer with the screw, the melt temperatures will gradually increase as the melt moves from the pushing flank to the trailing flank. The melt reaching the trailing flight flank moves up along the flight until it gets close to the barrel. At that point the cycle starts again. The outer region of the screw channel stays relatively cool because of the heat transfer from the melt to the barrel. The situation is quite different for the inner region of the screw channel.

The melt in the inner region remains in this region as it travels down the length of the screw channel – it will not come into close proximity of the barrel. The inner region is thermally insulated from the barrel by a relatively thick layer of polymer melt. As a result, the temperatures in this region increase higher than in any other part of the channel – it is a natural hot spot. In Figure 1.15 the temperatures in the hot spot are more than 100 °C above the barrel temperature.

When the screw discharges the melt, this melt will be non-uniform in temperature. A number of studies have analyzed the variation of melt temperatures in extruders. One such study was made by E. Brown and A. Kelly at Bradford University in England [232]. In this work, a fast response thermocouple (TC) mesh was used to measure melt temperature variation using three different extruder screws. The TC mesh was located in the breaker plate recess of the extruder barrel, see Figure 1.16.

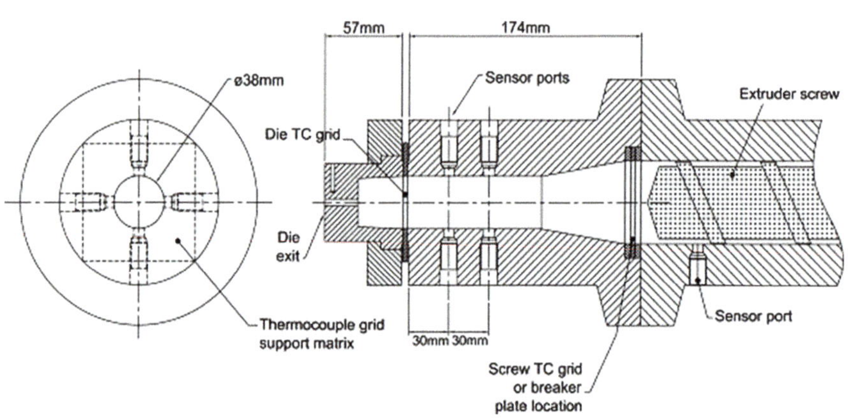

Figure 1.16 Experimental setup for melt temperature study

Melt temperature was measured at multiple locations, see Figure 1.17. It is clear from this figure that the melt temperatures across the channel are quite non-uniform with a low temperature of 190 °C and a high temperature over 215 °C.

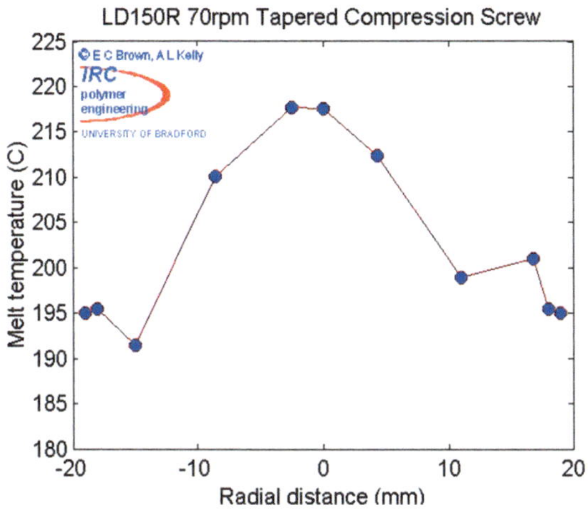

Figure 1.17 Melt temperatures versus radial distance from the center

The melt temperatures vary not only across the melt stream but also over time – this is shown in Figure 1.18.

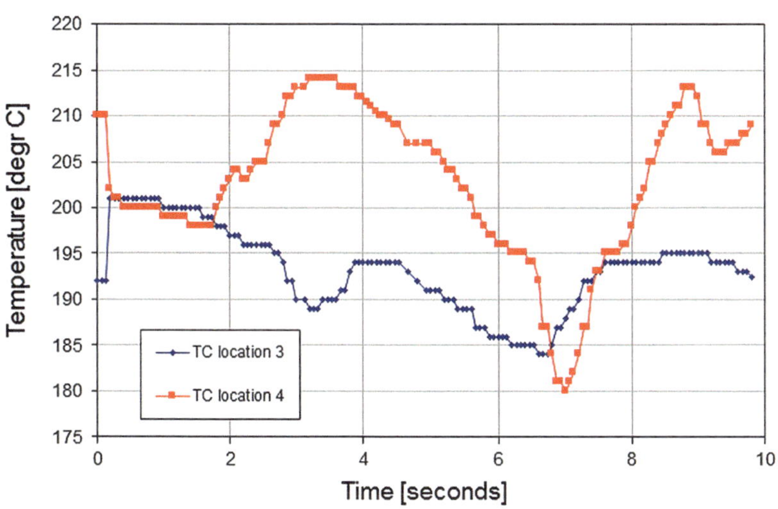

Figure 1.18 Melt temperature versus time

The melt at TC4 (radial location −10 mm in Figure 1.17) drops from 215 °C to 180 °C within four seconds and then increases again to almost 215 °C within two seconds. This is a rapid and substantial melt temperature variation! Interestingly, this melt temperature variation cannot be detected with an immersion melt temperature probe because of the slow response time.

In this study the melt temperatures were also measured with an infrared thermometer. The results are shown in Figure 1.19.

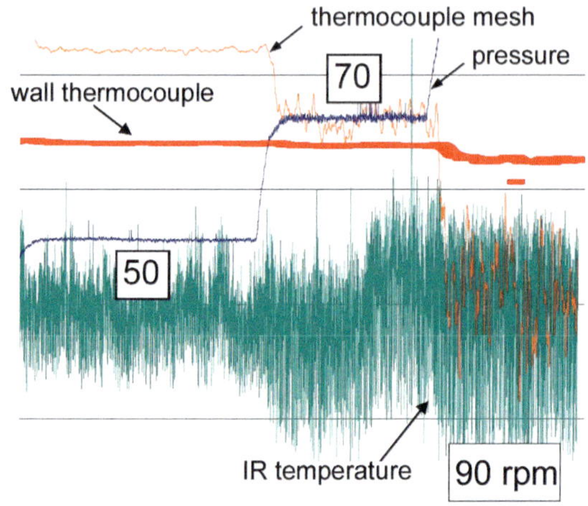

Figure 1.19 Comparison of TC mesh and IR temperatures

Figure 1.19 shows temperatures at three screw speeds: 50, 70, and 90 rpm. The temperature variation increases with screw speed. The variation in TC mesh temperatures is substantially smaller than in IR temperatures. Also, the IR temperatures show a much faster variation as a result of the short response time of the IR measurement.

Figure 1.20 shows infrared melt temperature measurements plotted versus time for three extruder screws running at 90 rpm [187].

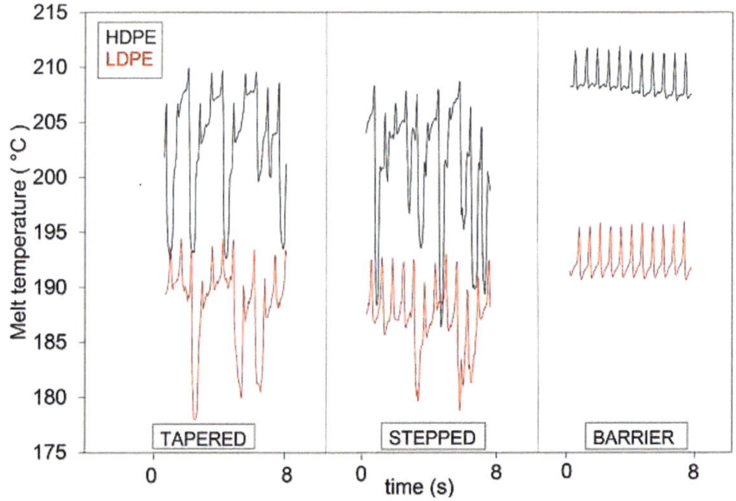

Figure 1.20 IR melt temperature measurements for three screws running at 90 rpm

The melt temperature variation for the tapered and stepped screw is quite large, 15–20 °C. The melt temperature variation for the barrier screw is smaller, about 5–7 °C, and displays a more regular pattern. The time between peaks corresponds closely with the screw speed – 90 rpm corresponds to 0.67 seconds per revolution. This means that the screw speed can be determined from the IR melt temperature measurement.

The reduced melt temperature variation in the barrier screw may be due to the barrier geometry. However, this screw was also equipped with a fluted mixing element in the metering section. Therefore, it is also possible that the reduced melt temperature variation is due to the presence of the fluted mixer. The tapered screw and the stepped screw had no mixing elements; therefore, these screws can be expected to result in a larger melt temperature variation.

These results clearly demonstrate that short-term melt temperature variations occur in polymer extrusion. In fact, these melt temperature variations are inherent in the extrusion process. They occur because polymers have very low thermal conductivity. As a result, conductive heat transfer is very slow. The time that is required for the melt temperatures to become uniform by conduction is very long – it can be measured in hours for production size extruders.

Typical residence times for the molten polymer in extruders are around 20–40 seconds. The residence times are usually much shorter, by orders of magnitude, than the times required to achieve uniform melt temperatures by conduction. Therefore, melt temperatures in extrusion will be non-uniform in most typical extrusion operations – especially when the extruder operates at high screw speed and when the melt temperatures are higher than the barrel temperature. These melt temperature non-uniformities are more pronounced on larger extruders, where heat transfer distances are greater.

When the temperatures of the melt exiting the extruder are non-uniform, the dimensions of the extrudate will become non-uniform as the extrudate cools down and solidifies. These dimensional variations result from non-uniform shrinkage. Saul et al. studied this problem in pipe extrusion [209]. The manifestation of this problem in pipe extrusion is the waviness of the internal diameter (ID), see Figure 1.21. The calibration process fixes the outside diameter of the pipe but the inside diameter will vary if the shrinkage is non-uniform.

Figure 1.21 Pipe with waviness of internal diameter

ID waviness is a common problem in pipe extrusion. However, the actual cause of this problem is not widely recognized. One approach to reduce the problem is the use of a large inventory die. Such a die will have long residence times in the die and this will reduce the melt temperature variation at the die exit. However, large inventory dies have several drawbacks:

- They are expensive
- They take up a large amount of space
- They take a long time to heat up
- Long residence times increase polymer degradation
- They increase change-over times
- They result in more scrap

A much more effective solution is to enhance the mixing capability of the extruder. With proper thermal mixing large inventory dies are not necessary and the ID waviness can be largely eliminated.

Data Collection Rate and Fast Fourier Transform

From the information presented earlier it is clear that rapid data collection is necessary for certain process variables such as melt pressure, melt temperature, and motor load. In order to capture short term process variation these variables have to be measured at least 50 times per second. Unfortunately, this capability does not exist on many extrusion lines.

A useful feature of a data acquisition system is the Fast Fourier Transform or FFT. This is a technique that determines the base frequencies making up a complex signal. FFT is very useful in troubleshooting because signals are often complex and difficult to interpret. This is because the extrusion process is influenced by many different factors (50 to 100). At any one time, 10 to 20 factors may have a measurable effect on the extrusion process. For that reason, process variables such as melt pressure, melt temperature, motor load, extrudate dimensions, etc. often show a complex pattern. Appendix 4 covers process signal analysis using Fast Fourier Transform.

1.8.2.3 Training

It is not enough to have good instrumentation and data acquisition capability. It is equally important to provide training so that operators and process engineers can take full advantage of the capabilities of the data acquisition system. This is an essential element in a successful extrusion operation. Some companies skip this critical step and, as a result, experience only limited benefits from a good data acquisition system.

1.8.3 Conclusions

Good instrumentation is critically important in extrusion. Without good instrumentation and data acquisition capability it is hard to:

- Know what is happening in the extruder
- Control the extrusion process
- Optimize the extrusion process
- Know when something is wrong
- Troubleshoot the extrusion process
- Have a profitable extrusion operation

Training is essential to derive maximum benefit from a good data acquisition system.

2 Tools for Troubleshooting

We will discuss a number of the important tools useful in the troubleshooting process.

2.1 Temperature Measurement Devices

One useful tool is a pyrometer with a surface contact probe and a melt probe (needle probe). The contact probe can be used to check for heater burnout and to check barrel and die temperatures as well as temperature distribution and variation. The melt probe can be used to check melt thermocouple accuracy and to measure the actual melt temperature as the extrusion material exits the die. The melt temperature at the die exit can be higher than the melt probe temperature at the end of the extruder barrel.

Another useful troubleshooting tool is the infrared thermometer. One example of a handheld infrared thermometer is shown in Figure 2.1. The non-contacting infrared (IR) thermometer allows temperature measurement in spots that are difficult to reach with a contacting thermometer. The IR thermometer also allows measurement of polymer melt temperature without damage to extruded product and the determination of melt temperature variations across the melt stream exiting a sheet die. Large melt temperature variations will generally create problems downstream.

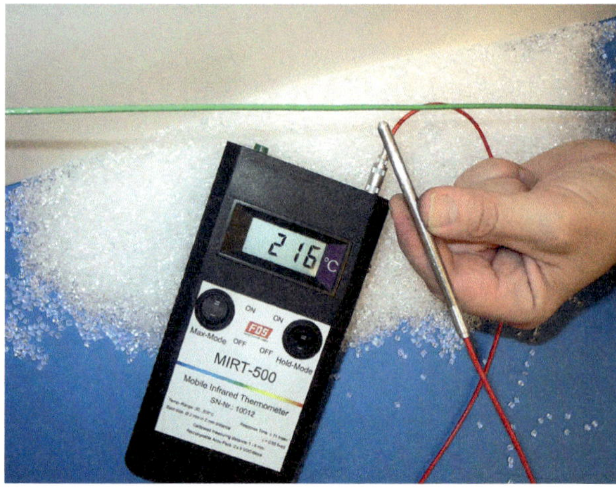

Figure 2.1 Example of a handheld infrared thermometer

2.2 Data Acquisition Systems

Data acquisition systems (DASs) are extremely useful in troubleshooting an extrusion process, because problems often occur when the operator is not watching the instrument panel of the extruder. Even when the operator is watching the instrument panel, he can only observe a limited number of variables at one time. A DAS that captures and saves important process data is indispensable in troubleshooting. When a problem occurs at 2:30 A. M., it is very difficult for a process engineer coming in at 7:00 A. M. to reconstruct the events at 2:30 A. M., if important process data were not recorded at that particular time.

A simple DAS is a chart recorder that can track important variables such as screw speed, die-head pressure, melt temperature, and motor amperage. More useful is a computer-based DAS. These come in two forms: portable data collectors/machine analyzers and fixed-station data acquisition systems. Today cloud-based DAS systems are commonly used. These are very convenient and relatively easy to implement. Several of these cloud-based systems are specifically designed for extrusion lines [e. g., 211–213, 215].

Today, there are process-monitoring systems that have the ability to determine the condition of machinery. One company active in this field is Prophecy Sensorlytics [216]. This system uses multi-function sensors placed on the machine. A data hub is plugged into an electrical outlet within 10 meters of the sensor and connected to the Wi-Fi network. The data travels to cloud-based servers that apply interpretive software and algorithms to track the condition of the machine.

The Prophecy Sensorlytics system has a self-learning mode that allows conversion to prescriptive maintenance after about six months of data logging. The system sends out e-mail or text messages when maintenance is required.

Extruder manufacturer Davis-Standard has developed DS Activ-Check for continuous extruder monitoring [214]. Activ-Check allows real-time predictive maintenance by providing early notifications of potential extruder failures. Machine operators are alerted to issues before they happen, reducing unplanned downtime while also collecting valuable data. Users receive notifications via e-mail or text, and continuous monitoring of production machine status is available on smart devices and remote PCs. Key parameters monitored include extruder reducer, lubrication system, motor characteristics, the drive power unit, barrel heating and cooling.

The system uses overview screens with graphic measurement information that provide a quick reference of monitor points and trend windows. Users can touch a location to view details, or use on-screen setup for e-mail or text notifications. Vibration sensors on the extruder reducer provide data regarding the condition of the gears, bearings, and lubrication system. The extruder motor is supplied with combination temperature sensors and vibration sensors. The reducer lubrication system includes pressure, flow, and temperature sensors to indicate system performance. Operators are also able to monitor key health indicator parameters in the drive power unit, providing an early indicator of potential power unit issues.

Appendix 4 covers the use of Fast Fourier Transform in data analysis. This appendix also covers the importance of the data capture rate. A low speed data capture rate does not allow analysis of high speed process variation. Even worse, it can distort process signals and cause a problem referred to as aliasing.

Data acquisition systems are also discussed in Section 4.25; this is a case study on gear pump speed variation in sheet extrusion. In this case study there was no data acquisition system. The problems that result from the absence of a DAS are discussed in some detail. Rapid changes, occurring within less than 1 second, happen in every extrusion process and they affect the vital signs of the extruder: melt pressure, melt temperature, and motor load. These changes occur fast enough that they cannot be captured by an operator trying to write down the data. As a result, a fast electronic data acquisition system is critical to a successful extrusion operation.

2.2.1 Portable Data Collectors/Machine Analyzers

Portable data collectors (PDCs) are similar to check sheets in that they can be easily moved around. In the process of injection molding, these devices are often referred to as portable machine analyzers (PMAs). Some of their important advantages are:

- They can record data in computer-readable form.
- They can take data directly from electronic sensors and gages, which makes PMAs fast and minimizes errors.
- Data can be analyzed internally to yield statistical information on mean, range, maximum value, minimum value, and standard deviation.
- PMAs are available with limit checking to actuate an alarm when data just taken is out of specification.
- PMAs can collect data on variables as well as attributes.

The use of PMAs has increased considerably as prices have come down to levels that are affordable even for small operations [3]. Several PMAs are presently available for less than $10,000. Most PMAs offer the user some flexibility in assigning inputs to the data acquisition channels; this even extends to auxiliary equipment (for example, dryers), and to external signals such as plant ambient temperature and relative humidity. Important inputs are:

- Melt pressure(s)
- Melt temperature(s)
- Screw speed
- Motor load
- Temperature of feed housing
- Barrel temperatures
- Die temperatures
- Line speed
- Extrudate dimensions
- Heating power at various temperature zones
- Cooling rate at various temperature zones

Other parameters to be monitored depend on the specifics of the operation. For instance, in vented extrusion it is often important to monitor the vacuum level at the vent port. For in-depth process analysis, a system capable of handling 32 channels or more should be used.

2.2.2 Fixed-Station Data Acquisition Systems

As the name implies, fixed-station data acquisition systems are fixed to one location because of size or because the wiring makes moving the unit very difficult. A fixed-station DAS can have a wide range of capabilities. A dedicated DAS may record data from only one extruder. A number of machine suppliers now offer extruders with integrated data acquisition and SPC capability.

A more sophisticated DAS may be capable of recording data from various sources, analyzing the data, presenting control charts, showing trend plots for different variables, and so on. These systems are often referred to as plant-wide monitoring systems. Central computer-based systems are now available which allow the user to view the operation of plant equipment and change the operating parameters on any selected equipment. A fixed-station DAS can take many different forms, depending on the application. A good DAS can be a very valuable tool in improving process control as well as in problem-solving. The following capabilities are useful [1] for application to extrusion:

- *Monitoring of many variables:* A typical extrusion process requires monitoring of 40 to 80 process parameters.
- *Variable sampling:* Data on slowly changing variables should be taken at least once per second; data for rapidly changing variables, such as melt pressure and hydraulic pressure, should be collected more frequently, typically 100 points per second. Some high-end systems sample up to 100,000 points per second.
- *Trending:* the capacity for displaying the variation of one or more process parameters over a particular time period. It is useful if the scales for these displays are adjustable. This capability is extremely helpful in troubleshooting and problem-solving.
- *Determination of statistical measurements:* (such as means, standard deviations, control limits) on process and product parameters.
- *Alarms for out-of-spec data:* and/or indications of assignable causes of variation.
- *Recipes:* the capacity for storing important process parameters for different products. This allows previous process conditions to be reproduced quickly and reliably.
- *Production summaries:* the capacity for following the amount of material being produced and presenting summaries for a shift, a day, a week, or other time period. This is a useful management tool for analyzing productivity on different production lines.

The following features must be considered:

- *User friendliness:* The system should be easy to use, intuitive, and should not require extensive training for operators to become accustomed to the system.
- *Accessibility:* Is access from a remote computer possible? With remote access, integration into a plant-wide information control system is possible.
- *Connectivity:* Can the DAS software work together with other software packages?
- *Upgradability:* Can new upgraded versions of software and/or hardware be readily implemented?
- *Cost:* A cost analysis should show that savings, in terms of improved production efficiency and quality, will outweigh the cost of a DAS.

The different methods of data collection each have their advantages and disadvantages. Check sheets are slow and prone to human errors and transcription errors, but they are also inexpensive, flexible, efficient for small amounts of data, and easy to use. Fixed-station DASs are fast and not prone to human and/or transcription errors, and they can handle large amounts of data. However, they tend to be expensive and more difficult to use. Thus, check sheets and fixed-station DASs have complementary areas of applications. Portable data collectors fall in between check sheets and fixed-station DASs. Table 2.1 shows a comparison of different data recording methods.

Table 2.1 Comparison of Data Recording Methods

	Check sheets	Portable data collectors	Fixed-station DAS
Speed	Slow	Fair	Fast
Human errors	Substantial	Small	Very small
Transcription errors	Substantial	No	No
Efficient for	Small amount of data	Large amount of data	Huge amount of data
Cost	Minor	Fair	Substantial
Flexibility	Very good	Fair	Good
User skill required	Low	Medium	High

For plant-wide monitoring systems, the extrusion process can be integrated with upstream and downstream operations. Bar coding can be used to achieve part traceability and can be integrated with automatic product tracking and warehousing. With the use of these systems, accurate data, such as throughputs cycle times, yields, efficiencies, scrap, labor allocation, and downtime, can be collected for job costing. Such information is essential for efficient plant management. Functions such as preventive maintenance, production scheduling, inventory, production reporting, order entry, job histories, real-time alarms, quality control, and SPC can be included in plant-wide monitoring systems.

Some plant-wide monitoring systems allow changes in such parameters as screw speed or barrel temperature to be made by remote control from a central terminal. Obviously, the results from these changes must be monitored carefully, because they can make the process either better or worse.

2.3 Light Microscopy

This analysis technique allows the observation of a polymeric sample to obtain information about its structure, processing, or manufacturing. Process failure or causes of damage may also be observed. Consequently, it is a very common technique for quality control, troubleshooting, and failure analysis.

A high-magnification microscope possesses magnification ranges of 50, 100, 200, and 500× with illumination from transmitted, reflected, or polarized light. A low magnification microscope will have magnification ranges from 7 to 80 times with the same light sources. There is also the possibility of a three-dimensional view. For sample preparation, a microtome with a range from 0 to 340 microns with 0.5 microns resolution and a polishing machine are required. An example of a light microscope is shown in Figure 2.2.

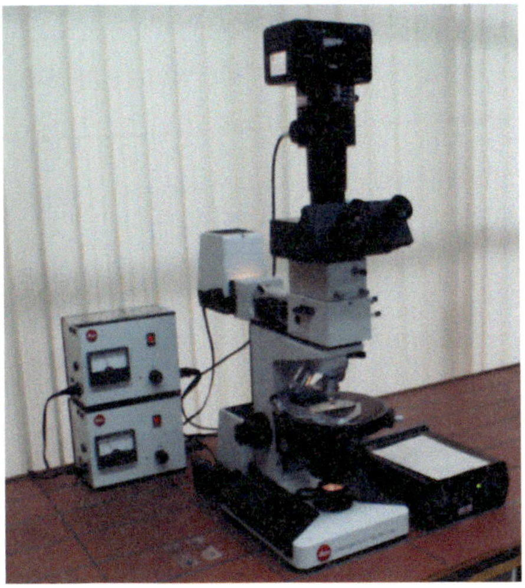

Figure 2.2
Light microscope

The application field for this technique includes:

- Analysis of defects in plastic products
- Observation of dispersion of a filler or a pigment in a polymeric matrix
- Visualization of residual stresses under polarized light
- Dimensional studies
- Detection of differences due to orientation
- Crystallization analysis
- Polymer reinforcement analysis
- Coextruded film analysis using colorimetry

The use of colorimetry involves the treatment of a sample to impart color to selected materials in the sample. For instance, in the absence of colorants, it can be difficult to distinguish the different layers in multi-layer film when natural polymers are used. Certain chemicals color some polymers but not others. For example, iodine will color nylon-6 red, ethylene-vinyl alcohol (EVOH) brown, typical tie layers gray, but will not change the color of polyethylene (PE) or polyethylene terephthalate (PET). Examples of iodine staining are discussed in Case Studies 4.7 and 4.14 of Chapter 4.

2.4 Thermochromic Materials

Thermochromic substances can be very useful in solving temperature-related problems. Because such materials change color irreversibly at a particular temperature, they are used in a number of applications. For example, thermochromic paints are used on heat-shrinkable sleeves in order for the installer, who is heating the sleeve with a torch, to see when the product has reached the proper installation temperature.

It is difficult to measure actual polymer melt temperatures inside an extruder. An immersion probe cannot be used because the rotating screw will shear it off. Flush-mounted temperature sensors do not give a good indication of the actual stock temperature. However, thermochromic materials can be added to the feedstock to determine whether the stock temperatures in the extruder exceed the color-transition temperature of the material.

Quite some time ago, Mennig [4] published a preliminary investigation into the use of thermochromic materials to indicate the temperature of materials in extrusion. The thermochromic materials used in the study were not clearly identified, and the study was flawed because the color-transition temperature was time-dependent. Obviously, this limits the usefulness of these particular thermochromic materials for accurate temperature indication.

Rauwendaal [5] employed thermochromic powders used commercially in thermochromic paints to study melt temperature in extruders. The study involved a small laboratory twin-screw extruder processing a highly filled high density polyethylene (HDPE). Though the barrel temperatures and exit melt temperature were quite reasonable, the material degraded inside the extruder.

The barrel temperatures were set at 200 °C and the exit melt temperature was measured at about 215 to 220 °C. These are normal temperatures for HDPE and should not cause degradation under normal circumstances, but it was suspected that the melt temperatures inside the extruder were much higher than the barrel temperature. Therefore, a thermochromic powder with a color-transition temperature of 250 °C was added to the feed. With this material, a very clear discoloration was noticed in the material exiting the die. Next, a thermochromic powder with a color-transition temperature of 300 °C was added. Again there was a very clear discoloration indicating that stock temperatures in the extruder exceeded 300 °C.

At this point the cause of degradation was quite clear. To determine the location of the high stock temperatures, the screws were pulled, and the color change along the screws was visually observed. The color change was found to occur in the screw section with high-restriction kneading blocks. Clearly, the shearing action in this part of the screw caused excessively high melt temperatures, resulting in degradation of the polymer. Based on these findings, the screw geometry was changed to provide less severe mixing and the problem disappeared. In this study, the thermochromic powders proved to be very useful because they provided information that was very difficult to obtain by any other means.

Thermochromic powders are not readily available; however, several companies produce thermochromic paints for specialty products such as heat-shrinkable sleeves. It would be helpful if a troubleshooting kit were available containing several thermochromic powders with different color transition temperatures.

2.5 Thermal Analysis, IR Spectroscopy, and Rheometry

In thermal characterization, a controlled amount of heat is applied to a sample, and its effect is measured and recorded. In isothermal operations, the effect is recorded as a function of time at constant temperature. In a programmed temperature operation, the temperature is changed in a predetermined fashion (at a certain rate, for example), and the effect is recorded as a function of temperature. General texts on thermal characterization have been written by Wendlandt [6], Daniels [7], Turi [8], Miller and Porter [9], and Mathot [10].

2.5.1 Differential Thermal Analysis and Differential Scanning Calorimetry

Differential thermal analysis (DTA) and differential scanning calorimetry (DSC) are similar techniques measuring change in the heat capacity of a sample. These techniques can be used to determine such information as various transition temperatures (T_m, T_g, T_α, T_β, etc.), specific heat, heat of fusion, percent crystallinity, onset-of-degradation temperature, induction time, reaction rate, and crystallization rate.

A DSC instrument operates by compensating electrically for a change in sample temperature. The power for heating is controlled in such a way that the temperature of the sample and the temperature of the reference are the same. A DSC curve displays the heat flow along the vertical axis against the temperature along the horizontal axis. A DTA instrument operates by measuring the change in sample temperature with respect to an inert reference. Newer DTA instruments with externally mounted thermocouples and reproducible heat paths have a precision comparable to DSC. Older DTA instruments with thermocouples placed in the sample are less accurate and their readings are less reproducible.

A DSC instrument (see Figure 2.3) measures the heat flow absorbed or produced by the plastic sample under a programmed temperature profile with a heat resolution of 3 µW and a temperature accuracy of ±0.1 °C. The heat flow is measured by monitoring the temperature difference between the sample and a reference material. This technique allows the acquisition of information about polymer thermal transitions, such as the glass transition temperature, the melt point, the crystallization temperature, and the degradation point. The plastic sample can be analyzed under an inert or oxidative atmosphere depending on the specific test or troubleshooting case. The typical temperature for this equipment ranges between −100 and 500 °C. The application field for this technique includes:

- Observation of first-order transitions according to International Organization for Standardization (ISO) and American Society for Testing and Materials (ASTM) standards. The first-order transitions are polymer melting and crystallization.
- Observation of second-order transitions according to ASTM standards. Second-order transitions are glass transition and residual stress relief.
- Thermal analysis together with other analytical techniques can be used for characterization of single polymers, blends, and coextruded films. Normally DSC is used to confirm or complete the analysis of Fourier transform infrared (FTIR), and light microscopy is used to observe its morphology.
- Observation of heat capacity as a function of temperature.
- Observation of thermal stability or oxygen induction time (OIT).
- Purity analysis of varied substances.
- Determination of reaction kinetics can also be analyzed using the heat of reaction.
- Determination of the kinetics of cure of thermosets.
- Evaluation of the compatibility of polymers and elastomers using the glass transition of the blend. A unique glass transition proves a perfect compatibility.

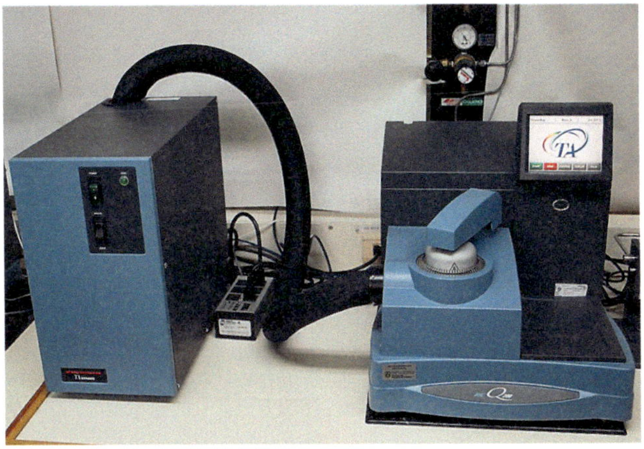

Figure 2.3 Differential scanning calorimeter

2.5.2 Thermogravimetric Analysis

A thermogravimetric analyzer (TGA) (see Figure 2.4) measures the weight changes of a plastic sample under a programmed temperature profile with a weight resolution of 0.1 µg and a weight accuracy of ±0.1%. This technique makes it possible to obtain important information about the degradation mechanisms of polymers including the temperature and kinetics of decomposition.

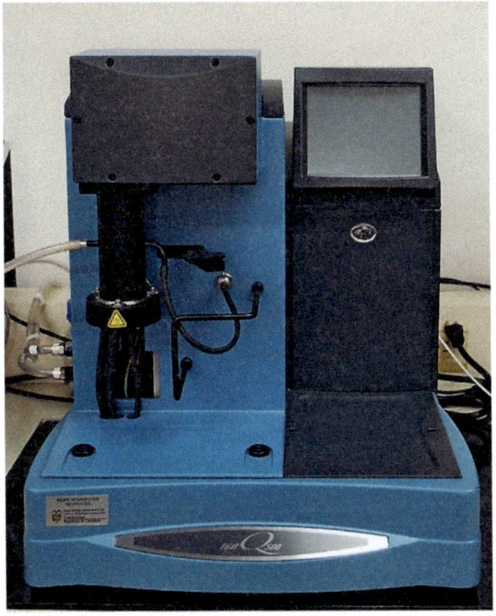

Figure 2.4 Thermogravimetric analyzer

The TGA also reveals significant details about the composition and the presence of additives in polymeric materials. The plastic sample can be analyzed under an inert or oxidative atmosphere, depending on the specific test or troubleshooting case. The typical temperature range of this equipment is between 25 and 1000 °C. The TGA is used to measure the amount and loss of moisture or diluent from polymers, as well as rates and temperatures of reactions. It is a convenient instrument for determining the polymer induction time. The sample size is usually less than one gram; thus, the amount of polymer required for characterization is minimal.

The application field for this technique includes:

- Thermogravimetric analysis under ISO and ASTM standards.
- The acquisition of useful information, such as the amount of volatile matter, polymer, inorganic filler, and reinforcement.
- The acquisition of not only the content, but also the type of filler or reinforcement present in an unknown sample using TGA coupled with FTIR.
- The analysis of evolved gases (also using TGA coupled with FTIR).
- Determination of the kinetics of decomposition of polymeric materials.

2.5.3 Fourier Transform Infrared Spectroscopy

All matter is composed of atoms or molecules that can be excited by infrared radiation. The spectrum of infrared absorption is a characteristic property of each substance. If the identification of an unknown plastic sample is required, FTIR spectroscopy can be used. An FTIR spectrophotometer is shown in Figure 2.5. Identification of additives, fillers, and reinforcements may be another important application of FTIR spectroscopy. The presence of specific vibrations in the infrared can be used for monitoring chemical reactions such as polymerization, crosslinking, or grafting, or even to suggest a reaction mechanism. The use of standards of known compositions makes it possible to calibrate the FTIR technique for quantitative analysis of polymers and additives.

The main characteristics of the FTIR equipment are:

- A wave number range of 7400 to 350 cm^{-1}.
- A wave number resolution of 0.5 cm^{-1}.
- The availability of accessories for special purposes. Several examples are attenuated total reflectance (ATR) for liquids and solid surface IR analysis (with less than 1 μm sample, nondestructive); transmissions of gases, solids, and liquids; diffuse reflectance of solids; and photoacoustic cell for small solid samples (nondestructive IR analysis).

Figure 2.5 Fourier transform infrared spectrophotometer

The application field of this technique includes:

- Identification of polymers and additives of unknown samples (depending on the sample complexity, additional analysis using DSC, TGA, or extraction could be required).
- Identification of elastomers and fillers in unknown rubber compounds (again, additional analysis using DSC, TGA, or extraction could be necessary).
- Characterization of coextruded films. This type of analysis also requires microscopy and DSC evaluation.
- Analysis of blooming problems in rubber compounds.
- Quantitative analysis of polymers and additives using appropriate standards of known composition.
- Troubleshooting of degradation problems in plastic and rubber materials. Degradation sometimes includes the appearance of molecular groups containing oxygen; such groups can be evaluated using the carbonyl index. Degradation of some polymers, including polyvinyl chloride (PVC), produces double bonds that can be observed in the IR spectrum.
- Evaluation of very complex compounds such as flexible PVC using FTIR with proper dissolution and extraction in specific solvents. With this technique it is possible to identify polymer modifiers, heat stabilizers, lubricants, fillers, and the *K*-value (a parameter related to viscosity) of the PVC resins.

2.5.4 Thermomechanical Analysis

Thermomechanical analysis (TMA) consists of measuring changes in the mechanical properties of extrusion materials as a function of temperature and/or time. A probe in contact with the sample moves as the sample undergoes dimensional changes. The movement of the probe is measured with a Linear Variable Displacement Transducer, LVDT. The sample deformations that can be measured are compression, penetration, extension, and flexure or bending.

2.5.5 Torque Rheometry

A torque rheometer (see Figure 2.6) measures the resistance of a polymeric sample to being mixed. This resistance is proportional to the sample's viscosity and is measured as the torque experienced by the rotors as a function of time. Although the torque rheometer is designed for optimal mixing, and at these conditions the response of the polymer melt and rubber compound are nonlinear and complex viscoelastic, its measurements approximate real-life process conditions.

The torque rheometer's measurements supply very important information regarding the compatibility of resins, the optimization of mixing compounds, the efficiency of additives in compounds, and polymer recycling. For thermoplastics, this information can include dynamic stability, mixing and processability of compounds, dispersion of filler and pigments, recycling studies, and the kinetics of decomposition or degradation of polymer.

Figure 2.6
Torque rheometer

The application field for this technique includes:

- Optimization studies of the mixing process of PVC and other plastic compounds. Two criteria must be considered: minimal power consumption and a high level of homogeneity of compounds.
- Evaluation of the efficiency of such additives as process aids, peptizers, processing and recycling stabilizers.
- Analysis of filler dispersion.

- Thermal stability measurements.
- First order decomposition kinetics.

The torque rheometer is a useful tool for developing new compounds to meet specific requirements. This device is wrongly denoted as rheometer, since the flow pattern inside is so complex that it is not possible to measure stress and deformation rate components individually.

2.5.6 High Pressure Capillary Rheometry

A high pressure capillary rheometer measures the flow properties of polymer melts; it comprises a capillary tube of specified diameter and length (e.g., circular capillary) for measuring differential pressures and flow rates. The results are presented as graphs of shear stress against shear rate at constant temperature or shear viscosity against shear rate at constant temperature. The flow through the capillary can be generated by controlled pressure or by controlled volume displacement inside a cylinder connected to the capillary. Some manufacturers install a small extruder to the cylinder to guarantee a permanent and homogeneous supply of material (Figure 2.7).

Figure 2.7
High pressure capillary rheometer

In general, rheometers are built to produce simple isothermal flows or simple deformation rate patterns. Therefore, viscosity can be expressed with very simple mathematical expressions. For the construction of rheometers and for their use it is pursued to have only one component for stress and deformation rate. With these rates the "shear viscosity" can be calculated easily. Capillary rheometry deals with the experimental description of the material flow through a channel with a geometrically well defined cross section (circular, rectangular, or annular), where the characteristic dimensions (diameter, width, height) are small enough compared with the channel or capillary length.

The flow in a capillary rheometer is a steady-state flow. These devices allow the study of shear viscosity as a function of temperature and shear rate in a wide range, usually from 10^1 to 10^4 s^{-1}.

A viscosity value obtained from the equation $\tau_{wap} = \eta_{ap} \cdot \dot{\gamma}_{wap}$ is only valid for Newtonian fluids; where τ_{wap} is the apparent shear stress at the wall, η_{ap} is the apparent shear viscosity and $\dot{\gamma}_{wap}$ is the apparent shear rate at the wall. We use the term apparent shear viscosity when applied to non-Newtonian fluids such as polymers. To obtain the actual shear viscosity, the apparent viscosity should be corrected. The Weissenberg-Rabinowitsch correction is the most used method for it. To apply this correction method it is necessary to run several measurements under different shear rates, using the same capillary and keeping a constant polymer temperature for all runs.

Rectangular channels are used instead of circular channels to obtain low shear rates. The bigger dimension of those capillaries make it possible to measure the gradient pressure in the channel region where the entrance effects are not present.

The application field for this technique includes:

- Viscosity curves of thermoplastic materials at several constant temperatures. This information is used as an important input for designing extrusion screws and extrusion dies, and to predict polymer processability.
- Viscosity curves deliver important information about molecular structure of polymers, for example, estimation of average molecular weight and molecular weight distribution (MWD).
- A rheological model of polymer materials can be easily obtained for simplicity of data handling, in particular for computer simulations and screw and die design. Possible rheological models to be used are: Carreau, Carreau-Yasuda, Power law, Vinogradov, and others.
- Flow instabilities or anomalies can be observed at a specific range of shear stresses. This information is used as an important input for designing extrusion screws and extrusion dies.
- Rheological data can be used to evaluate polymer compatibility between a thermoplastic resin and a masterbatch and to select an adequate resin for co-extrusion applications.
- Analysis of polymer viscoelasticity is possible by measuring the first normal stress difference.
- The ratio between diameter of capillary and strand, called die swell ratio, is used to characterize elastic behavior of polymers.
- Evaluation of processing aids in thermoplastic and rubber compounds.

The capillary rheometer is a useful tool for troubleshooting, process monitoring, and developing new screws and dies to meet specific requirements.

2.5.7 Rotational Rheometry

A rotational rheometer measures the flow properties of a polymer melt that is deformed between two bodies, where one of them remains still and the other one rotates. The bodies can be two concentric cylinders (known as a Couette Rheometer), where generally only the internal cylinder rotates, or two flat, horizontal discs, where the bottom disc is stationary.

The most used rotational rheometers for characterization of polymer melts use discs. One of the discs, the rotating one, may be a circular flat plate or a cone. The stationary disc is flat. The combination of cone and plate has the benefit of producing a constant shear rate in the whole region between the two discs. Both discs are placed in an open space inside an oven (Figure 2.8). This limits the rotation speed, especially for substances with low viscosity at test temperature.

Figure 2.8 Rotational rheometer

The rotational rheometers do not have restrictions for low shear rates (only limited by type of transducers and control unit). They deliver more information about material rheological behavior than capillary rheometers do, for example, information on the visco-elastic behavior of the polymer melt.

This rheological information is important for structural rheology, which deals with the relationship between molecular structure of polymers (for example, molecular weight, molecular weight distribution, MWD, molecular branches) and their rheological behavior.

Because of the viscoelastic properties of polymer melts during the shear deformation of the material between the discs it appears a normal force, which tries to separate both discs (known as Weissenberg effect). This normal force to the discs creates primary and secondary normal stresses than can be expressed like shear viscosity as a function of shear rate and temperature.

There are two basic types of rotational rheometers: rheometers with controlled stress are called controlled stress rheometer CSR (i. e., controlled torque) and rheometers with controlled rate of deformation are known as controlled rate rheometer CRR (i. e., controlled rotational speed). In most cases, existing rotational rheometers can operate in both modes.

Rotational rheometers can carry out stationary, oscillatory, and transient tests, depending on what is controlled and measured: torque (stresses) or angular velocity (deformations).

The rotational rheometers under stationary mode for steady-state measurements are suitable to obtain the shear viscosity for very low shear rates – typical measurements from 10^{-2} to 10^{2} s^{-1} – and are an excellent complement to measurements with capillary rheometers.

Shear viscosity can be obtained as the quotient of shear stress and shear rate. For a cone and plate rheometer, where one disc is rotating with a constant angular velocity Ω and the cone angle θ_0 is very small (< 5°), the shear rate $\dot{\gamma}$ is constant in the region between discs and is given by the expression

$$\dot{\gamma} = \frac{\Omega}{\theta_0} \tag{2.1}$$

The shear stress can also be considered as constant in the region between discs and can be obtained from the measured torque T using the equation

$$\tau = \frac{3T}{2\pi R^3} \tag{2.2}$$

For a parallel plate rheometer, where one disc, with radius R, is rotating with a constant angular velocity Ω and the distance between plates is h, the shear rate $\dot{\gamma}$ can be obtained using the following equation and assuming a Newtonian fluid.

$$\dot{\gamma} = \frac{\Omega \cdot R}{h} \tag{2.3}$$

The shear stress is given by the expression

$$\tau = \frac{2T}{\pi \cdot R^3} \tag{2.4}$$

The Weissenberg-Rabinowitsch correction for the viscosity is required because the equations above are only valid for Newtonian fluids. Parallel plate rheometers have the disadvantage to need the Weissenberg-Rabinowitsch correction, but they are more suitable to measure viscous materials at low shear rates.

Figure 2.9 shows the use of several rheometrical techniques (rotational, rectangular, and circular capillaries) to obtain a complete viscosity curve of a polymeric material or polymeric formulation.

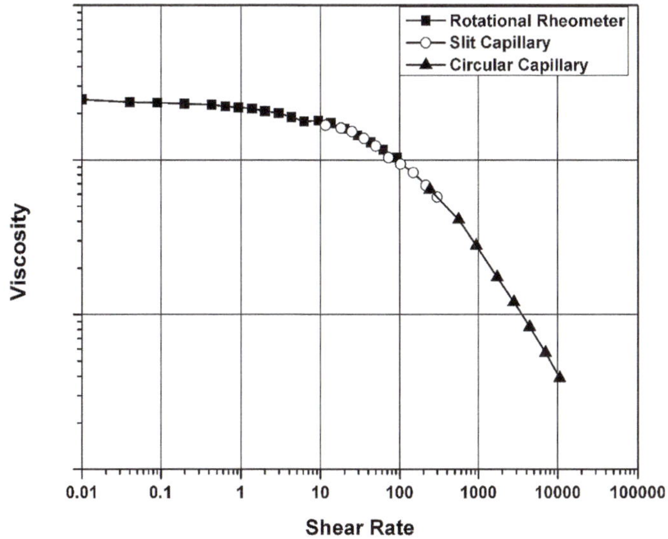

Figure 2.9 Viscosity curve of a polyolefin using rotational and capillary rheometry

The measurements from rotational rheometers as a complement of the measurements from capillary rheometers are a useful tool for troubleshooting the different extrusion processes, monitoring and process control, improving actual screws or designing new screws and dies to meet specific necessities or solve problems.

2.5.8 Other Thermal Characterization Techniques

While TMA refers to the measurement of a static mechanical property, there are also techniques that employ dynamic measurement. In torsional braid analysis (TBA), a sample is subjected to free torsional oscillation. The natural frequency and the decay of oscillations are measured, providing information about the viscoelastic behavior of materials. However, the procedures for these measurements are elaborate and time-consuming. In dynamic mechanical analysis (DMA), a sample is exposed to forced oscillations. Many useful properties can be measured by this technique.

Thermal optical analysis (TOA) – also known as thermal depolarization analysis (TDA) – and the depolarization light intensity (DLI) method measure the conversion of plane polarized light to elliptically polarized light in semi-crystalline polymers. The intensity of the depolarized light transmitted through a sample is a function of the level of crystallinity. Melting and recrystallization phenomena can be analyzed, but the technique does not appear to be sensitive to glass transitions [11].

2.6 Miscellaneous Tools

Many basic tools are useful for troubleshooting. A tape measure can be used to measure distances from 20 cm to several meters, and dial calipers are useful to measure such items as extruded products, dimensions of the extruder screws, and extrusion tooling with greater accuracy. A stopwatch is an indispensable tool for measuring screw speed or line speed, for blender calibration, and for other rate measurements. A scale can be used to measure the output of an extruder. A small voltmeter is very useful in making sure that voltage and resistance levels of various components are at their required values, and a millivolt source can be used to verify thermocouple inputs and to check controller response and line continuity.

A digital still camera is a valuable tool in troubleshooting for accurate visual documentation of the information such as configuration of the extrusion line, details of the screw geometry, die geometry, wear regions on the screw, contamination in the product, or unusual buildup on the screen pack. It can also be used for preparation of clear documentation of operating procedures and other processes, and the digital photos can easily be transferred to a personal computer and used in reports, manuals, presentations, and such.

A digital video camera is also useful in troubleshooting to capture motion allowing for instant determination of the frequency of pulsing of the extrudate coming out of the die. It can record sound as well as images, and the soundtrack can be analyzed separately. For instance, the soundtrack of a high-frequency vibration can be captured on video, downloaded to a personal computer, and used to determine the frequency of the vibration. Digital still cameras capture a static event with better quality, but a video camera is an excellent tool for capturing dynamic events. In troubleshooting, the dynamic events are often the most important.

2.6.1 Infrared Thermography

Infrared thermography or thermal imaging is an effective tool to capture the invisible infrared energy naturally emitted from all objects. Infrared radiation is part of the electromagnetic spectrum extending from 0.7 microns to 1000 microns, which includes radio waves, microwaves, visible light, ultraviolet (UV), gamma, and X-rays. In fact, the 0.7 to 14 micron band is used for IR temperature measurement.

A thermographic camera (see Figure 2.10) detects radiation in the infrared range and produces images of that radiation, so-called thermograms. The emissivity depends on the material to be measured and its surface characteristics. Therefore, the IR camera requires in-situ calibration.

The portable IR camera can easily determine extruder overheating problems, random shutdown of heating bands, uniformity of extrudate melt temperature, extrudate calibration and cooling, among others.

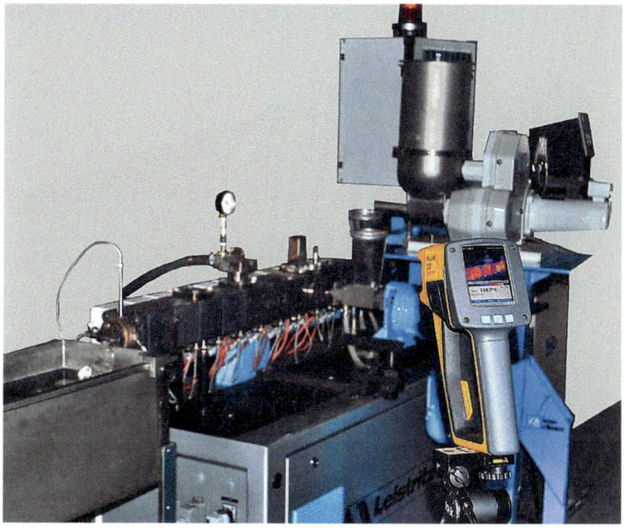

Figure 2.10 IR thermographic camera

2.6.2 The Smartphone

The smartphone is ubiquitous today but it has not been around very long. Smartphones did not really take off until around the year 2000. Today it is hard to imagine life without a smartphone. It turns out that the smart phone is one of the most useful tools in troubleshooting because it is truly a multi-functional device. Here is a partial list of functions that smartphones can perform in troubleshooting:

- Photo camera
- Video camera
- Timer
- Stopwatch
- Calculator
- Sound recording
- Remote access
- Flashlight
- Microscope
- Profile dimensions
- IR camera
- Strobe tachometer

Some of these capabilities require a plug-in device. Figure 2.11 shows the ProScope microscope attachment that turns an iPhone into a microscope with a magnification range from 20 × to 80 ×.

Figure 2.11 Microscope attachment for an iPhone

The smartphone can become an infrared camera by plugging in an infrared unit. This is a very useful and powerful tool for troubleshooting extrusion problems. Figure 2.12 shows an infrared unit made by FLIR that plugs into an iPhone.

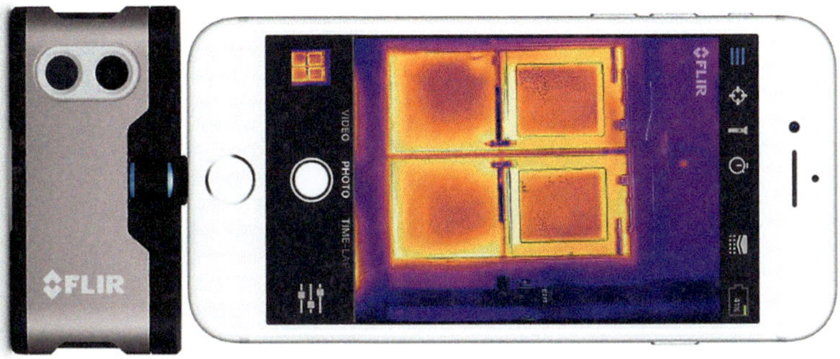

Figure 2.12 FLIR One for iOS personal thermal imager

The strobe tachometer measures the magnetic field from a spinning magnet. The magnetic field is measured by the iPhone's magnetometer. This allows measurement of rotational speed from 6 rpm to 60,000 rpm.

The smartphone can also be used to determine dimensions of extruded products. Perdikoulias presented a paper [178] at the Extrusion 2017 conference on software developed that allows smartphones to measure dimensions of extruded profiles.

2.6.3 Power Measurements

Power measurements, such as voltage, amperage, and frequency, are useful to verify thermocouple and energy inputs. Power measurements over time are important to check that energy parameters are at their required values.

The application field for this tool includes:

- Voltage and amperage measurements per phase (AC motors)
- Detection of power harmonics
- Identification of power quality problems
- Determination of power factor
- Monitor efficiency improvements based on energy consumption
- Monitor demand at defined periods

A portable tester for power measurements is shown in Figure 2.13.

Figure 2.13 Portable tester for power measurements (courtesy Fluke)

3 Systematic Troubleshooting

3.1 Upsets versus Development Problems

This chapter will primarily focus on upsets, problems that occur in an existing extrusion line for an unknown reason. If the extrusion line had been running fine for a considerable period of time, then it is clear that there must be a solution to the problem. Thus, the objective of troubleshooting is to find the cause of the upset and eliminate it. On the other hand, there may be no solution to a development problem. Solving a development problem involves establishing a condition that has not been achieved before. If it is physically impossible to establish the desired condition, then, clearly, there is no solution to the problem. A functional analysis of the process should make it possible to determine the bounds of the conditions that can be realized in practice.

3.2 Machine-Related Problems

In machine-related problems, mechanical changes in the extruder cause a change in extrusion behavior. These changes can affect the drive system, the heating and cooling system, the feed system, the forming system, or the actual geometry of the screw and the barrel. The main components of the drive are the motor, the reducer, and the thrust-bearing assembly. Drive problems manifest themselves as variations in rotational speed and/or the inability to generate the required torque. Problems in the reducer and thrust bearings are often associated with clear audible signals of mechanical failure. If the problem is suspected to be the drive, make sure that the load conditions do not exceed the drive capacity.

3.2.1 The Drive System

Older motor drive systems generally consist of a direct current (DC) brush motor, a power conversion unit (PCU), and operator controls. A frequent problem with the motor itself is worn brushes; these should be replaced at regular intervals as recommended by the manufacturer. The manufacturer's recommendations should also be followed in troubleshooting an extruder drive. A typical troubleshooting guide for a DC motor is shown in Table 3.1.

Table 3.1 Troubleshooting Guide for DC Motor

Problem	Possible cause	Action
Motor will not start	Low armature voltage	Make sure motor is connected to proper voltage
	Weak field	Check for resistance in the shunt field circuit
	Open circuit in armature or field	Check for open circuit
	Short circuit in armature or field	Check for short circuit
Motor runs too slow	Low armature voltage	Check for resistance in armature circuit
	Overload	Reduce load or use larger motor
	Brushes ahead of neutral	Determine proper neutral position for brush location
Motor runs too fast	High armature voltage	Reduce armature voltage
	Weak field	Check for resistance in shunt field circuit
	Brushes behind neutral	Determine proper neutral position for brush location
Brushes sparking	Brushes worn	Replace
	Brushes not seated properly	Reseat brushes
	Incorrect brush pressure	Measure brush pressure and correct
	Brushes stuck in holder	Free brushes, make sure brushes are of proper size
	Commutator dirty	Clear commutator
	Commutator rough or eccentric	Resurface commutator
	Brushes off neutral	Determine proper neutral position for brush location
	Short circuit in commutator	Check for shorted commutator, and check for metallic particles between commutator segment
	Overload	Reduce load or use larger motor
	Excessive vibration	Check driven machine for balance
Brush chatter	Incorrect brush pressure	Measure and correct
	High mica	Undercut mica
	Incorrect brush size	Replace with proper size
Bearings hot	Belt too tight	Reduce belt tension
	Misaligned	Check alignment and correct
	Bent shaft	Straighten shaft
	Bearing damage	Inspect and replace

3.2.2 The Feed System

The most important component of the feed system in a flood-fed extruder is the feed hopper and its stirrer and/or discharge screw. A mechanical malfunction of this system can be determined by visual inspection. If the feed hopper is equipped with a discharge screw (crammer feed), the speed of the discharge screw should be checked for unusual variation. For proper functioning, the drive of a crammer feeder should have a torque feedback control to ensure constant feeding and to avoid overfeeding.

Many extruders have square feed hoppers with rapid compression in the converging region. Extruder manufacturers often choose this geometry because of the ease of manufacture, but this hopper geometry does not promote steady flow. When flow instabilities occur in a feed hopper, the extruder operator will often hit the hopper with a heavy object to get the flow going again. As a result, hoppers that cause flow problems often show signs of abuse such as surface damage, dents, scrapes. Such damage is a strong indication of poor feed hopper design.

3.2.3 The Heating and Cooling System

The heating and cooling system exercises a certain degree of control of the polymer melt temperature. However, stock temperature deviations do not necessarily indicate a heating or cooling problem because heat is transferred directly to (or removed from) the barrel and only indirectly to (or from) the polymer. It is actually the barrel temperature that is controlled. The local barrel temperature, as measured with a temperature sensor, determines the amount of barrel heating or cooling.

The stock temperature is generally controlled by changing the setpoint of the temperature zones along the extruder. However, due to the slow response of the melt temperature to changes in heat input, only very gradual stock temperature changes can be effectively controlled by setpoint changes. Rapid stock temperature fluctuations (cycle times less than about five minutes) usually cannot be reduced via a melt temperature control system. Such fluctuations are indicative of conveying instabilities in the extrusion process and can only be effectively reduced by eliminating the cause of the conveying instability.

The heating system can be checked by changing the setting to a much higher temperature, for instance 50 °C above the regular setting. The heater should turn on at 100% power, and the measured barrel temperature should begin rising in about one to two minutes. If the heater does not turn on at full power, the barrel temperature measurement is in error, or there is a problem with the electronic circuit of the temperature controller. If the heater turns on at full power but the temperature does not start to rise within two to four minutes, either the barrel temperature measurement is incorrect, or there is poor contact between the heater and barrel. The cooling system can be checked similarly by changing the setting to a much lower temperature, for instance 50 °C below the regular setpoint. If the cooling does not turn on at full capacity, the barrel temperature measurement is in error, or there is a problem in the circuit of the temperature controller. If the cooling turns full on at full capacity but the temperature does not start to drop within two to four minutes, either the barrel temperature is incorrect or the cooling device is inoperable. This checkout procedure is summarized in Table 3.2.

Table 3.2 Heating and Cooling System Check

Heating system: Increase setpoint of temperature zone by 50 °C	
Heater turns on full blast and the barrel temperature rises in about 2 minutes	Heating system normal
Heater turns on full blast but the barrel temperature does not change	Poor contact of heater to barrel, insufficient heating capacity, temperature sensor failure
Heater output does not change	Heater failure, controller bad
Cooling system: Reduce setpoint of temperature zone by 50 °C	
Cooling on full blast and the barrel temperature drops in about 2 minutes	Cooling system normal
Cooling on full blast but the barrel temperature does not change	Temperature sensor failure, insufficient cooling capacity, cooling system not functioning at all
Cooling output does not change	Cooling system bad, controller bad

If a substantial amount of cooling is required to maintain the desired stock temperature, this is generally a strong indication of excessive internal heat generation by frictional and viscous dissipation. Internal heat generation can be reduced by lowering the screw speed or by changing the screw design. The main screw design variable that affects viscous heating is the channel depth. Increasing the channel depth will reduce shear rates and viscous heating. Mechanical changes in the forming system are tied to the extrusion die and downstream equipment. These elements can be subjected to simple visual inspection to detect mechanical changes. Changes in the geometry of the screw and/or the barrel are often caused by wear. Because wear is a very important element in the performance of extrusion machinery, it will be discussed in detail in Section 3.2.5.

3.2.4 How Screw Design Can Affect Extruder Performance

One possible cause of a number of extrusion problems is poor screw design. It is helpful to have some rudimentary knowledge of screw design so that a problem caused by poor screw design can be recognized. Screw design is a very large topic and we will not go into the details of screw design. For a detailed discussion on screw design, the reader is referred to the Polymer Extrusion book [176]. However, it is important to have an understanding of some of the main screw design issues.

The screw is the heart of the extruder. The screw performs the following functions:

- Feeding of the plastic particles (usually pellets)
- Conveying of the plastic
- Heating of the plastic
- Melting of the plastic
- Mixing of the plastic
- Degassing of the plastic (in vented extruders)
- Pressure development

Functional Zones of the Extruder

The main functional zones of the extruder are feeding, solids conveying, melting, and melt conveying in the extruder and melt conveying in the die. These zones have to be in balance for the extruder to achieve stable extrusion conditions. The balancing of the functional zones occurs by the pressures that develop over the length of the extruder. For instance, if the solids-conveying rate is higher than the melting rate, the pressure at the end of the solids-conveying zone will increase as shown in Figure 3.1.

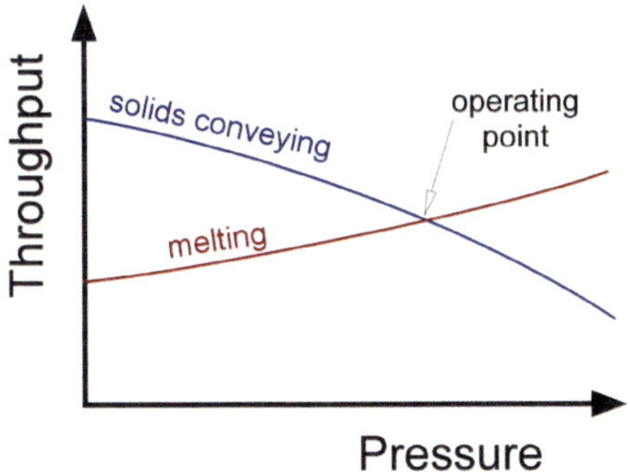

Figure 3.1 Throughput versus pressure solids conveying and melting – stable conditions

The pressure shown in Figure 3.1 is the pressure at the end of the solids-conveying zone; this is also the pressure at the start of the melting zone. An increase in this pressure will reduce the solids-conveying rate and increase the melting rate. When the solids-conveying rate equals the melting rate, the two zones are balanced. This is where the solids-conveying characteristic curve intersects with the melting characteristic curve. This point is called the operating point.

We have unstable conditions when solids conveying cannot reduce to the point where the solids-conveying rate equals the melting rate; this situation is shown in

Figure 3.2. This condition is not unusual in grooved feed extruders. With a grooved feed the solids-conveying rate can be higher than the melting rate because the grooves produce very effective solids conveying. This is why standard compression screws generally do not work in grooved feed extruders and special decompression screws are used. These screws have a very shallow feed section and a metering section that is actually deeper than the feed section.

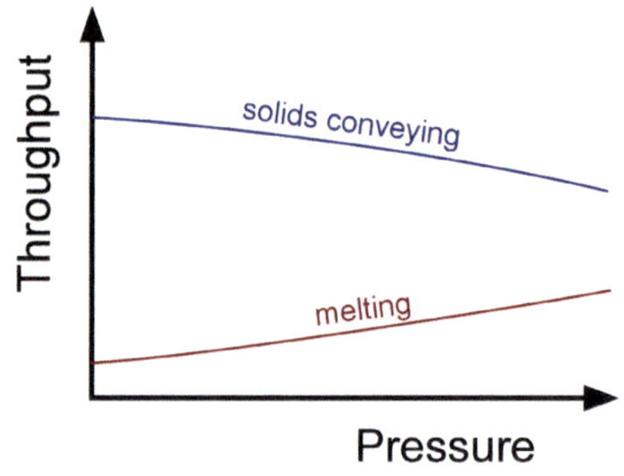

Figure 3.2 Throughput versus pressure solids conveying and melting – unstable conditions

If the solids-conveying rate exceeds the melting rate, we can approach the problem two ways. We can make process changes that will reduce the solids-conveying rate. Alternatively, we can make process changes that will increase the melting rate. Some options to reduce the solids-conveying rate are:

- Starve feeding
- Change barrel temperatures in feed section
- Change screw temperature in feed section
- Change temperature of feed housing

The effect of starve feeding is illustrated in Figure 3.3.

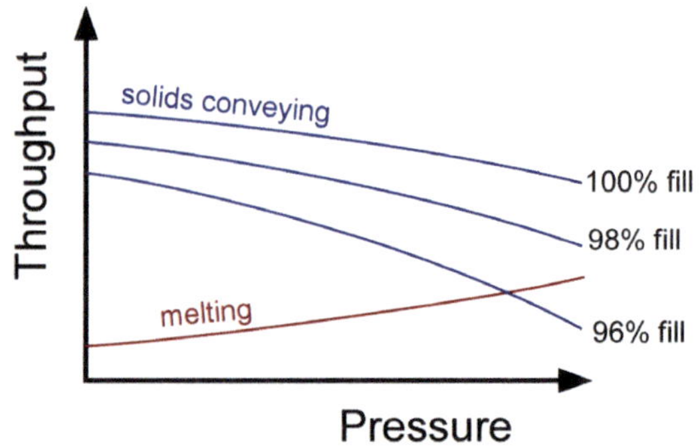

Figure 3.3 Effect of starve feeding on solids-conveying characteristic curve

Figure 3.3 shows that a 4% starvation (96% fill) reduces pressure development in solids conveying to the point where the solids-conveying rate can match the melting rate. Some options to increase the melting rate are:

- Preheating the plastic
- Changing barrel temperatures
- Changing screw temperatures

Figure 3.4 shows the effect of preheating.

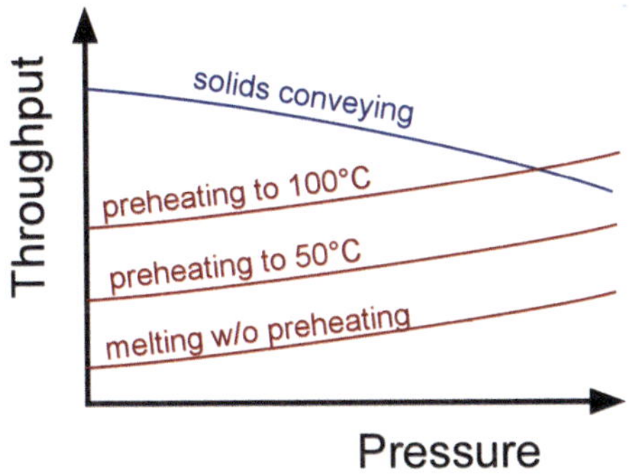

Figure 3.4 Effect of preheating on solids melting characteristic curve

Figure 3.4 shows that preheating to 100 °C increases the melting rate to the point where it matches the solids-conveying rate. This also shows that generally there are several options available to solve an extrusion problem. We can produce similar characteristic curves for melt flow in the extruder and in the die, see Figure 3.5.

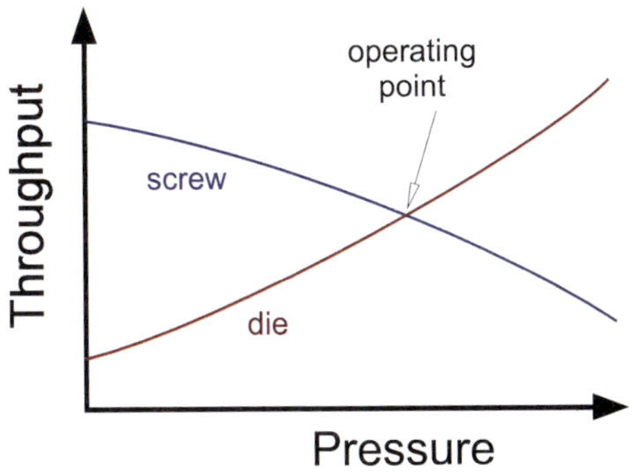

Figure 3.5 Screw and die characteristic curves

In Figure 3.5 the pressure is the die inlet pressure; this is also the extruder discharge pressure. If we have a high restriction die, the screw will have to develop high pressure. This is a common situation in blown film extrusion. If the screw is incapable of generating the diehead pressure, unstable conditions may result and the output may need to be reduced. There are a number of options to remedy this situation:

- Increase die temperatures
- Modify die design
- Change screw design
- Use a gear pump
- Reduce plastic viscosity
- Reduce barrel temperatures in metering section
- Increase screw temperature in metering section

Changing die temperature or barrel temperature is easy. However, such simple changes may not eliminate the problem. In that case, more significant changes may be required.

Screw Design Considerations

A good screw design provides for good solids conveying, melting, mixing, degassing, melt conveying, and pressure development. Degassing is only required in vented extruders. Figure 3.6 shows a standard extruder screw. This type of screw can be referred to as a simple conveying screw.

The screw has a single flight wrapped around the core of the screw. The screw has three geometrical sections, the feed section, the compression section, and the metering section. Some screws have more than three geometrical sections; for example, vented screws have six or more geometrical sections. The flighted length of the screw can vary from 20*D* to 40*D* (*D* is the screw diameter) or longer. Most extruder screws have a flighted length between 24*D* and 35*D*. A screw with a single compression section is referred to as a single stage screw. A screw with two compression sections is referred to as a two-stage screw; these screws are used in vented extruders.

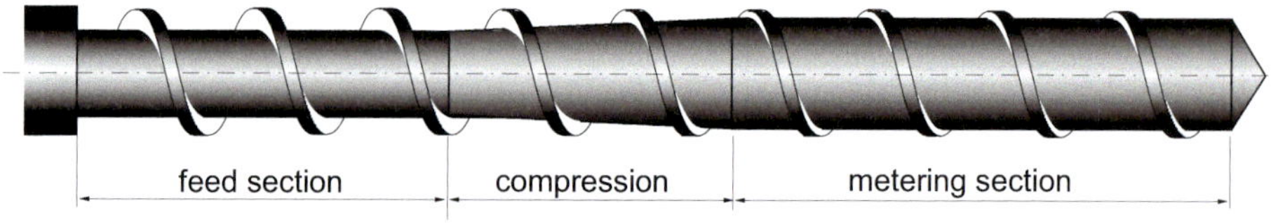

Figure 3.6 Standard extruder screw

The feed section has a large channel depth, while the metering section has a smaller channel depth. The ratio of the feed depth to the metering depth is usually referred to as the compression ratio even though it is actually a channel depth ratio. In the compression section the channel depth gradually reduces from the feed depth to the metering depth.

Some of the important screw design variables are:

1. Screw diameter
2. Screw length
3. Length feed section
4. Number of flights in feed section
5. Flight helix angle in feed section
6. Flight width in feed section
7. Flight flank radius in feed section
8. Length of compression section
9. Number of flights in compression section
10. Flight helix angle in compression section
11. Flight width in compression section
12. Flight flank radius in compression section
13. Length of the metering section
14. Number of flights in metering section
15. Flight helix angle in metering section
16. Flight width in metering section
17. Flight flank radius in metering section

If we use a simple conveying screw with a single-flighted geometry and constant flight pitch, flight width, and flight flank radius, the number of screw design variables is reduced as follows:

1. Screw diameter
2. Screw length
3. Length feed section
4. Flight helix angle
5. Flight width
6. Flight flank radius
7. Length of compression section
8. Length of the metering section

The design of a simple conveying screw is not too complicated. These screws were very common up until the 1970s. Unfortunately, simple conveying screws have limited melting and mixing capability. Therefore, simple conveying screws tend to have poor process stability and low output capacity. As a result, many modifications were made to the simple conveying screw to improve melting and mixing.

Some important developments are listed below:

- The barrier screw by Maillefer (1959)
- The fluted mixer by LeRoy and Maddock (1967)
- The Saxton mixer by Saxton (1961)
- The grooved feed extruder (1960s)
- The grooved barrel extruder by Grünschloss (1999)
- The high speed single-screw extruder (2005)
- The high speed grooved barrel extruder (2015)

This list can be extended significantly because there have been many more important developments. The list shows that important developments did not stop in the 1960s. There have been very significant developments since the year 2000 and we can expect these developments to continue.

Another change that has occurred is the length of the extruder. In the 1950s the typical flighted length was about 20*D*. In the 1970s this increased to 24*D* and in the 1990s the length increased to 30*D*. Today (2019) many single-screw extruders have a flighted length of 30*D* to 35*D*. Very high speed extruders have a length from 35*D* to 45*D*. The increase in length has coincided with significant increases in output capacity. Table 3.3 lists typical output of a 75-mm extruder with various lengths values.

Table 3.3 Output Capacity versus Flighted Length

Flighted length	Typical output capacity [kg/h]
20*D*	100
24*D*	150
30*D*	250
35*D*	1000*
40*D*	2000*

* high speed

The increases in output have been made possible by increasing length of the extruder in combination with increases in screw speed. Modern high speed single-screw extruders run at screw speeds in the range of 1000 to 2000 rpm. It can be expected that in the future screw speeds over 2000 rpm will become reality.

Today it is possible to run at production rates with a small high speed extruder. In 2015 Oliver Kast presented data on a 35-mm 34*D* high speed grooved barrel extruder at the PPS meeting in Graz, Austria. At a screw speed of 1200 rpm the extruder output ranged from 300 to 400 kg/h dependent on the polymer extruded.

In the past an output of 400 kg/h required a 120-mm conventional extruder. Today this same output can be achieved with a 35-mm extruder. Such small extruders with high output capacity provide significant advantages in terms of machine cost, floor space, energy consumption, operating cost, noise level, etc.

Screw Design Features and Extruder Performance

We will now discuss design features that improve extruder performance and those that tend to be detrimental to extruder performance. Most single-screw extruders have a single-flighted design although there may be a barrier section along the melting zone. In the barrier section there is a main flight and a barrier flight. The barrier flight separates the unmelted plastic (the solid bed) from the melted plastic (the melt pool) as shown in Figure 3.7.

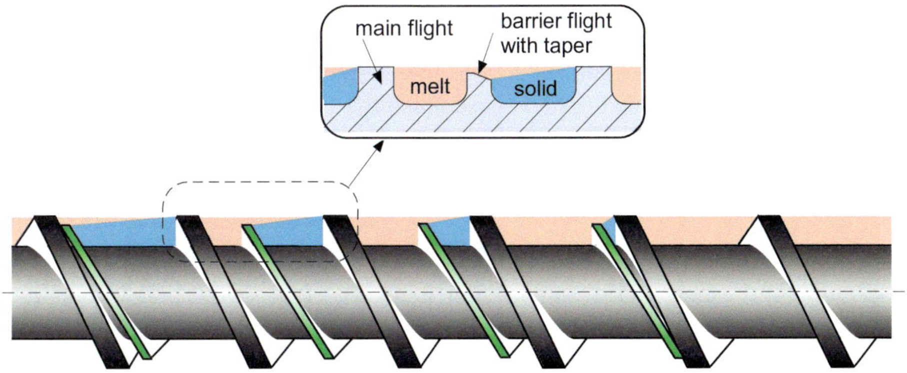

Figure 3.7 Illustration of a barrier screw

This barrier screw has a barrier flight with a tapered section on the top of the barrier flight. This taper is used to create elongational flow as the melt passes from the melt film above the solid bed to the melt pool. This is a patented design [138] developed by Rauwendaal Extrusion Engineering.

The length of the feed section is generally 5*D*–10*D*. The feed depth for a smooth bore extruder ranges from 0.15*D* to 0.20*D*. With grooved feed extruders the feed depth is generally less than 0.10*D*. A very common flight lead is 1.0*D*; this is referred to as a square pitch screw. This lead corresponds to a flight helix angle of 17.67 degrees. The flight flank radius should be equal to or greater than the flight height, see Figure 3.8.

Many screws are manufactured with a small flight flank radius (r_1<H, Figure 3.8 top left). A small radius can cause solids-conveying problems and will cause hangup of the molten polymer. Such a screw design should be avoided especially for thermally sensitive polymers.

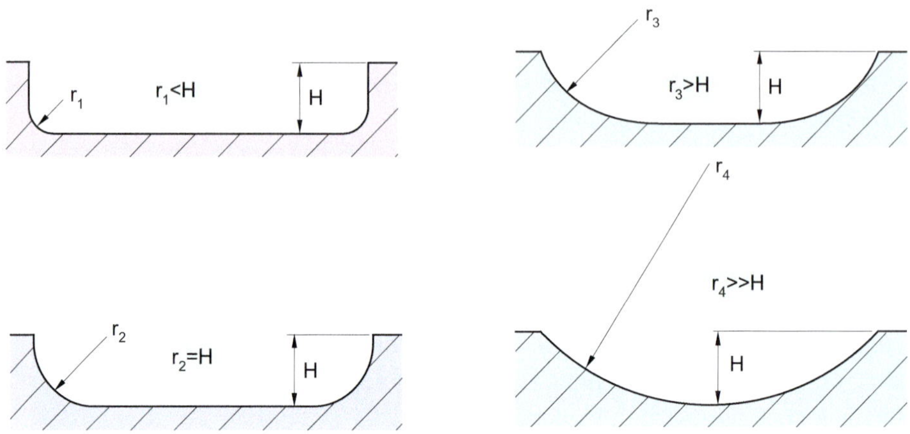

Figure 3.8 Flight flank radius

The length of the compression section is often about 10*D*. Very short compression sections should be avoided because these make the process susceptible to surging. Some resin producers recommend rapid compression screws for certain polymers, e.g., fluoropolymers. However, processors generally do not achieve good results with rapid compression screws. Very long compression sections (>15*D*) should also be avoided because these tend to cause unstable process conditions.

The length of the metering section is often about 10*D*. It is beneficial to incorporate a distributive mixing section into the last 2*D*–3*D* of the metering section. This will improve the homogeneity of the melt discharged by the screw into the die. Figure 3.9 shows a distributive mixing section.

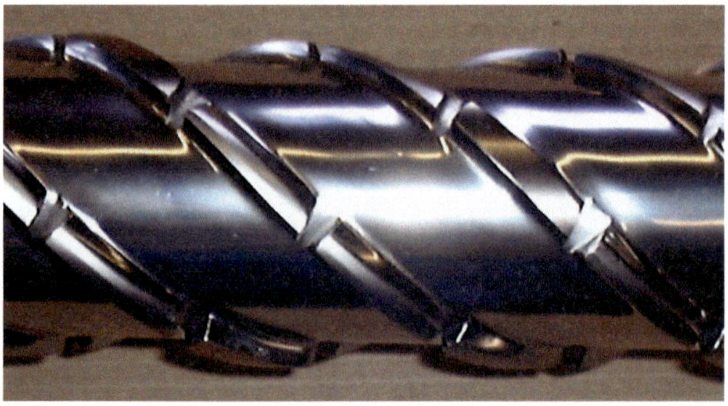

Figure 3.9 Example of a distributive mixing section

This is a multi-flighted mixing section with tapered slots in the flights. The tapered slots produce elongational flow; this mixing section is covered by US Patent 6,136,246 [139].

It is also beneficial to incorporate a dispersive mixing section in the early part of the metering section. Dispersive mixing sections are often designed as a fluted mixer, see Figure 3.10. In this CRD-Z mixer the barrier flight helix angle is greater than the wiping flight helix angle. This results in a highly streamlined geometry. The barrier flight in this mixer has a tapered geometry so that elongational flow occurs as the melt flows over the barrier flight.

Figure 3.10 Example of dispersive mixing element

One advantage of a dispersive mixer is that unmelted pellets cannot travel beyond the mixing section. This is one way to avoid unmelt in the extruded product.

General features to look for in mixing devices:

- Forward mixing capability – this will reduce pressure drop
- Barrel should be completely wiped
- Mixer should have a streamlined geometry to minimize hangup
- Elongational mixing is preferred over shear mixing

When the barrel is not completely wiped by the mixer a stagnant layer will form on the barrel surface. This layer will result in poor heat transfer and increase polymer degradation. An example of a mixing section with poor streamlining and large non-wiped sections is the blockhead mixer shown in Figure 3.11.

Figure 3.11 Blockhead mixer

For about 75% of the length of the mixer the barrel is not wiped by the screw. This makes degradation likely to occur. Also, the square mixing pins create a highly non-streamlined geometry. Despite these obvious drawbacks, the blockhead mixer is still commonly used.

Elongational mixing is preferred over shear mixing. Elongational mixing offers the following advantages over shear mixing:

- Less viscous dissipation and reduced energy consumption
- Lower melt temperatures
- Better dispersive mixing

The flight depth profile of the screw has to be designed using the rheological properties of the polymer. It is important that the amount of viscous heating by the screw is correct. Generally, we would like for about 80% of the heat to be supplied by the screw and 20% by the barrel heaters. Problems will occur when the screw supplies more than 100% of the heat needed to increase the temperature of the polymer to the desired process temperature. This type of screw is called a hyperactive screw.

A hyperactive screw will result in excessive energy consumption and melt temperatures. In turn, excessive melt temperatures will cause polymer degradation. In a hyperactive screw the melt temperatures will be higher than the barrel temperatures. This is one way to identify a hyperactive screw. Another way to identify a hyperactive screw is to determine the specific energy consumption (SEC) by the screw. The SEC can be determined by dividing the motor power by the throughput. If the motor power is expressed in kW and the throughput in kg/h, the SEC is expressed in kWh/kg.

The amount of energy needed to bring the polymer up to process temperature is the enthalpy rise. The enthalpy rise is determined by measuring the specific heat of the polymer from room temperature to the desired process temperature – it is a thermal property of the polymer. The enthalpy rise is typically measured using a differential scanning calorimeter (DSC). For polyethylene the enthalpy rise is around 0.16 kWh/kg. If the SEC is greater than the enthalpy rise we are dealing with a hyperactive screw, especially when the SEC is substantially greater. For this reason, the SEC is an important process parameter in the extrusion process.

In order to design a screw correctly, the melt flow properties of the polymer have to be taken into account. It is not possible to design a screw without information on the melt flow properties. When melt flow properties are available, the amount of viscous heating by the screw can be calculated. This allows a design of the screw such that the viscous heating matches the enthalpy rise of the polymer.

For low viscosity (high melt index) polymers the screw should have shallow flights so that enough viscous heat can be generated. For high viscosity (low melt index) polymers the screw should be deep-flighted to avoid excessive viscous heating.

This is the reason why screws used for rubber extrusion have very deep flights. The viscosity of rubbers is much higher than the viscosity of regular plastics.

It should be clear from the discussion above that it is not possible to design a screw that can process all plastics because plastics have a wide range of viscosities. If an extruder is required to run plastics with large differences in viscosity it may be necessary to have two or more screws so that the different plastics can be processed appropriately.

3.2.5 Wear Problems

Wear occurs in all machinery with moving parts and, unfortunately, extruders are no exception. Extruder wear problems can take many shapes and forms. A good general text on wear in polymer processing is the book by Mennig [12].

Extruder wear generally results in an increase in the clearance between screw flight and barrel. Wear often occurs toward the end of the compression section. This type of wear is more likely to occur when the screw has a high compression ratio. Wear in the compression section of the screw reduces the melting capacity and will lead to temperature non-uniformities and pressure fluctuations. Wear in the metering section of the screw will reduce the pumping capacity; however, the reduction in pumping capacity is generally quite small, as long as the wear does not exceed two to three times the design clearance. An increased flight clearance will also reduce the effectiveness of heat transfer from the barrel to the polymer melt and vice versa; this may contribute to temperature non-uniformities in the polymer melt.

Wear can only be detected by disassembling the extruder and inspecting the screw and barrel. If the wear is serious enough to affect the extruder performance, it will often be noticeable to the naked eye. However, measuring the ID of the barrel and the OD of the screw over the length of the machine is recommended. If this is done regularly, it can readily be determined how fast wear is progressing with time. By extrapolating to the maximum allowable wear, the troubleshooter can determine when the screw and/or barrel should be replaced or rebuilt.

If replacement as a result of wear is necessary after several years of operation, the easiest solution to a wear problem is to simply replace the worn parts. However, if replacement as a result of wear becomes necessary within a short period of time, for instance several months, then simple replacement may not provide an acceptable solution. For frequently occurring wear problems, the cost of downtime and replacement parts can easily become unacceptable, and a solution has to be found to reduce the actual wear rate instead of simply replacing the worn parts. To reduce the wear rate, the troubleshooter must understand the wear mechanism(s) in order to determine the most effective way to reduce wear.

3.2.5.1 Wear Mechanisms

Five mechanisms of wear can be distinguished:

- Adhesive wear
- Abrasive wear
- Laminar wear
- Surface-fatigue wear
- Corrosive wear

When wear occurs, often more than one mechanism is at work. For example, *adhesive wear* may occur during high-stress metal-to-metal contact. Because the actual contact area is much smaller than the apparent contact area, local welds can form at points of contact. This phenomenon is referred to as cold welding. The sliding motion causes a rupture in the weld region, and small fragments of the weld region are carried away with one member of the sliding system. Usually fragments of the softer material are transferred to the harder material. The attrition rate depends on the shear strength of the adhesive junctions. Adhesive wear is generally more severe with sliding contact of similar metals, where it is often referred to as galling. In sliding motion between dissimilar metals, the adhesive junction will contain a spectrum of compositions. Adhesive wear can be significantly reduced when the spectrum of compositions in a junction contains brittle intermetallic compounds that fracture easily. Oxide layers formed at the interface will also reduce adhesive wear because oxides will not bond. Because adhesive junctions are only formed between clean surfaces, lubricants are often used to reduce the chance of adhesive wear.

Abrasive wear is a micro-cutting process. In two-body adhesive wear, the asperities of the harder member penetrate the softer one and remove material as a result of the sliding motion. In three-body abrasive wear, hard particles are pre-embedded in the material of at least one member of the wear system. The hardness ratio of the materials has been found to be the most important material characteristic in abrasive wear, although fracture toughness also seems to play a role [13]. Krushchov [14] found that the wear resistances of pure metals and annealed steels increase proportionally with hardness. Strain hardening or precipitation hardening does not result in improved resistance to abrasive wear because the micro-cutting process already yields maximum local strain hardening.

Laminar wear occurs when the shear strength in the heterogeneous portion of the interfacial layer is higher than the shear strength of the homogeneous portion of the interfacial layer. Laminar wear takes place only at the thin outer layers of the interface and is sustained only if the outer layer of the heterogeneous interface continuously regenerates as an oxide or other reactive layer. The formation of reactive layers can be controlled by additives in the lubricant. Laminar wear is, to a certain extent, a mild form of corrosive wear. A mild corrosive or oxidative action affects only a thin layer of the newly generated metallic surface. When the new surface layer is formed, the reaction will stop.

Surface-fatigue wear consists of a separation of microscopic and macroscopic material particles from the surface, which is caused by fatigue crazing, cracking, and breakup under specific mechanical, thermal, and chemical load, occurring during rolling contact between two surfaces. Fatigue cracking is initiated by alternating thermal or mechanical load and can occur even with direct metal contact. Surface-fatigue wear is characterized by considerable induction times and relatively large penetration depths. A familiar example is the pitting of roller bearings and gears.

In *corrosive wear*, a chemical reaction attacks at least one of the sliding surfaces. During extrusion, corrosive wear usually occurs in combination with one or more of the other wear mechanisms. The combined chemical and mechanical attack of the sliding surface can cause wear rates far in excess of what would be expected based on the individual contributions of the two wear types. The most important mechanisms for wear occurring during extrusion are adhesive, abrasive, and corrosive wear.

3.2.5.2 Test Methods for Wear

There are basically two ways to test the wear characteristics of materials involved in the extrusion process. One method is to run the actual machine under normal operating conditions and measure the progress of wear at regular intervals. This approach is time-consuming and expensive, but it does yield accurate and representative results. However, it does not allow a simple analysis of the parameters that influenced the wear process.

An interesting technique for relatively quick wear studies on actual extruders was developed at the Institut für Kunststoff Technologie (IKT) in Stuttgart, Germany by Fritz and coworkers [15]. A reference surface of the machine is made radioactive by proton and neutron bombardment to a depth of 30 to 80 μm. The impulse rate from the measured isotope diminishes linearly with activation depth. This allows accurate measurement of wear over short time periods. The wear process can be accurately characterized in approximately one to three hours. The particular study mentioned [15] measured wear in the feed section of a screw. The extruder was equipped with a grooved barrel section. The abrasive filler was titanium dioxide, which was added to the virgin polymer as a masterbatch. Considerable wear occurred when the virgin polymer was in pellet form, but no wear could be detected when the virgin polymer was in powder form.

Another method of testing the wear characteristics of extrusion system materials involves the use of model systems. A test specimen is subjected to certain load conditions to simulate actual service conditions. Such wear testers allow tribologically relevant loads to be preselected.

Tribological parameters, such as temperature and coefficient of friction, can be measured and recorded continuously. This method provides a relatively quick and inexpensive determination of wear characteristics. However, information obtained from a wear tester can only be transferred to practice if the wear conditions in the model system are essentially the same as those in the real system, that is, the extruder itself.

Many mistakes have been made in transferring information from a short wear test to actual extruders simply because of differences in the tribological conditions of the wear process. It is often not realized that the tribological parameters, friction and wear, are not material properties, but properties of a complex system. Thus the transfer from a model system to an actual extruder must be made cautiously. The same measurements must be taken while the actual extruder is in use to ensure that results from the model system are also valid for the real system. A good review and analysis of methods for testing wear in the polymer processing industry was provided by Mennig and Volz [16]. Because there are many different types of wear in polymer processing, there is, unfortunately, no universal wear tester.

Mennig and Volz distinguish four types of testing: metal-liquid wear, metal-solid polymer wear, corrosive wear, and metal-to-metal wear. As early as 1944, a test device was proposed by Mehdorn [11] to measure metal-liquid wear. This test simulated wear conditions in a press used for injection of thermosets. The test geometry is shown in Figure 3.12.

A molten or liquid mass of polymer is forced against a test specimen, from which the mass is deflected; the material exits through a small clearance of 0.4 mm. This test device provides relatively quick results. The disadvantages of this test method are the complex geometry of the clearance, non-uniform flow conditions at the specimen, and the fact that increased wear changes the resistance to flow and thus the wear conditions.

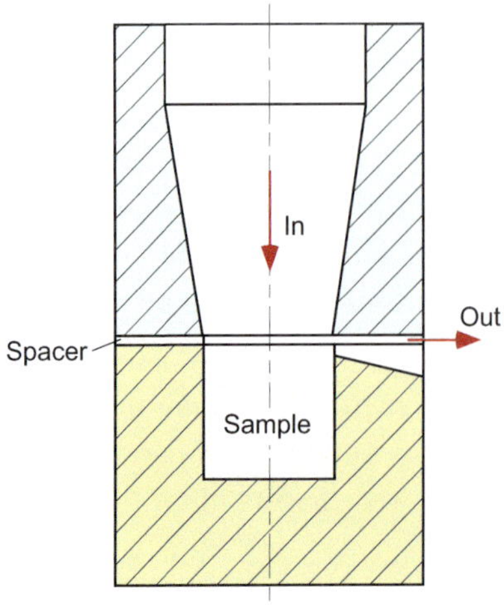

Figure 3.12
Wear test device proposed by Mehdorn [11]

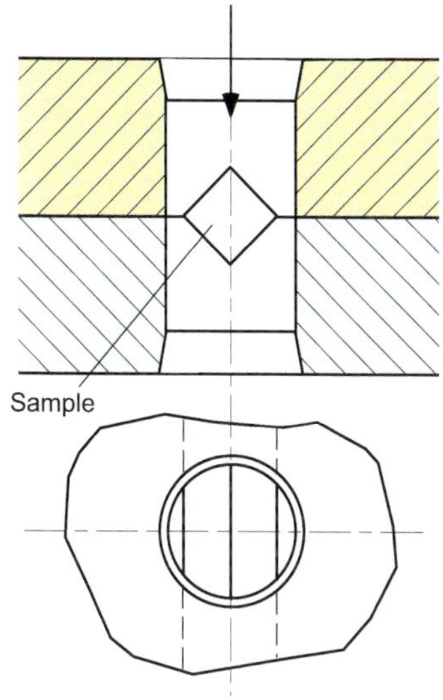

Figure 3.13
Wear test proposed by Bauer et al. [17]

Another method was developed by Bauer, Eichler, and John [17] in 1967. Figure 3.13 shows the geometry of their test apparatus.

The specimen is a diamond-shaped obstruction in the center of a flow channel. A commercial wear test apparatus based on this geometry is the Tribotest from Brabender OHG (Duisburg, Germany). This test is often referred to as the Siemens-Method wear test. Eichler and Frank [18] modified this test to make it more suitable for injection molding.

An entirely different test geometry was developed at the DKI (Deutsches Kunststoff Institut), utilizing a flat plate geometry as shown in Figure 3.14. The rectangular test gap has a length of 12 mm, a width of 10 mm, and a height that is adjustable from 0.1 to 1.0 mm. This geometry has been used for studies with thermoplastics [19] as well as thermosets [20]. A modification of the flat plate wear tester is the BASF wear tester, which simulates the wear process in a molding machine.

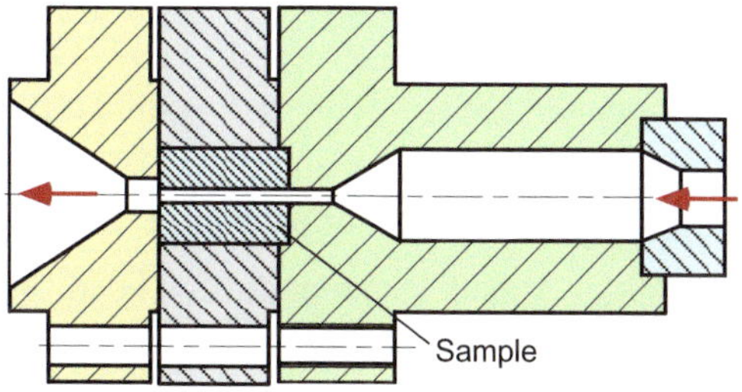

Figure 3.14 DKI flat plate wear test apparatus

Another test apparatus developed at the DKI is the ring method (see Figure 3.15), which simulates conditions occurring in the annular space between the tip of the screw flight and the extruder barrel [19, 21].

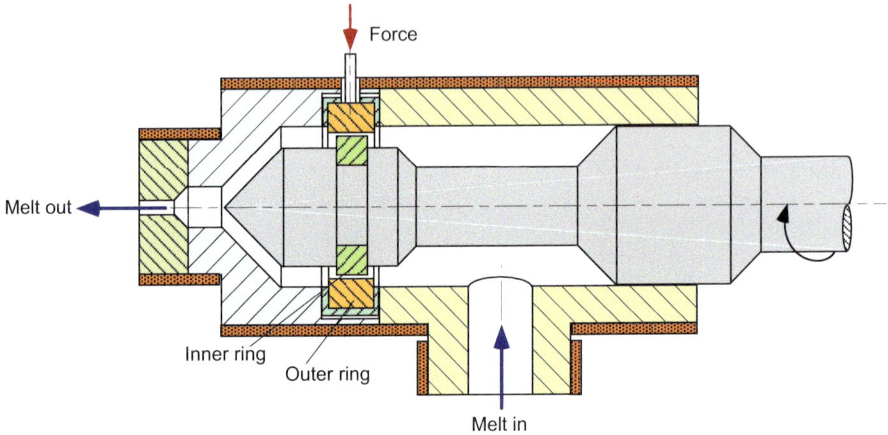

Figure 3.15 DKI ring wear test apparatus

Plumb and Glaeser [22] developed a test method for filled elastomers based on a capillary rheometer. The specimen in this test is a cone-shaped torpedo in the flow channel. The flow conditions change in axial direction and with it, the local wear rate.

A test developed at Georgia Marble Company, a supplier of calcium carbonate, uses an aluminum breaker plate at the end of a screw extruder to evaluate abrasive wear of mineral fillers. The amount of wear is determined by measuring the weight loss of the breaker plate over a certain run time [23].

Metal-to-solid polymer wear occurs in the solids-conveying zone of the extruder. The introduction of the grooved barrel extruder has significantly increased interest in and concern about wear in this portion of the extruder. Grooved barrel sections substantially increase the shear and normal stresses between the polymer solid bed and the metal surfaces. As a result, grooved barrel sections are much more susceptible to wear than smooth barrels. The first systematic study of wear in the solids-conveying section of extruders was made by Fritz [24], using a diamond-shaped specimen that protruded into the screw channel.

The universal disk tribometer, a model system developed at the DKI in the form of a disk wear tester, was discussed by Volz [25]. The concept of the disk tribometer is partially based on a modified friction tester developed at Enka Glanzstoff [26]. The geometry of the disk tribometer is schematically shown in Figure 3.16.

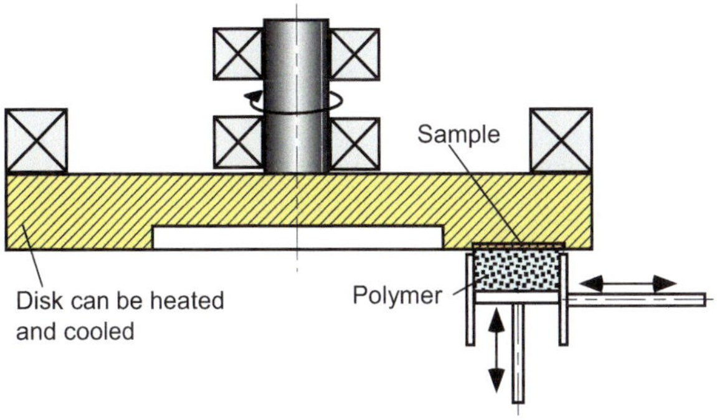

Figure 3.16 Universal disk tribometer

The disk can be heated and cooled. The annular groove at the bottom surface of the disk contains the metal specimen. The disk tribometer allows measurement of frictional force, normal force, and wear at contact pressures of up to 150 MPa.

A test for corrosive wear was first proposed by Calloway, Morrison, and Williams [27]. They used vacuum pyrolysis at normal process temperature. A metal specimen is suspended in the extracted volatiles for 24 hours at room temperature. A similar apparatus was used by Mahler [19] and Braun and Maelhammar [28] to test for corrosion at elevated temperatures and pressures. One disadvantage of vacuum pyrolysis is that the corrosion conditions are different from those existing in the actual extrusion process. This drawback is reduced in the test procedure used by Knappe and Mahler [21].

Other tests have been described by Moslé et al. [29] and Maelhammar [30]. The latter test measures a combination of metal-liquid wear and corrosive wear. The apparatus was modified by Volz [31] for thermosets. The volatiles were extracted from the polymer melt, which had been prepared in an injection molding machine and sheared through a test gap. By taking electrochemical measurements, Volz proved that there were significant differences in the corrosive action of volatiles separated from injection-molded samples of thermosetting polymers depending on the size of the test gap.

Tests for metal-to-metal wear can utilize the standard test methods, provided the proper intermediate material can be introduced between the metallic surfaces. Broszeit used the cylinder-disk apparatus to study metal-to-metal wear [32]. Saltzman et al. [33–35] used the Alpha LFW-1 test apparatus (see Figure 3.17). A stationary block is forced against a rotating ring by a dead-weight load. The bottom part of the ring is immersed in a water-oil emulsion. The presence of the water-oil emulsion is a drawback for this test because the actual wear behavior, in an extruder with a polymer melt as the intermediate material between screw and barrel, is bound to be substantially different from the wear behavior in the LFW-1 test apparatus. No data have been published on metal-to-metal wear with a polymer melt as the intermediate material.

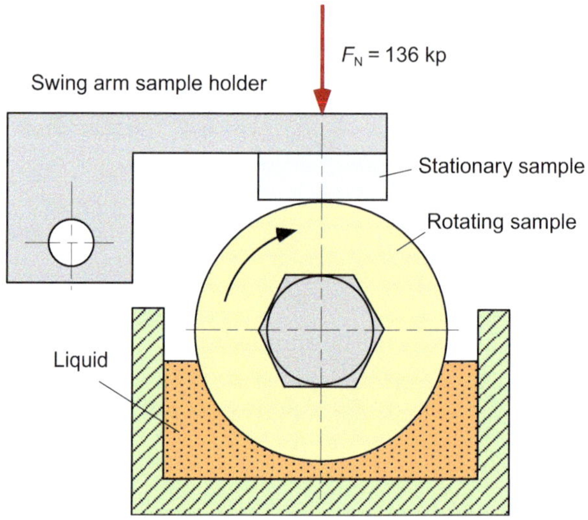

Figure 3.17 Alpha LFW-1 wear test apparatus

3.2.5.3 Causes of Wear

In polymer-metal wear, the main causes of wear are abrasion and corrosion. Abrasive wear is generally due to abrasive components in the polymer matrix. Whether these components are classified as fillers, reinforcements, or additives, they can cause significant wear when the filler is hard and present in significant amounts. Factors affecting the wear are particle hardness, particle size, particle shape, and loading [36]. One indication of the ability of a filler to cause abrasive wear is its ranking on the Mohs scale. This scale ranks a material from one to ten according to its ability to scratch another material or to be scratched by it. A very soft material, such as talc, is ranked at the bottom of the scale, rank one, and a very hard material, such as diamond, is ranked at the top of the scale, rank ten. The Mohs scale ranking of several fillers is given in Table 3.4. However, the filler hardness only partially determines the wear characteristics of a filled compound. The particle size is another important parameter. Generally, the severity of wear is less with small particle size. In glass-fiber-reinforced compounds, there is less wear with shorter fiber lengths [25]. This is believed to be due to the greater mobility and reduced kinetic energy of the smaller glass fibers. Particle shape also has a strong influence on wear characteristics. Mahler's experiments [37] with glass-fiber-reinforced nylon and glass-bead-reinforced nylon showed the wearing intensity of the fiber-reinforced compound to be fourteen times higher than that of the compound reinforced with glass spheres. The ability to cause wear is more pronounced when the particle has sharp corners and a large aspect ratio. A spherical particle shape minimizes wear. Unfortunately, a spherical shape is often undesirable with respect to mechanical properties, electrical properties, and others.

Table 3.4 Mohs Scale Ranking of Various Fillers

Calcined kaolin	7
Silica	6.5
Glass	6
Perlite	5.5
Wollastonite	5.5
Mica	3
Calcium carbonate	3
Kaolin	2
Alumina trihydrate	1
Talc	1

In a study on glass-reinforced polymers, Mahler [19] found that some polymers can cause corrosive wear as well as abrasive wear and thus have much higher wear intensity than other polymers. He found that the wear intensity of glass-fiber-reinforced nylon 6,6 against 9S20K steel was about thirteen times higher than reinforced styrene acrylonitrile (SAN) and polycarbonate (PC). Olmsted [38] reported similar findings with injection molding of glass-fiber-reinforced nylon 6. Severe wear occurred on both the screw and the barrel. Primarily corrosive-type wear occurred, caused by a silane wetting agent on the fiber. The decomposition temperature of the wetting agent was lower than the process temperature, and hence degradation occurred, resulting in a corrosive attack on the screw and barrel. However, Mahler's similar work [19] using nylon 6,6 filled with glass fibers coated with an aminosilane coupling agent did not reveal any corrosive wear resulting from the coupling agent.

Surface treatment of fillers can reduce wear. A comparison between coated and uncoated calcium carbonate showed the wear intensity of a rigid PVC compound with coated filler to be three to nine times lower than the same compound with uncoated filler [3]. The wear intensity was measured using the aluminum breaker plate test discussed earlier. It is good practice to incorporate abrasive fillers, such as glass fibers, at a point in the extrusion process when the polymer is already molten. This allows the melt to coat the filler and reduce the wear intensity. When abrasive fillers are added with solid polymer particles, the abrasive action is much more severe, and rapid wear will occur in the solids-conveying section of the extruder. This is why glass fiber is generally added in a downstream barrel opening or, in the case of a tandem extrusion setup, in a downstream extruder.

Corrosive wear also occurs in non-filled polymers. Well-known examples are fluoropolymers and chlorine containing polymers, such as PFA, and PVC. Fluoropolymers have a tendency to form hydrofluoric acid at high temperatures in combination with air and moisture. PVC tends to generate hydrochloric acids at elevated temperatures. The corrosion problem is generally more severe with rigid PVC than with flexible PVC. When such polymers are extruded, the metal parts in contact with the polymer should be made out of a corrosion-resistant metal, such as Hastelloy, 17-4 PH, 15-5 PH.

Corrosion can also occur with hygroscopic polymers, such as acrylonitrile-butadiene-styrene (ABS), polyamide (nylon, PA), PET, PMMA, when moisture is released under high temperature and pressure, forming high pressure steam. Braun and Maelhammar [28] found that PA 6,6 splits into various corrosive components. Calloway et al. [27] found that the corrosive attack of high impact polystyrene (HIPS) is dependent on the chlorine and sulphur content of the carbon blacks. Moslé et al. [29] found that degradation products of ABS can cause corrosive attack in extruders.

Some abrasive components in the compound are foreign objects resulting from contamination or human error. Hard foreign objects, such as wrenches, bolts, and knives, can cause severe wear in a very short period of time. Magnetic traps are available to catch metallic objects, but there is no simple method to successfully remove all foreign objects from the feed stock. Good housekeeping procedures and conscientious personnel will go a long way in reducing the chance of foreign objects ending up in an extruder. One word of caution is in order for barrier-type screws. If small particles that do not melt at operating temperatures are present in the polymer feed stock, they will get trapped at the end of the barrier section, if the particle size is larger than the barrier clearance.

Another important cause of wear in extrusion equipment is metal-to-metal wear. Unfortunately, relatively little is known about this type of wear. A number of circumstances can cause metal-to-metal contact between screw and barrel. Metal-to-metal contact will necessarily occur at startup and can also occur as a result of misalignment, a warped barrel, or a warped screw. Metal-to-metal contact can also occur at the feed opening of an extruder as described by Luelsdorf [39], particularly when the feed opening is offset and when it forms a sharply tapered angle with the circumference of the screw.

The author (CR) has also experienced cases where metal-to-metal wear occurred in the feed throat of an extruder as a result of improper feed opening geometry. Radiographic analysis of wear particles indicates [39] that temperatures in the contact zone between screw and barrel may exceed 800 °C. This is confirmed by personal observations of screws subjected to severe metal-to-metal wear. A very noticeable purple discoloration indicating exposure to temperatures over 500 °C occurred in the metal in the wear region.

Metal-to-metal wear can also occur in intermeshing twin-screw extruders. Counter-rotating twin-screw extruders are more susceptible to metal-to-metal wear than co-rotating twin-screw extruders, and therefore generally operate at rather low screw speeds. Unfortunately, serious metal-to-metal wear problems can also occur in co-rotating twin-screw extruders.

Lai Fook and Worth [40] proposed a modified flight geometry to increase the centering force on the screw in order to reduce metal-to-metal contact. The two flight geometries they proposed based on theoretical calculations are shown in Figure 3.18.

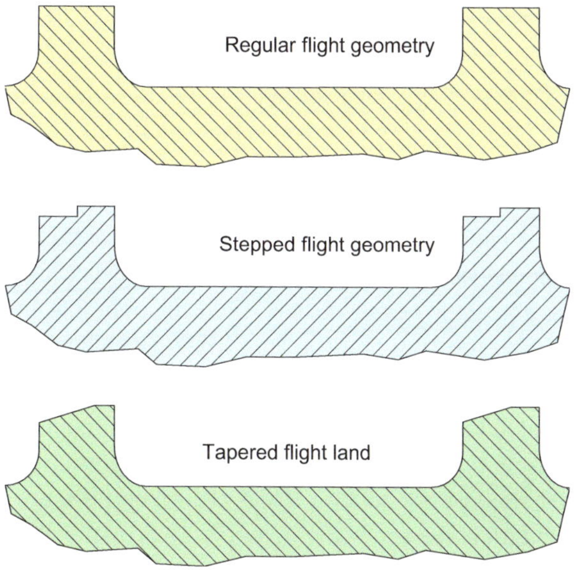

Figure 3.18 Flight geometry to reduce the chance of metal-to-metal wear

Actual measurements of the tangential pressure profile differed from predicted values by a factor of about ten. This indicated that the analysis employed was not entirely realistic. In particular, neglecting side leakage results in large errors in the predicted values. No data were presented on the actual centering force acting on the screw. Winter [41] analyzed the non-isothermal flow of a power law fluid in the flight clearance, and obtained solutions using a numerical procedure. Very high temperature increases in the polymer melt were found to occur in the clearance, and these temperature changes affected the velocity profile. Since the mass flow rate could not change, Winter adjusted the pressure gradient along the gap to satisfy the continuity equation. This led to a prediction of large negative pressure gradients at the leading edge of the flight and large positive pressure gradients at the trailing edge of the flight, as shown qualitatively in Figure 3.19. The actual pressure reached extreme values at about one third of the way from the leading edge. The calculated values ranged from 5 to 20 MPa.

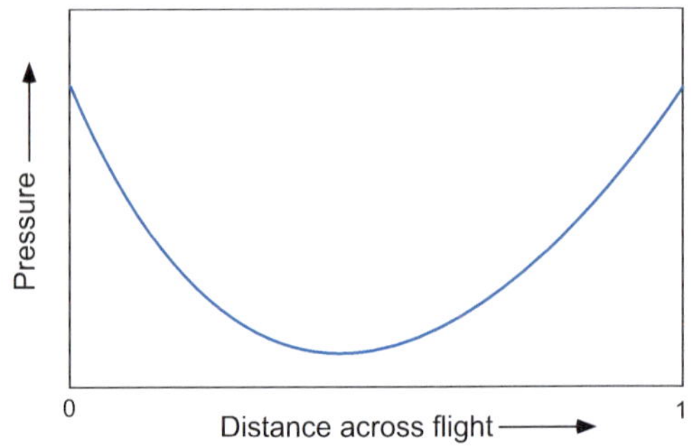

Figure 3.19 Pressure profile in the flight clearance as predicted by Winter [41]

The predicted pressure profile is obviously a direct result of the assumptions made in the calculations. Winter [41] assumed isothermal conditions at the barrel wall and adiabatic conditions at the flight tip. With stock temperature increases on the order of 100 °C or more, it is unlikely that the isothermal boundary condition assumption is valid for the barrel. For the same reasons, it is unlikely that the adiabatic boundary condition assumption is valid for the flight tip, particularly because the temperature of the rest of the screw would be much lower. Unfortunately, it is difficult to measure actual temperature and pressure profiles. Thus, the predicted temperature and pressure profiles have not been compared to experimental results.

Winter [41] postulated that the pressure minimum in the middle of the clearance can cause the screw to be pushed against the barrel by pressure on the other side of the screw. This would only be true if the clearance pressure profiles change along the helical length of the screw flight, and if the pressure profiles in the screw channel itself do not play a significant role. Most likely, the actual situation will be considerably more complex. Following Lai Fook and Worth [40], Winter recommended beveling the flight tip to create a tapered gap between flight and barrel. Another recommendation was to alter the thermal boundary conditions, for instance by heating the barrel above the temperature of the screw flight. It is interesting to note that according to Winter's analysis, the chance of metal-to-metal contact is reduced when the flight clearance and helix angle are increased, and when the flight width is decreased. If this is true, these measures will have a dual benefit because the power consumption will also be reduced. However, these measures do not work in practice.

One disadvantage of the stepped or beveled flight geometry is that, when the extruder is running empty or partially empty, the apparent contact area between screw and barrel will be considerably reduced. As a result, there will be significantly greater stresses at the actual contact area, and wear, particularly adhesive wear, is more likely to occur. Though Volz [25] reports that flight lands with a hydrodynamic slide bearing function are used by a number of extruder manufacturers, it does not appear that this flight geometry is widely used in the industry.

Metal-to-metal contact between screw and barrel will occur when the extruder is empty because of gravitational sag of the screw. In theory, this type of wear will be most pronounced at the very end of the screw and diminish in the upstream direction, but in practice maximum wear generally occurs toward the end of the compression section of the screw. This indicates that the screw is supported by the polymer melt at the end of the screw and is subjected to a substantial lateral force in the compression section of the screw. It is unlikely that this lateral force is caused by temperature-induced pressure gradients in the flight clearance, because the flow process in the flight clearance will not change significantly from the point where melting begins to the very end of the screw. Therefore, it seems more likely that the lateral force is caused by the conveying process in the screw channel.

In the melting zone of the extruder, there is a continuous deformation of the solid bed. In the compression section of the screw, the solid bed is compressed between the root of the screw and the barrel. Very large pressure can build up in this screw section, particularly when the compression ratio is high. These rapid pressure changes along the screw can easily cause an imbalance of the lateral forces acting on the screw. The imbalance seems to be the most likely cause of metal-to-metal contact between screw and barrel. Considering how frequently this wear problem occurs, it is surprising how little attention this problem has received in the open technical literature.

3.2.5.4 Solutions to Wear Problems

The key to finding the best solution to a wear problem is to identify both the cause and the mechanism of wear. For example, if the screw OD is worn but not the flight flanks and the root diameter, then the problem is probably caused by contact between screw and barrel. A hard-facing can be put on the tip of the flight, but this will not eliminate the cause of the problem, though it may reduce the magnitude of the problem. The problem can actually be eliminated by a change in the screw design or by altering the pressure profile along the screw by another method, for example, by starve feeding the extruder.

Corrosive wear can usually be identified by a pitted, worn surface. The best solution to corrosive wear is to eliminate the corrosive component from the compound. However, when this is not possible, corrosion-resistant materials, such as stainless steel, Inconel, or Hastelloy, must be used in the extruder. To select the best material of construction, the troubleshooter should identify the chemical species that are causing the corrosive attack. Various metal handbooks contain information about the chemical resistance of many metals against a number of chemical species. Figure 3.20 shows a flow chart that will aid in the systematic troubleshooting of wear problems.

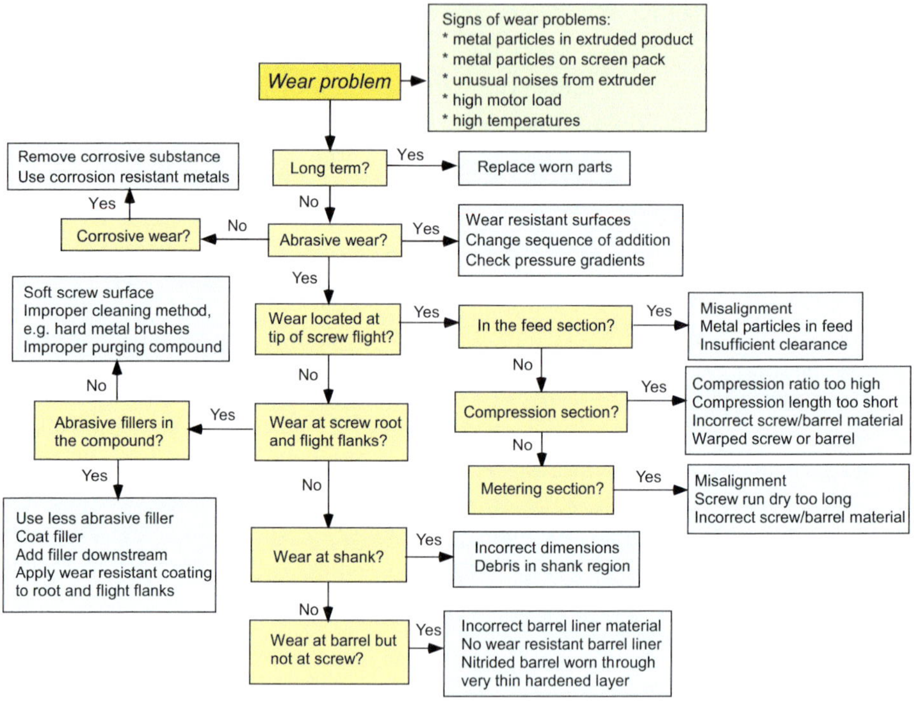

Figure 3.20 Flow chart for troubleshooting wear problems

A large number of materials are available for the screw and barrel. Most extruder barrels in the United States have a liner, which is centrifugally cast into the barrel. The barrel liner is made of a wear-resistant material, often boron-stabilized white irons with a Rockwell C hardness of approx. 65, containing iron chromium boron carbides. Bimetallic barrels provide better wear resistance than nitrided barrels (O'Brien [42] and Thursfield [43], among others). The liner material can be formulated to give good abrasion resistance, good corrosion resistance, or a combined abrasion and corrosion resistance. The correct choice of screw material will depend to some extent on the liner material, particularly if metal-to-metal contact takes place.

Recommended screw materials for several commercial barrel liners are shown in Table 3.5. The recommendations are based on metal-to-metal wear tests on the Alpha LFW-1 machine. As discussed earlier, this test does not fully simulate actual conditions in an extruder, but results from more realistic tests are not available in the open literature.

Table 3.5 Recommended Screw Flight Materials (Courtesy [35])

Barrel liner material		
Xaloy 101	Xaloy 306	Xaloy 800
Screw material		
Colmonoy 56	Colmonoy 6	Colmonoy 6
Colmonoy 6	Colmonoy 56	Colmonoy 56
Haynes 711	Colmonoy 63	Xaloy 008
Colmonoy 5	Stellite 1	Nye-Carb
Colmonoy 63	Nye-Carb	Xaloy 830
Stellite 1	Stellite 6	Ferro-Tic
Ferro-Tic (iron)	Stellite 6H (severe wear)	Stellite 6H (severe wear)
HC-250		
Colmonoy 84		
Triballoy T-700		

The most common screw material is 4140 steel, which has the advantages of low price, good machinability, and the ability to be used with hard-facing and chrome plating. A disadvantage of 4140 steel is its relatively poor wear resistance, as discussed, for instance, by Hoffmann [44].

As a result, 4140 screws are often flame-hardened, plated, or hard-faced when used in more demanding applications. Chrome plating is often used on extruder screws. This produces a hard wear-resistant layer, up to 70 R_c in hardness, which is usually quite thin, approx. 25 to 75 μm, and does not form a hermetic seal to most corrosive substances. Thus, because of its porosity, chrome plating does not substantially improve the corrosion resistance of the screw. It is quite resistant to abrasive wear, but because of its limited thickness, does not provide much protection.

Nickel plating is also used quite frequently on extruder screws. It is applied in a thickness similar to chrome plating. The surface hardness of nickel plating tends to be somewhat lower than chrome plating. However, the thickness of the nickel coating is generally more uniform, and the coating is much less porous than chrome plating. Therefore, nickel plating usually offers better corrosion protection than chrome plating.

A very large number of proprietary plating materials and plating processes are available today. Without exception, the claims made by the supplier of the proprietary plating are quite impressive. Unfortunately, these claims are rarely based on long-term extrusion tests, so the troubleshooter must be quite cautious in selecting a proprietary plating. An interesting proprietary plating is the Poly-Ond plating. It is basically an electroless nickel plating impregnated with a fluoropolymer. This yields a moderately hard, corrosion-resistant layer with a low coefficient of friction. The low coefficient of friction makes it attractive for use in injection molds and on extruder screws. Luker, from Killion Extruders, reporting [45] on tests with Poly-Ond plated extruder screws, found that output increased from 5 to 36% for a number of different polymers. A Teflon-impregnated nickel plating (Nedox) was discussed by Levy [46], and a Teflon-impregnated chrome plating was discussed by Trompler [47].

Metal coatings, such as chrome plating and nickel plating, are generally more effective in reducing wear than hardening of the base material. The various methods used for hardening, such as flame hardening, nitriding, carburizing, and induction hardening, are all basically case-hardening processes with limited depth of penetration and the ability to provide limited improvement in wear characteristics. Further, heat-treated steels have reduced hardness and wear resistance at elevated temperatures.

Hard-Facing Materials

Another technique used for improving the wear resistance of screws and barrels is hard-facing or hard-surfacing. Hard-facing materials are generally nickel- or cobalt-based and contain various metal carbides, such as chromium carbide and tungsten carbide. The most common alloys are Stellite[1] (St. Louis, MO) and Colmonoy[2] (Madison Heights, MI), but many other materials are available today. Hard-facing materials are applied by welding, spraying, or casting in a thickness ranging from 1 to 3 mm.

The steps involved in the hard-facing process are shown in Figure 3.21. On a worn screw, the flights are ground to a uniform undersize, the hard-facing material is welded onto the flights, the hard-facing is ground to the original OD, and the sides of the flights are machined to provide a smooth transition from the flight flank to the hard-facing material.

A critical step in the application of the hard-facing is the preheating of the screw before application and, more importantly, the gradual cooling of the screw after the hard-facing has been applied. If the screw cools too rapidly, thermal stresses will develop in the hard-facing, and cracks will form. Because hard-facing materials are hard and brittle, it is quite easy for small cracks to form. In fact, small hairline cracks cannot be completely avoided in many hard-facing materials. In general, the higher the carbon content of the hard-facing material, the greater its tendency to crack and the better its wear characteristics.

---

[1] Stellite is a trademark of Stoody Deloro Stellite (previously Cabot Stellite), a division of Thermodyne, St. Louis, MO

[2] Colmonoy is a trademark of Wall Colmonoy Corporation, Madison Heights, Michigan

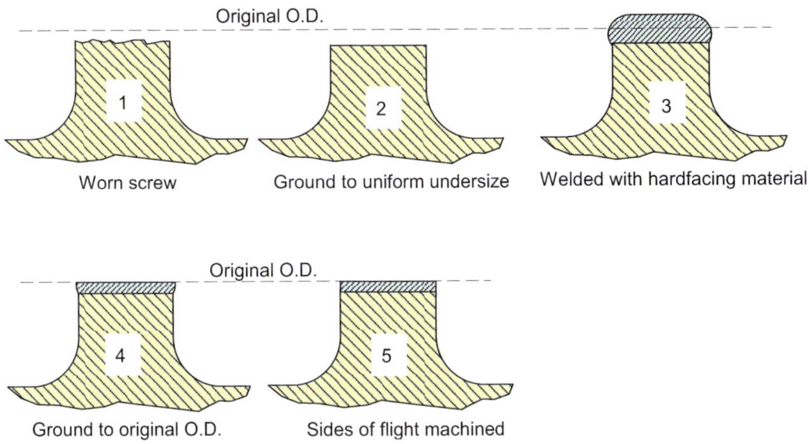

Figure 3.21 Steps involved in applying hard-facing to worn screw

Screws in which high-carbon, highly wear-resistant hard-facing materials have been used will generally show some degree of small hairline cracks. If such hard-facing does not show any hairline cracks, the hard-facing material may be suspect. It is possible that softer, less wear-resistant hard-facing has actually been used. Alternatively, the hard-facing material has been diluted too much with the base screw material, which will result in a softer hard-facing material that will wear rapidly. For this reason, it is a good idea to measure the hardness of the hard-facing after the screw is manufactured. If the hardness is substantially below the normal value, it is likely that the hard-facing is overdiluted, or the wrong hard-facing was applied.

The cracks that are cause for concern are the ones that traverse completely through the hard-facing material and then turn in a circumferential direction. When such cracks occur, parts of the hard-facing may come loose. This may happen when the hard-facing is not properly applied, when the screw is not handled properly before installation, or when the screw is exposed to excessive stresses during the extrusion process.

Most hard-facing materials do not plate well. This results in a wavy line at the intersection of the welded material, a problem that can be eliminated by the use of an inlay as shown in Figure 3.22. This figure shows two basic hard-facing geometries, one in which the hard-facing is applied to the full width of the flight, and the other in which the hard-facing has been inlaid. Inlays can only be applied to new extruder screws.

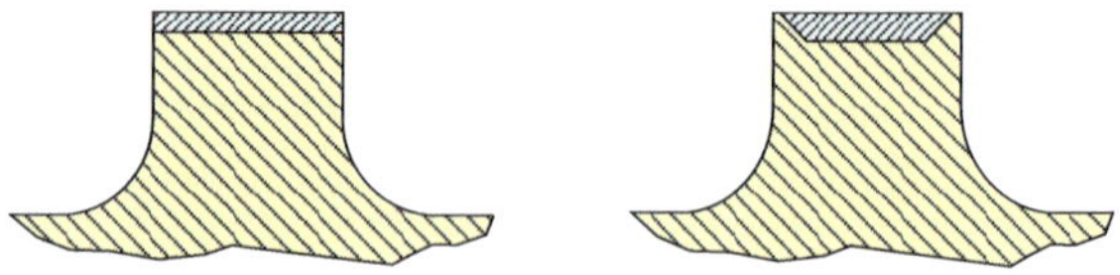

Figure 3.22 Full width and inlay hard-facing

Applying hard-facing to small screws (diameter less than 40 mm) is more difficult than applying hard-facing to large screws. For this reason, smaller screws are often manufactured without hard-facing. For small screws it is economical to make the screw out of a wear-resistant tool steel, such as D2 or CPM, because the weight of the screw is small and the material cost relatively low. For larger screws the material cost is a more important issue and it is more important to use a low-cost steel.

Hard-facing can offer substantial improvements in wear resistance of heat-treated steels. Lucius reported increases in service life by a factor of 2 to 25 using hard-facing on a screw used in extrusion of glass fiber reinforced nylon 6,6 [48]. Two common hard-facing materials are Stellite and Colmonoy. Colmonoy is a nickel-based alloy containing chromium, iron, boron, and silicon. Stellite is a cobalt-based alloy containing chromium and tungsten. Other hard-facing materials are Haynes 711, HC-250, Triballoy T-700, Ferro-Tic HT6 and M6, Nye-Carb, and Xaloy 008, 830, 101, 306, and 800. Properties of various hard-facing materials are given in Table 3.6 [49]. Metal-to-metal wear tests of these materials were described by McCandles and Maddy [35]. These materials are applied by welded using tungsten inert gas (TIG), transferred arc plasma, and oxyacetylene.

Table 3.6 Properties of Hard-Facing Materials

Product	Base material	Hardness, R_c	Cracking tendency	% Carbon	% Chromium	% Tungsten	% Boron	Cost/ lb [$]
Stellite 1	Cobalt	48-54	High	2.5	30.0	12	–	25-40
Stellite 6	Cobalt	37-42	Medium	1.1	28.0	4	–	25-40
Stellite 12	Cobalt	41-47	Medium	1.4	29.0	8	–	25-40
Colmonoy 5	Nickel	45-50	Medium	0.65	11.5	–	2.5	15-25
Colmonoy 56	Nickel	50-55	High	0.70	12.5	–	2.7	15-25
Colmonoy 6	Nickel	56-61	High	0.75	13.5	–	3.0	15-25
Colmonoy 83	Nickel	50-55	High	2.0	20.0	34	1.0	40-50
N-45	Nickel	30-40	Medium	0.3	11.0	–	2.2	15-25
N-50	Nickel	40-45	Medium	0.4	12.0	–	2.4	15-25
N-56	Nickel	45-50	High	0.6	13.5	–	2.8	15-25

Molybdenum-based hard alloys have also been used on extruder screws. These materials are relatively soft, approximately 40 Rockwell C, but have good lubricity. In some instances, molybdenum hard-facing improved wear resistance approximately 500% over the more common, but harder, hard-facing compounds. Molybdenum-based hard-facing alloys are used primarily with nitrided barrels, not only in single-screw extruders but also in twin-screw extruders. The use of molybdenum hard-facing in bimetallic barrels often results in rapid wear of the screw. Since most extruders in the U.S. have bimetallic barrels, molybdenum-based hard-facing for extruder screws has not found widespread use.

In some cases, improvements in wear resistance can be obtained by diffusion coating hard-facing alloys. Panzera and Saltzman [50] tested treated and untreated hard-facing alloys against carburized SAE 4620 rings using the Alpha Model LFW-1 wear tester. Three casehardening processes were selected: aluminum diffusion coating, boron diffusion coating, and ion nitriding. Case-hardened cobalt-based hard-facing alloys exhibited significantly improved wear resistance against carburized SAE 4620 steel, but nickel-based hard-facing alloys were unaffected by ion nitriding. The wear resistance of aluminized cobalt-based alloys improved after a porous outer layer was ground off. Boriding reduces the wear resistance of both nickel- and cobalt-based facing alloys. Removing the porous outer layer improved the wear properties of the borided nickel-based alloy.

Panzera and Saltzman [50] also investigated the effect of work hardening on hard-facing alloys using shot peening. The cobalt-based alloy hardened to a depth of about 250 μm, but the nickel-based alloy did not work harden. The cobalt-based alloy's wear resistance against carburized SAE4620 steel (as measured on the LFW-1 wear tests) did not improve with work hardening.

Another process that has been used to surface-harden extruder screws is chemical vapor deposition (CVD). This process and some of its applications have been described by Bonetti [51]. This process has been used to apply a thin layer (approximately 4 microns) of a very hard titanium nitride coating to the screw surface. Hardness values of about 110 Rockwell C can be achieved. Obviously, with a screw of such extreme hardness, it is imperative that the barrel material is compatible with the screw material.

3.2.5.5 Rebuilding Worn Screws and Barrels

In a correctly designed extruder, the majority of the wear should be concentrated on the screw because the screw can be replaced and rebuilt more easily than the barrel. In fact, the rebuilding of extruder screws has become so common that the rebuilding business has become a major segment of the extrusion industry. There are more than 70 companies in the U.S. involved in the rebuilding of extrusion equipment. For a number of these companies, screw rebuilding constitutes the major part of their business.

One reason for the popularity of screw rebuilding is that rebuilding is usually considerably less expensive than replacement. Rebuilding is usually done with hard-facing materials. With the proper choice of hard-facing material, the rebuilt screw can be better than the original screw. It usually makes no economic sense to rebuild extruder screws having diameters of less than 40 mm because the cost of rebuilding may be the same or higher than the manufacture of a new screw. Also, applying hard-facing to a worn, small-diameter screw is difficult, and the results are often less than satisfactory. However, larger diameter screws can be hard-faced without difficulty and generally can be rebuilt numerous times. Properties of several hard-facing materials were listed in Table 3.6 [49].

The steps involved in rebuilding a screw are [49]:

1. The screw is set up in a lathe and the center is found. At this time the screw is checked for straightness and concentricity;
2. The screw is polished and prepped for stripping of the existing chrome;
3. The entire screw is submerged in an acid bath to remove the chrome plating;
4. The screw is moved to the grinder where it is ground undersize;
5. The screw flights are welded with a hard-facing material such as Colmonoy 56 or Stellite 12;
6. The screw is again moved to the grinder for rough grinding after welding, where it is also checked for straightness;
7. The flight grinder is used to trim the sides of the flight;
8. The screw is moved to the polishing booth for a rough polish;
9. The screw is inspected and buffed for chrome plating if needed;
10. Chrome plating is applied to the entire root and bearing surface;
11. The screw is buffed after plating;
12. The screw is ground to the final OD specification;
13. Final polishing and buffing is applied to the screw as needed;
14. Front surface of the screw is ground, size registered, and the OD bored;
15. Final inspection.

Application of Hard-Facing Materials

There are four commonly used hard-facing techniques in the industry [52]: oxy-acetylene, tungsten inert-gas (TIG), plasma transfer arc (PTA), and metal inert-gas (MIG), each of which has certain advantages and disadvantages, discussed later. Sometimes a layer of stainless steel is applied to the flight before applying the hard-facing material to control the dilution of the hard-facing material with the screw base material, to improve the bond, and to reduce the cracking of the hard-facing. Table 3.7 provides a comparison of the various methods.

Table 3.7 Comparison of Different Welding Methods

Method	Speed of application	Metal dilution	Integrity of the weld	Ease of automation
Oxyacetylene	Poor	Good	Fair	Poor
Tungsten inert-gas	Good	Fair	Good	Fair
Plasma transfer arc	Fair	Fair	Good	Good
Metal inert-gas	Excellent	Poor	Good	Excellent
Laser	Excellent	Excellent	Excellent	Excellent

Oxyacetylene Welding

In this process, an intense flame is produced by burning a controlled mixture of oxygen and acetylene gas (Figure 3.23). The gases are drawn from separate sources through pressure regulators, introduced into a torch for mixing, and then exit the welding nozzle where they are ignited. The flame intensity depends on the flow rate of the gases, the gas mixture ratio, the properties of the fuel gas selected, and the type of nozzle used. Welds are formed from the weld puddle created by contact of flame, work piece, and welding rod. Oxyacetylene welding requires a high degree of skill to obtain high quality deposits, and the process is slow. The benefit of oxyacetylene welding is that it provides the least base metal dilution of any non-laser method. A one-layer deposit is usually sufficient to reach the desired hardness.

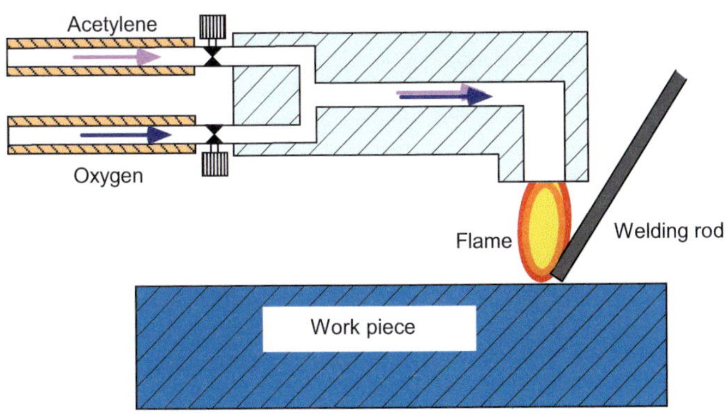

Figure 3.23 Oxyacetylene welding

Tungsten Inert-Gas Welding

Tungsten inert-gas welding is an arc fusion welding process in which intense heat is produced by an electric arc between a non-consumable, torch-held tungsten electrode and a work piece (Figure 3.24). An inert shielding gas, generally argon, is introduced through the torch to protect the weld zone from atmospheric contamination. TIG welding is the method most commonly used in the manufacturing and rebuilding of extruder screws. The localized intense heat of TIG results in some base metal dilution, and as a result, it may be necessary to apply a second layer to achieve full hardness of the hard-facing material.

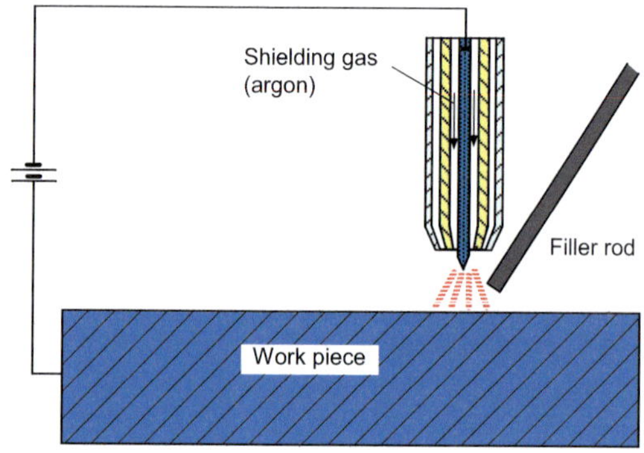

Figure 3.24 Tungsten inert-gas welding process

Plasma Transfer Arc Welding

A PTA torch consists of an electrode in the center, surrounded by a double-walled tube which carries the powdered metal. Argon gas passes through this annulus and is circulated around the welding zone to provide a shield around the arc region, while metal powder is metered through the holes in the inside wall of the tube. Both exit onto the work piece through an arc struck between the electrode and the work piece (Figure 3.25).

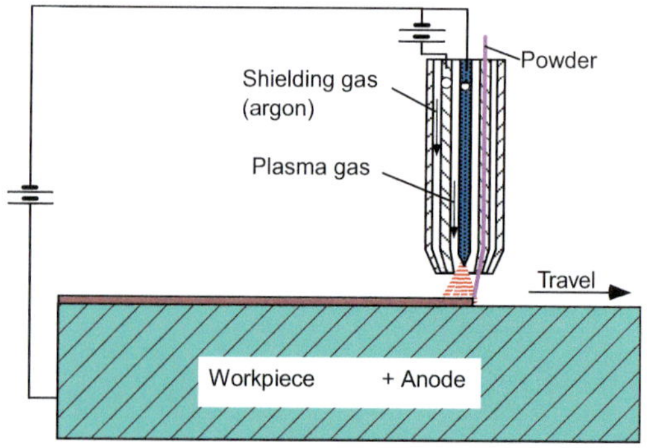

Figure 3.25 Plasma transfer arc welding process

Metal Inert-Gas Welding

In the MIG welding process, an electric arc is established between the work piece and a wire electrode, which is continuously fed through a torch by a wire feeder. The arc continuously melts to form the weld puddle, and an appropriate gas or gas mixture shields the weld area from atmospheric contamination. The MIG process has the advantages of high deposition rates, speed, and excellent weld quality, but results in more base metal dilution than other processes. As a result, application of a second layer of weld may be necessary to achieve the desired hardness.

Laser Hard-Facing

Another method of applying hard-facing is laser hard-facing on the flight lands. The common hard-facing materials listed in Table 3.6 can be applied by laser hard-facing along with tungsten carbide composites. The tungsten carbide particle can be spherical or angular in shape. Important benefits of the laser process are the low heat input, the minimal dilution of the deposited alloy, the large variety of (powder fed) hard-facing compositions, and the fact that overlays are metallurgically bonded and impervious. These characteristics lead to high quality overlays without cracks, minimal porosity, and high hardness values. For instance, Colmonoy 56 PTA powder can be laser deposited with resulting hardness values in the low 60s R_C. A final thickness of 0.4 to 0.15 mm can be achieved for crack-sensitive materials. The final thickness achievable for less crack-sensitive materials such as Stellite 6 is not limited because multi-layer deposits can be applied. This method can always be used on new screws; however, its use in rebuilding screws is generally evaluated on a case-by-case basis.

Rebuilding of Extruder Barrels

Rebuilding barrels is usually considerably more difficult than rebuilding screws. If the barrel wear does not exceed 0.5 mm, the whole barrel can be honed to a larger diameter and an oversized screw can be placed in the machine. The obvious disadvantage of this procedure is the fact that nonstandard barrel and screw dimensions result. Thus, screws from other machines can no longer be used in the nonstandard extruder. If barrel wear occurs near the end of the barrel, a sleeve can be placed in the barrel. In most cases, however, the barrel wear is such that replacement of the barrel makes more sense than sleeving or increasing ID by honing.

3.2.6 Screw Binding

One particular problem in extruders occurs when the screw suddenly stops rotating and gets stuck in the extruder barrel and/or feed housing. This problem does not occur very often, but when it does, it wreaks havoc with the machine. The most frequent cause of this problem is a difference in thermal expansion between the screw and barrel. Corrosion-resistant screws commonly used in the extrusion of fluoropolymers are particularly susceptible to screw binding. This usually results in considerable damage to the machine and substantial downtime. We will analyze the mechanism of screw binding and give recommendations on how to prevent this problem [53].

3.2.6.1 Extrusion of Fluoropolymers

In the extrusion of fluoropolymers, the extruder screw and die must generally be made out of a highly corrosion-resistant material. Common materials used for this purpose are Hastelloy, Monel, Inconel, and Duranickel. These materials exhibit much better corrosion resistance than the typical 4140 steel used for most extruder screws, but they also have other properties that are quite different from 4140 steel. These other properties make these corrosion-resistant screws much more susceptible to screw binding than screws made of 4140 steel. Since screw binding usually results in considerable damage to the extruder with associated downtime and cost, not to mention aggravation, it is important for processors to be aware of the pitfalls of using screws of highly corrosion-resistant materials.

3.2.6.2 The Mechanics of Screw Binding

When a screw is installed in an extruder, the typical radial clearance between the screw and the barrel is $0.001D$, where D is the inner diameter of the extruder barrel. This is the clearance at room temperature. When the machine is in operation, the actual clearance between the screw and the barrel can be quite different. The two main causes for the change in clearance under actual processing conditions are the processing temperature and compressive load on the screw. When the processing temperature is much higher than room temperature, the clearance can change

- when the screw and barrel have different coefficients of thermal expansion (CTEs) and
- when the temperature of the screw is different from that of the barrel.

3.2.6.3 Changes in Clearance Due to Temperature Differences

When the temperature of the screw and barrel increases by the same amount, both the screw and barrel will increase in diameter due to thermal expansion. If the CTE of the screw is greater than the one of the barrel, the clearance between the screw and barrel will decrease with increasing temperature. Values of the CTE for several materials are shown in Table 3.8 together with data on thermal conductivity and elastic modulus.

Table 3.8 Thermal Properties and Elastic Modulus for Several Screw Materials

Material	Coefficient of thermal expansion [/°C]	Thermal conductivity [J/m·s·K]	Elastic modulus [MPa]
Hastelloy C276	11.16×10^{-6}	11.25	2.00×10^{5}
Inconel 718	12.96×10^{-6}	11.42	2.00×10^{5}
Inconel 625	12.78×10^{-6}	9.86	2.07×10^{5}
Monel 400	13.86×10^{-6}	21.80	1.79×10^{5}
Monel 500	13.68×10^{-6}	17.47	1.79×10^{5}
4140 steel	11.34×10^{-6}	42.56	2.00×10^{5}
4340 steel	11.34×10^{-6}	42.21	2.00×10^{5}
17-4 stainless	10.44×10^{-6}	17.82	2.00×10^{5}
316 stainless	18.54×10^{-6}	16.09	2.00×10^{5}
304 stainless	18.72×10^{-6}	16.26	2.00×10^{5}

For a 25.40-mm barrel running at 333.3 °C above room temperature, the increase in ID is 0.0965×10^{-2} mm when the CTE is 11.34×10^{-6}/°C. For a 25.3492-mm Monel screw with a CTE of 13.86×10^{-6}/°C running at 333.3 °C above room temperature, the increase in screw diameter will be 0.1168 mm. Thus, the difference between the thermal expansion of the screw and the thermal expansion of the barrel diameter is approx. 0.02 mm, or 0.01 mm based on the radius. If the radial clearance is 0.0254 mm, the clearance will decrease to 0.01524 mm due to the differential thermal expansion. Thus, the clearance is decreased but still greater than zero, provided both the screw and barrel are at the same temperature.

In an operating extruder, however, it is not likely that the screw and barrel will be at the same temperature. The largest difference between the screw and barrel temperature is likely to occur in the feed throat. The feed throat of most extruders is water cooled and, therefore, close to room temperature in many cases. The screw temperature in the feed section, however, can be (and in many cases will be) much higher because thermal conduction of the high temperatures in the compression and metering section will cause the temperature of the feed section to rise.

If the feed throat is maintained at room temperature and the screw temperature in the feed section is 166.7 °C higher, the screw diameter in that section will increase from 25.3492 to 25.4076 mm, if the CTE = 13.86×10^{-6}/°C. This corresponds to 0.0023 mm per mm of screw diameter. The screw diameter now is larger than the diameter of the feed throat and the screw will bind! The question can be asked, is this temperature difference not equally likely to occur in a screw made out of 4140 steel? The answer is no, and the reason has to do with the thermal conductivity of these materials. The thermal conductivities of corrosion-resistant metals tend to be considerably lower than that of steel, by a factor of three to five (see Table 3.8).

The lower thermal conductivity of corrosion-resistant materials will reduce the amount of heat that can be transferred from the screw shank to the reducer. As a result, the shank and feed section of the screw will be at a higher temperature than a high conductivity screw. Figure 3.26 shows the thermal expansion graphed against the temperature difference between the screw and barrel. The graph shows several values of the coefficient of thermal expansion. From Figure 3.26, it is clear that it takes a temperature difference of 165 to 220 °C to cause the screw to lock up in the barrel, considering that the coefficient of thermal expansion is in the range of 10×10^{-6} to 17×10^{-6}/°C. It is clear that if a polymer is processed at 370 °C, it is quite possible that the screw temperature will be more than 165 °C above the barrel temperature.

When viscous heating is significant, the screw temperature will tend to be higher than the barrel temperature, at least with a neutral screw. Janssen et al. [54] found that in extruders without screw cooling, the screw temperature gives a much better indication of the mean polymer temperature than does the barrel temperature. Finite element analysis of non-isothermal, non-Newtonian flow in extruders [55] showed that screw temperatures tend to be higher than barrel temperatures when viscous heating is significant. As a result, the screw temperature in the metering section of the screw may be significantly higher than the barrel temperature and, therefore, the temperature difference between screw and barrel in the feed section may be greater than what might be assumed based on measured barrel temperatures.

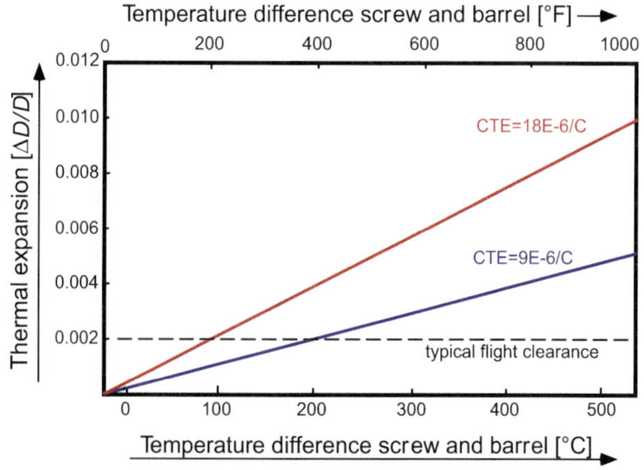

Figure 3.26 Thermal expansion versus temperature difference

There are few publications available to provide data on screw and plastic temperatures along an extruder. The publication by Marshall et al. [56] provides some interesting experimental data. Their research confirms that the screw temperature in the metering section is higher than the barrel temperature. Further, screw temperatures in the feed section were found to be in the range of 115 to 127 °C, while barrel temperatures were at 190 °C. When barrel temperatures are around 370 °C, it can be expected that the temperature of the feed section of the screw will be in the range of 200 to 260 °C, if not higher.

3.2.6.4 Analysis of Temperature Distribution in Extruder Screws

In order to determine whether the mechanism for screw binding proposed in the previous section is correct, predictions of the temperature distribution in extruder screw processing FEP were made using finite element analysis. The FEHT program developed at the University of Wisconsin-Madison was used [57]. Figure 3.27 shows the thermal boundary conditions that were used in the analysis; 697 nodes were used with 1280 triangular elements. The feed throat temperature was set at 15.6 °C; the barrel temperatures were set at 288, 371, 371, and 371 °C; the screw shank was set at 93 °C; the screw tip was at 371 °C; and the heat flux at the screw centerline was zero in radial direction. The program does not take into account viscous dissipation or convection; the heat transfer is purely by conduction. The thermal conductivity of the FEP is taken as 0.246 J/m·s·K. Figure 3.28 shows the predicted temperature distribution in a screw made out of 4140 steel. Figure 3.29 shows the temperature distribution in a Monel screw.

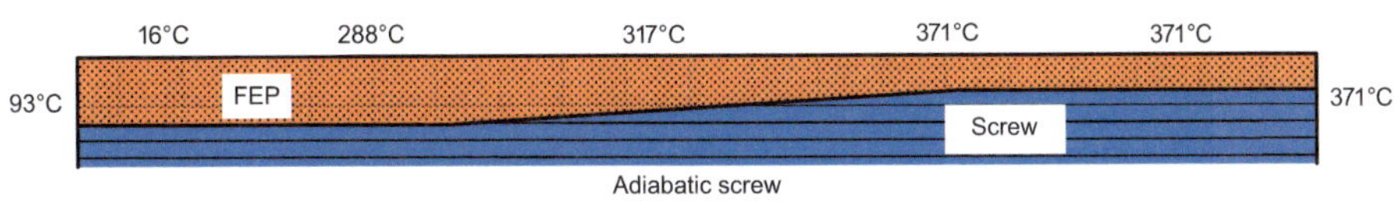

Figure 3.27 Schematic of thermal boundary conditions for FEA

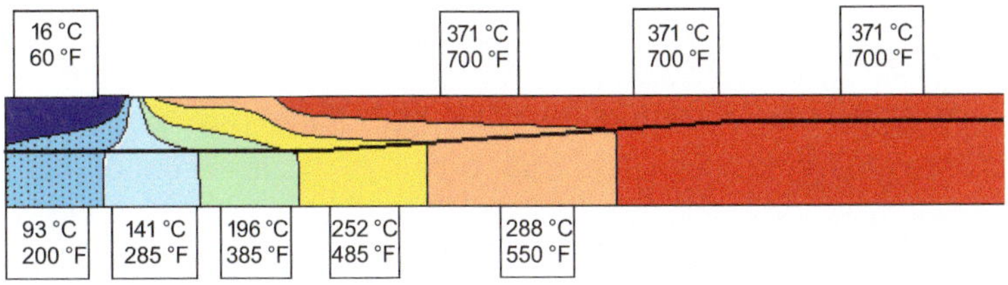

Figure 3.28 Predicted temperature distribution in 4140 screw

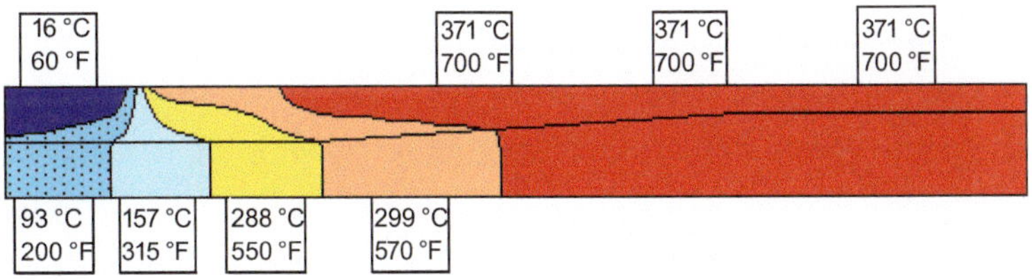

Figure 3.29 Predicted temperature distribution in Monel screw

When Figure 3.28 and Figure 3.29 are compared, it is clear that higher temperatures occur in the feed section of the Monel screw. This must be due to the thermal conductivity because that is the only difference between the two cases. These studies predict that a screw made from a material of low thermal conductivity can develop higher temperatures in the feed section of the screw during the extrusion process. In the case of the Monel screw, the screw temperatures in the feed section ranged from about 150 to 260 °C.

3.2.6.5 Change in Clearance Due to Compressive Load

When an extruder screw presses upon the plastic melt to force it through a die, the pressure at the end of the screw will cause a compressive thrust load on the screw. As a result, the length of the screw will decrease, and the diameter of the screw will increase. The relative increase in the screw diameter can be expressed as:

$$\frac{\Delta D}{D} = \frac{\Delta L}{2L} = \frac{P}{2E} \tag{3.1}$$

where D is diameter, ΔD the increase in diameter, L the length, ΔL the increase in length, P the pressure, and E the elastic modulus.

Values of the elastic modulus for several materials are shown in Table 3.8. If the pressure is 34.5 MPa (5000 psi) and the modulus 2.07×10^5 MPa, the $\Delta D/D$ = 8.3×10^{-5}. Thus, for a 25.4-mm screw, the increase in diameter will be 0.0021 mm. Therefore, the increase in diameter due to compressive load is quite small compared to the effect of differential thermal expansion and as a result, radial expansion of a screw due to compressive load is likely to be only a minor factor in the chance of the screw locking up in the barrel.

3.2.6.6 Results from Analysis

The analysis above confirms that corrosion-resistant screws do indeed have a greater chance of locking up in an extruder than screws made out of regular 4140 steel. The main reason for this appears to be the low thermal conductivity of highly corrosion-resistant metals, which causes a large temperature difference to develop between the feed throat and the feed section of the screw during the extrusion process. The higher screw temperature will cause the screw to expand more than the feed throat and the barrel, possibly causing the screw to bind. Finite element analysis results confirm that the lower thermal conductivity of a corrosion-resistant screw leads to a higher temperature in the feed section of the screw.

The reason that screw-binding problems often occur with highly corrosion-resistant materials is that these screws are typically used for fluoropolymers processed at high temperatures (approx. 370 °C or 700 °F). In this case, there are several factors that make screw binding more likely. First, at such high process temperatures there will be a higher temperature difference between the screw and the barrel in the feed section. Second, this temperature difference will be greater in a screw made of a highly corrosion-resistant material versus one of 4140 steel, due to its lower thermal conductivity. Third, because such highly corrosion-resistant materials have a higher coefficient of thermal expansion than 4140 steel, a screw of such a material will expand more.

Several measures that can be taken to reduce the chance of screw binding include:

- Cooling of the screw in the feed section
- Increasing the temperature of the feed throat
- Decreasing the temperatures in the transition section
- Decreasing the temperatures in the metering section
- Increasing the clearance in the feed throat region

In most cases, the best way to avoid binding problems is to reduce the screw diameter in the feed section by at least 0.002 mm per mm of screw diameter. Since most plastics are fed in pellet form, increasing the flight clearance in the early part of the feed section is not likely to affect the performance of the extruder, but this will substantially reduce the chance of the screw locking up in the extruder barrel or feed throat.

3.3 Polymer Degradation

Polymer degradation is a frequent problem in extrusion. Degradation usually manifests itself as discoloration, release of volatile components (smoking), or changes in mechanical properties. Five types of degradation can be distinguished according to their mode of initiation: thermal, chemical, mechanical, radiation, and biological. Degradation processes are generally quite complex, often involving more than one type of degradation. Thermo-oxidative degradation and thermo-mechanical degradation are examples of dual types of degradation. This situation is quite similar to wear in extruders, in which usually more than one wear mechanism is operational at any one time.

3.3.1 Types of Degradation

In extrusion, the first three types of degradation, thermal, mechanical, and chemical degradation, are the most important.

3.3.1.1 Thermal Degradation

Thermal degradation occurs when a polymer is exposed to an elevated temperature in an inert atmosphere in the absence of other compounds. The resistance against such degradation depends on the nature and the inherent thermal stability of the polymer backbone. There are three main types of thermal degradation: depolymerization, random chain scission, and unzipping of substituent groups.

Depolymerization and unzipping involve a reduction in the length of the main chain by sequential elimination of monomer units. Polymers such as polymethylmethacrylate, polyformaldehyde, and polystyrene degrade by this mechanism. Polystyrene unzips to some extent during degradation, although only approx. 40% is converted to monomer. Unzipping of substituent groups is an important thermal degradation mechanism because it is the primary breakdown process for polyvinylchloride. Random scission occurs in many polyolefins because of their simple carbon chain backbone.

It is often difficult to distinguish between thermal and thermo-chemical degradation because polymers are rarely chemically pure. Impurities and additives can react with the polymeric matrix at sufficiently high temperatures.

3.3.1.2 Mechanical Degradation

Mechanical degradation refers to molecular scission induced by mechanical stresses. These stresses can be shear stresses or elongational stresses or a combination of the two. Mechanical degradation of polymers can occur in the solid state, in the molten state, and in solution. An extensive review of the field of mechanically induced reactions in polymers was published by Casale and Porter [58]. In an extruder, mechanical stresses are applied mostly to the molten polymer.

Various theoretical approaches have been developed to describe mechanical degradation. One of the earlier studies was made by Frenkel [59] and Kauzmann and Eyring [60], who proposed that in a shear field, linear macromolecules are extended in the direction of motion. The strain on the molecule is primarily concentrated at the middle of the chain. No degradation is expected when the degree of polymerization is below a certain critical value. Bueche [61] predicted that entanglements also produce preferential tension in the mid-section of macromolecules. Thus, chain scission is more likely to occur in the center of the chain. He also predicted that main chain rupture would increase dramatically with increasing molecular weight.

These theoretical considerations suggest that mechanical degradation in polymer melts or solutions is a non-random process, producing new low molecular weight species with molecular weights of one-half, one-fourth, one-eighth (and so forth) of the original molecular weight. Because of the elevated temperature of polymer melts, mechanical degradation is essentially always combined with thermal degradation, and possibly with chemical degradation. When a polymer melt is exposed to intense mechanical deformation, and the rate of deformation is non-uniform, local temperatures can rise substantially above the bulk temperature. Thus, bulk temperature measurements may not properly reflect actual stock temperatures. This is the case in screw extruders and in high-intensity internal mixers, where very high local temperatures can occur. In such devices, pure mechanical degradation is unlikely to occur. Therefore, degradation processes in polymer melts involving mechanical stresses tend to be rather complex.

Some workers have reported that degradation at processing conditions is almost exclusively thermal [62, 63], while others conclude that degradation is mainly mechanical [64, 65]. Most workers, however, deduce that, though the nature of degradation is basically thermal, there is a distinct reduction in the temperature necessary for reaction due to mechanical energy stored within the polymer chains as a result of mechanical deformation. This corresponds to a shear-induced change in the potential energy function for thermal bond rupture as proposed by Arisawa and Porter [66]. In practice this means that if the polymer is exposed to a mechanical deformation, the polymer induction time determined under quiescent conditions will be longer than the actual induction time.

Because of the aforementioned complications with mechanical degradation in polymer melts, mechanical degradation can be more easily studied in polymer solutions. Casale and Porter [58] reviewed most of the work in this area up to 1978. Work in this area published between 1978 and 1984 is summarized in a later publication [67]. A more recent publication by Odell, Keller, and Miles [68] describes an elegant technique for continuously monitoring the molecular weight distribution (MWD) of a polymer solution undergoing mechanical deformation. A cross-slot device was used in this study to apply a pure elongational flow field to dilute solutions of narrow MWD atactic polystyrene. Information on the MWD of this polymer was obtained by measuring birefringence. Repeated breakage of the stretched molecules at their centers was observed, as shown by the MWD before and after mechanical deformation of the polymer (Figure 3.30).

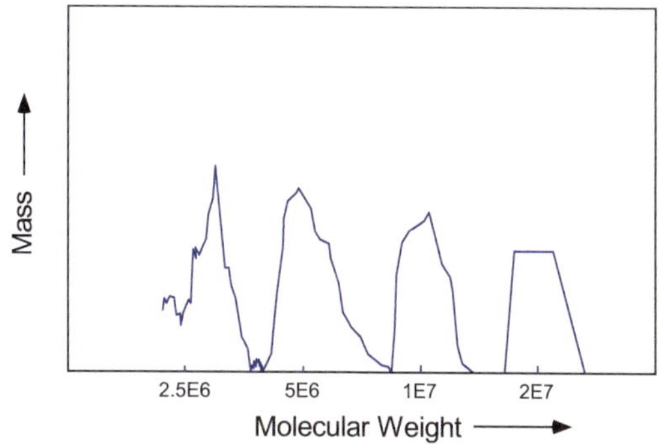

Figure 3.30 Molecular weight distribution after mechanical deformation

3.3.1.3 Chemical Degradation

Chemical degradation refers to processes that are induced by chemicals in contact with a polymer. These chemicals can be acids, bases, solvents, reactive gasses, and so on. In many cases, a significant conversion is observed only at elevated temperatures because the activation energy for these processes is high. Two important types of chemical degradation are solvolysis and oxidation. Solvolysis reactions involve the breaking of C–X bonds, where X represents a non-carbon atom. Hydrolysis is an important type of solvolysis. The reaction basically occurs as follows:

$$-\overset{|}{\underset{|}{C}}-X-\overset{|}{\underset{|}{C}}- \; + \; H_2O \longrightarrow -\overset{|}{\underset{|}{C}}-X-H \; + \; HO-\overset{|}{\underset{|}{C}}-$$

This type of degradation occurs in polyesters, polyethers, polyamides, polyurethanes, and polydialkylsiloxanes. Polymers that tend to absorb water are the most likely to undergo hydrolysis. For instance, during the extrusion of polyester or polyamide, it is very important that the polymer be properly dried beforehand. The stability of polymers against solvolytic agents is important in many applications. Such important polymers as PVC, PMMA, PA, PC, PETP, polyurethane (PU), polyacrylonitrile (PAN), and polyoxymethylene (POM) have poor stability against acids and bases at room temperature. Polyolefins and fluoropolymers tend to exhibit good stability against these solvolytic agents.

Oxidative degradation is a very common type of degradation occurring in polymers during extrusion at elevated temperatures. Thus, the degradation becomes a thermo-oxidative degradation. Polymer degradation starts with the initiation of free radicals. Free radicals have a high affinity for reacting with oxygen to form unstable peroxy radicals. The new peroxy radicals will abduct neighboring labile hydrogens, producing unstable hydroperoxides and more free radicals that will start the same process again. This results in an autocatalytic process, that is, a process that self-propagates once it has started. Under continuous initiation, the reaction rate is accelerated, resulting in an exponential increase in conversion with reaction time. The process will stop when a reacting chemical species is depleted or when the propagation is inhibited by reaction products.

Oxidative degradation in polymers is generally combated with the addition of antioxidants. The purpose of the antioxidant is to intercept radicals or prevent radical initiation during the various phases of a polymer's life: polymerization, processing, storage, and end use. According to their functionality, antioxidants can be classified as primary or secondary antioxidants. Primary antioxidants, or chain terminators (also referred to as free-radical scavengers), interrupt chain reactions by tying up free radicals. Secondary antioxidants, or preventive antioxidants (also referred to as peroxide decomposers), destroy hydroperoxides. Primary antioxidants are primarily hindered phenols and aromatic amines that tie up polymeric peroxy radicals through hydrogen donation, forming polymeric hydroperoxide groups and relatively stable antioxidant species. Secondary antioxidants are various phosphorous or sulfur containing compounds, particularly phosphites and thioesters that reduce hydroperoxides to inert products, thus preventing the proliferation of alkoxy and hydroxy radicals. Selecting an effective antioxidant package is a key factor to the success of a plastic product. Some of the factors that should be considered in the selection of an antioxidant are toxicity, volatility, color, extractability, odor, compatibility, supply, cost, and performance.

3.3.2 Degradation in Extrusion

Degradation during the extrusion process will often be a combination of thermal, mechanical, and chemical degradation. Factors that are important in determining the rate of the various types of degradation are:

- Residence time and residence time distribution (RTD)
- Stock temperature and distribution of stock temperatures
- Deformation rate and deformation rate distribution
- Presence of solvolytic agents, oxygen, or other degradation promoting agents
- Presence of antioxidants and other stabilizers

The first three factors are strongly influenced by the machine geometry and by operating conditions. The presence of solvolytic agents or oxygen can also be influenced by operating conditions. For example, oxygen can be eliminated from the extruder by putting the feed hopper under a nitrogen blanket. The presence of antioxidants and other stabilizers is part of the material selection process. Proper selection of a stabilizer package is very important; however, the details of how a troubleshooter can determine the right stabilizer package are outside the scope of this book.

3.3.2.1 Residence Time Distribution

The RTD of an extruder is a valuable piece of information concerning the details of the conveying process in the machine. The RTD is directly determined by the velocity profiles within the machine. Thus, if the velocity profiles are known, the RTD can be calculated. Various workers have made theoretical calculations of the RTD in single-screw extruders [69–71] using velocity profiles within the machine. Obviously, the predicted RTD is only as accurate as the velocity profiles that formed the basis of the calculations. In single-screw extruders, the velocity profiles can be determined reasonably well, although usually a substantial number of simplifying assumptions are made. Calculation of velocity profiles for other screw extruders, for example, twin-screw extruders, is rather complex, and thus, prediction of the RTD more difficult.

Experimental determination of the RTD of an extruder yields information about the conveying process in the extruder that is useful in a number of other areas as well as in the analysis of the risk of degradation in the machines. The RTD can be used to analyze the mixing process in an extruder. When an extruder is used as a continuous chemical reactor, the RTD provides important information for process design and process analysis. The RTD also serves as a good selection criterion; for instance, an extruder used in profile extrusion should have a narrow RTD and short residence time. Experimental studies of RTD in single-screw extruders have been reported by a number of investigators [70, 72–76]. Experimental studies of RTD in twin-screw extruders have been reported by Todd [77], Janssen et al. [78, 79], Rauwendaal [74], Walk [80], and Nichols et al. [81].

The RTD is determined by measuring the output response after a change in input. This is referred to as the stimulus response method as discussed by Levenspiel [82] and Himmelblau and Bischoff [83]. The system is disturbed by a stimulus and the response of the system to the stimulus is measured. Two common stimulus response protocols are the step input response protocol and the pulse input response protocol (Figure 3.31).

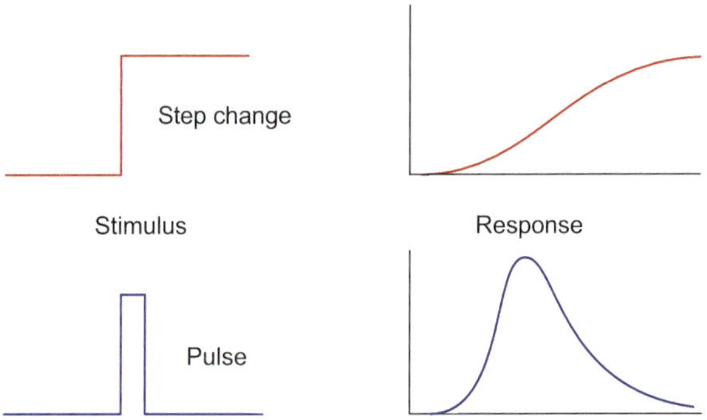

Figure 3.31 Response to step change and pulse input

The response to a step input is an S-shaped curve (Figure 3.31, top). Other stimuli that can be used are random input and sinusoidal input. The response to a pulse input is a bell-shaped curve (Figure 3.31, bottom). The ideal pulse input is of an infinitely short duration; such an input is called a delta function or impulse. The normalized response to a delta function is called the C-curve and the total area under the curve is assigned a value of one.

The definitions of RTD functions were developed by Danckwerts [84]. The internal RTD function, $g(t)dt$, is defined as the fraction of fluid volume in the system with a residence time between t and $t + dt$. The external RTD function, $f(t)dt$, is defined as the fraction of the exiting material with a residence time between t and $t + dt$. The cumulative internal RTD function, $G(t)$, is defined as:

$$G(t) = \int_0^t g(t)dt \tag{3.2}$$

$G(t)$ represents the fraction of the fluid volume in the system with a residence time between 0 and t. The cumulative external RTD function is defined as:

$$F(t) = \int_{t_0}^t f(t)dt \tag{3.3}$$

where t_0 is the minimum residence time. $F(t)$ represents the fraction of exiting flow rate with a residence equal to or shorter than t. For very long times t, both functions G and F become equal to unity:

$$G(\infty) = F(\infty) = 1 \tag{3.4}$$

The mean residence time $\bar{t}$ is given by the following expression:

$$\bar{t} = \int_0^\infty t f(t)dt \tag{3.5}$$

The mean residence time is determined by the volume of the machine V, the degree of fill X of the machine, and the volumetric flow rate $\dot{V}$:

$$\bar{t} = \frac{XV}{\dot{V}} \tag{3.6}$$

The relationship between the internal RTD function and the cumulative external RTD function is given by:

$$g(t) = \frac{1 - F(t)}{\bar{t}} \tag{3.7}$$

In the flow of a Newtonian fluid through a pipe, the RTD can be calculated rather easily using the expression for the velocity profile given in [161]. The external RTD function is:

$$f(t) = \frac{2t_0^2}{t^3} \quad \text{for } T \geq t_0 \tag{3.8}$$

where the minimum residence time is:

$$t_0 = \frac{4\mu L^2}{\Delta P R^2} \tag{3.9}$$

Figure 3.32 shows a typical cumulative RTD curve for a single-screw extruder as determined experimentally [74]. The curve is for a 25-mm extruder running at 20 rpm with an output of 2.3 kg/h. The mean residence time in this example is

5.9 minutes. This type of information is useful because one can easily tell the fraction of the material spending a certain time in the machine. For instance, in the situation covered by Figure 3.32, more than 1% of the material is inside the extruder for a residence time of 17.7 minutes, three times the mean time. If the induction time for degradation of the material at the process temperature is less than 17.7 minutes, 1% of the material can be expected to be degraded.

Figure 3.33 shows several cumulative RTD curves for an intermeshing counter-rotating twin-screw extruder [74]. The shape of the curve changes substantially when the processing conditions are changed. The narrowest RTD is obtained by running the extruder at low speed and high output. Figure 3.34 shows several cumulative RTD curves for an intermeshing co-rotating twin-screw extruder [74]. Clearly, the curves show considerable deviations from positive conveying characteristics for the co-rotating twin-screw extruder.

A major advantage of these normalized RTD curves is that conveying characteristics of different extruders can be directly compared. Together, Figure 3.32, Figure 3.33, and Figure 3.34 make clear that the conveying characteristics of the single-screw extruder are quite positive compared to the two twin-screw extruders. This is partially due to the plug flow of the solid bed in the single-screw extruder. The solid bed in a twin-screw extruder is not continuous and generally does not extend over a large proportion of the length of the machine.

The RTD is strongly dependent on the screw design and the operating conditions. Kemblowski and Sek [76] discussed the influence of these factors on the RTD of single-screw extruders in some detail, and the author (CR) [74] has done the same for twin-screw extruders.

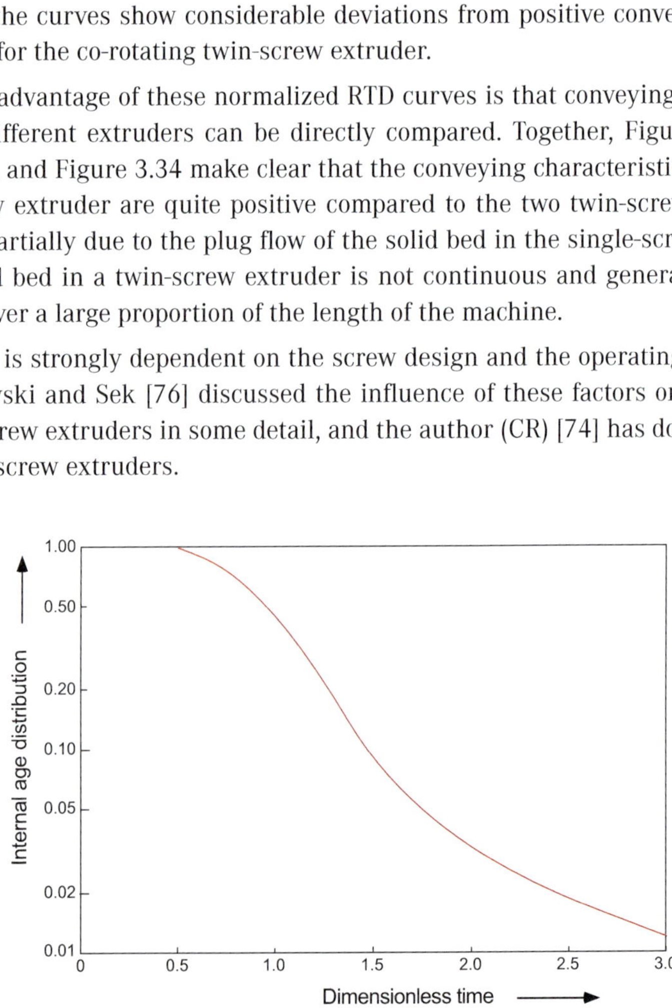

Figure 3.32 Residence time distribution of a single-screw extruder

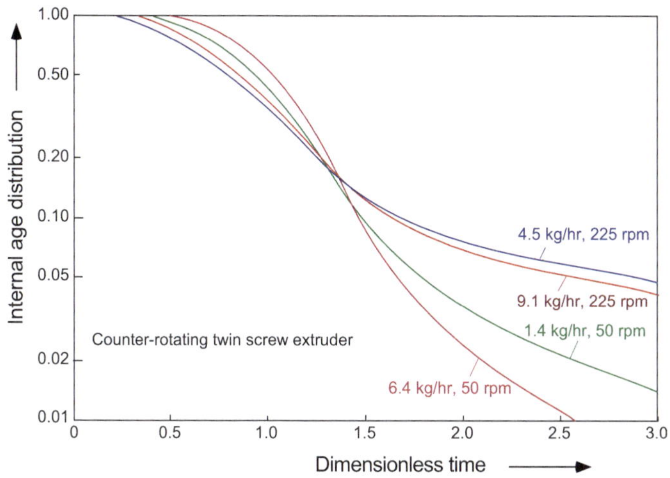

Figure 3.33 Residence time distribution of a counter-rotating twin-screw extruder

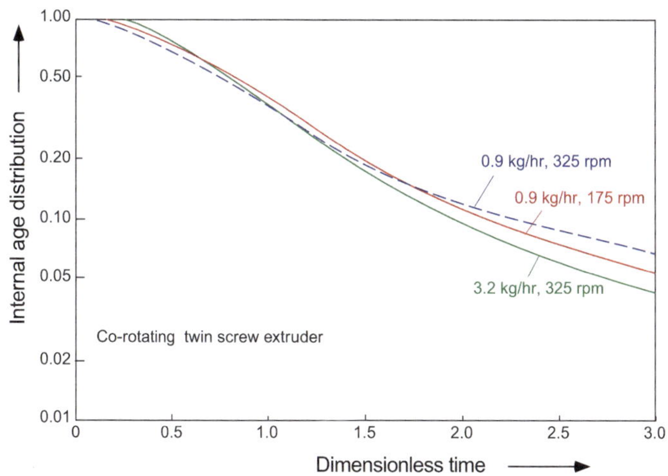

Figure 3.34 Residence time distribution of a co-rotating twin-screw extruder

3.3.2.2 Temperature Distribution Simple Calculations

Obviously, the residence time and its distribution only partially determine the chance of degradation in an extruder. Other factors that play an important role are the actual stock temperatures and the strain rates to which the polymer is exposed. These two factors are closely related. In the extruder, there are two major areas of concern, the screw channel and the flight clearance. Janssen, Noomen, and Smith [85] studied temperature distribution of the polymer melt in the screw channel. Temperature distribution of the polymer right after the end of the screw was measured, for instance, by Anders, Brunner, and Panhaus [86]. The temperature variations in the screw channel at the end of the screw were reported to be less than 5 to 10 °C and relatively close to the barrel temperature. More recently, Noriega et al. [87] measured melt temperature distribution with a thermocomb and found temperature variations as high as 20 to 30 °C.

The situation in the screw clearance is substantially different from the screw channel. The strain rates in the screw channel are relatively low and the temperature variations are also relatively low. In the screw clearance, however, the strain rates are very high and the stock temperature increase can also be very high. This can be verified by the following simple analysis. The shear rate in the clearance is approximately the Couette shear rate:

$$\dot{\gamma}_{cl} = \frac{\pi DN}{\delta} \tag{3.10}$$

where D is the diameter, N the notational speed, and δ the gap in the flight clearance. The corresponding viscous heat generation per unit volume for a power law fluid is:

$$\dot{E}_{cl} = m\left(\frac{\pi DN}{\delta}\right)^{1+n} \tag{3.11}$$

where m is the consistency index and n the power law index.

If it is assumed that there is no exchange of heat between the polymer melt and the screw and between the polymer melt and the barrel, the average increase in adiabatic temperature can be determined from:

$$\Delta\bar{T}_a = \frac{\dot{E}_{cl}\bar{t}_R}{C_p\varrho} \tag{3.12}$$

where $\bar{t}_R$ is the average residence time of the polymer melt in the flight clearance, C_p the specific heat, and ϱ the melt density. The average residence time in the flight clearance is approximately:

$$\bar{t}_R = \frac{2w}{\pi DN \sin\varphi} \tag{3.13}$$

where w is the flight width and φ the flight helix angle.

Combining Equation 3.10, Equation 3.11, and Equation 3.12, the average rise adiabatic temperature in the clearance can be written as:

$$\Delta\bar{T}_a = \frac{2mw(\pi DN)^n}{\varrho \sin\varphi C_p \delta^{1+n}} \tag{3.14}$$

The average temperature increase is directly proportional to the consistency index, m, and the tangential flight width, $w/\sin\varphi$, and is strongly dependent on the radial clearance, δ, the power law index of the polymer melt, n, and the screw speed, N. Figure 3.35 shows the effect of flight clearance, δ, and the power law index, n, for a 114-mm (4.5-inch) extruder running at 100 rpm; the specific heat is 2250 J/kg·°C, the melt density is 900 kg/m^3, and the consistency index is 10^4 Pa·s^n.

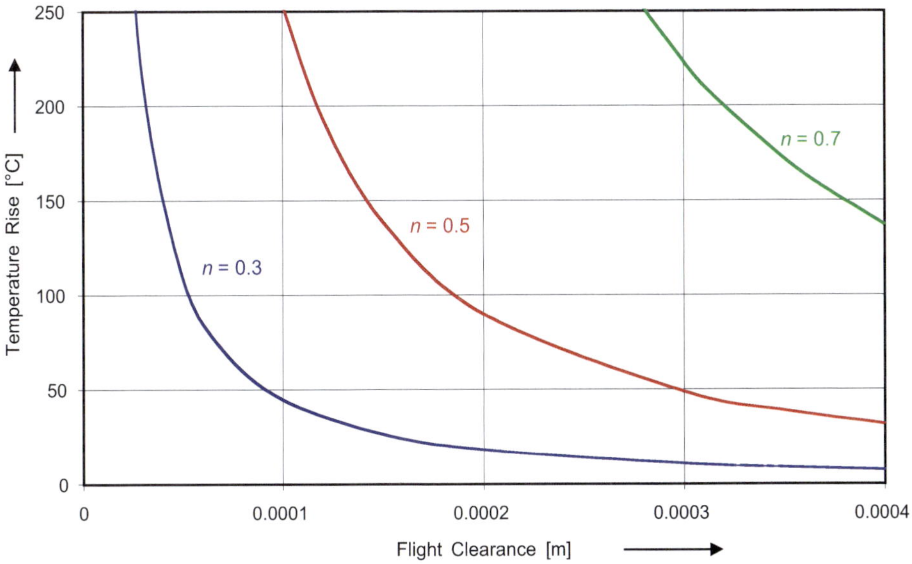

Figure 3.35 Adiabatic temperature rise versus flight clearance

It is clear that the adiabatic temperature rise in the flight clearance can be surprisingly high, especially considering the very short residence time of approximately one-tenth second. However, the actual temperature increase will be less than the adiabatic temperature rise because there will be heat transfer to the screw and to the barrel. In reality, the thermal boundary conditions at the barrel and the flight land will be somewhere between adiabatic and isothermal.

The temperature increase in the clearance can be substantially reduced by simply decreasing the flight width and increasing the flight helix angle. These same measures will also substantially reduce the power consumption of the extruder. Thus, proper design of the screw flight is of great importance in reducing power consumption and reducing the chance of degradation in the extruder.

Another reason that the flight clearance is so important in degradation processes occurring in the extruder is the fact that, in addition to high stock temperatures, the polymer melt is exposed to very high strain rates, both elongational and shear. As discussed earlier, this causes a flow-induced change in the potential energy function for thermal bond rupture. Thus, the degradation will be more severe than it would be based on just the effect of temperature. Obviously, another important point is to eliminate dead spots in the screw design and in the die design. Hangup of material can be very detrimental and should be avoided if at all possible. For instance, fluted mixing sections with a 90° helix angle should not be used with polymers that have a tendency to degrade because such mixing sections have stagnating regions.

The lowest temperature rise will occur in the extreme case that both screw flight surface and barrel surface can be maintained at constant temperature, i. e., isothermal boundary conditions. This situation was analyzed by Meijer, Ingen-Housz, and Gorissen [88] with the primary purpose to determine the thermal development length. They assumed that the clearance flow is dominated by drag flow in tangential direction. The thermal development length for the Newtonian case was found to be approximately 0.36 N_{pe}, where N_{pe} is the Peclet number. Thus, the length required for thermal development can be written as:

$$L_t = \frac{0.36 v_b \delta^2}{\alpha} \tag{3.15}$$

where α is the thermal diffusivity, v_b is the barrel velocity, and δ is radial clearance.

The thermal development length is directly proportional to the barrel velocity v_b (and thus the screw speed) and to the radial clearance squared. With thermal diffusivity values of about 10^{-7} m²/s, the thermal entrance length will be the same order of magnitude as the tangential flight width when the clearance has the normal design value ($\delta \cong 0.001D$). Thus, the temperature profile at the exit of the flight clearance will be very close to the fully developed temperature profile. The fully developed temperature profile for the isothermal case can be written as [88]:

$$T_t = \frac{\mu v_b^2}{2k}\left(\frac{y}{\delta} - \frac{y^2}{\delta^2}\right) + \frac{(T_b - T_s)}{\delta} y + T_s \tag{3.16}$$

where μ is the viscosity, v_b the barrel velocity, k the thermal conductivity, y the normal distance, δ the flight clearance, T_b the barrel temperature, and T_s the screw temperature.

The maximum temperature T_{max} that can develop in the isothermal case is:

$$T_{max} = \frac{\mu v_b^2}{8k} + \frac{1}{2} T_b + \frac{1}{2} T_s \tag{3.17}$$

The first term to the right of the equal sign represents the viscous temperature rise. If the viscosity in the flight clearance is written for a power law fluid, the viscous temperature rise can be written as:

$$\Delta T_{\text{viscous}} = \frac{m v_b^{n+1}}{8k\delta^{n-1}} \tag{3.18}$$

The viscous temperature rise in isothermal conditions is plotted against the flight clearance in Figure 3.36. The data for the isothermal viscous temperature increase were calculated for a 114-mm extruder running at 100 rpm with a polymer melt having a consistency index of $m = 10^4$ Pa · s^n and a thermal conductivity $k = 0.25$ J/m · s · °C. It is interesting to see that the increase in the isothermal viscous temperature becomes greater with a larger clearance, while the increase in the adiabatic temperature decreases with the clearance. Under both thermal boundary conditions, the increase in temperature increases strongly with the power law index of the polymer melt. This indicates that the increase in melt temperature in the flight clearance will be greater for weakly shear-thinning polymers than for highly shear-thinning polymers, which have lower power law indices.

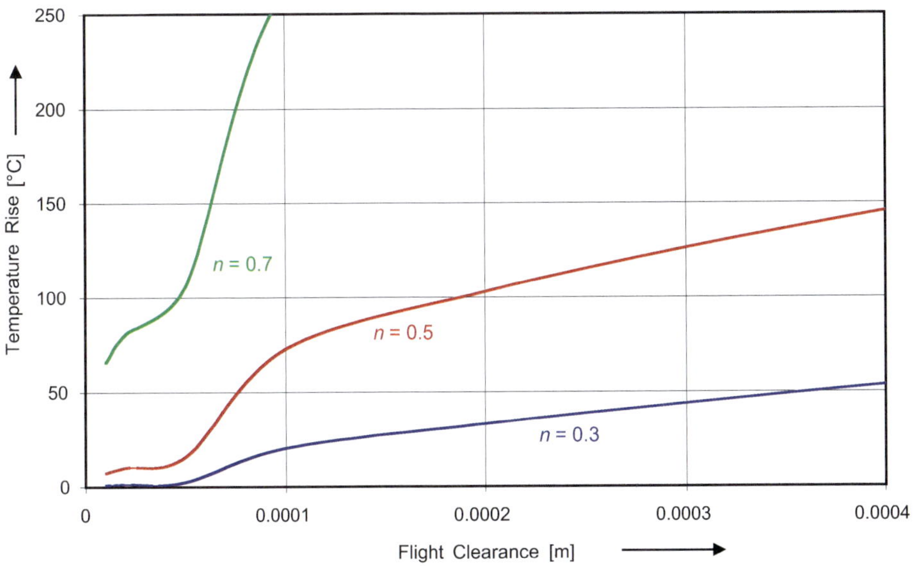

Figure 3.36 Isothermal viscous temperature increase versus flight clearance

In reality, true isothermal conditions may not be achieved because the high heat fluxes required at the screw and barrel interface to maintain isothermal conditions may not be physically possible. Thus, the actual maximum stock temperature in the clearance will be somewhere between the adiabatic and the isothermal case. The expressions for the increase in melt temperature are only valid for Newtonian fluids with a temperature-independent viscosity in pure drag flow. Obviously, for a temperature-dependent fluid, the increase in melt temperature will be less than that of the temperature-independent fluid. The pressure gradient in the flight clearance will also affect the velocities and temperatures. The pressure gradient in the flight clearance will usually be negative, decreasing from the pushing to the trailing flight flank. This will decrease the melt temperature increase in the flight clearance relative to the case of pure drag flow.

The values for the increase in melt temperature in the flight clearance for both adiabatic and isothermal conditions are shown in Figure 3.37. When the values of the flight clearance are small, the adiabatic temperature rise is much greater than the isothermal temperature rise. However, at a certain clearance value, the adiabatic and isothermal curves intersect. At clearance values higher than the intersection value, the isothermal temperature rise is actually greater than the adiabatic temperature rise. For low values of the power law index, the crossover clearance value (where $\Delta T_{adiabatic} = \Delta T_{isothermal}$) is about 0.001*D*, a typical flight clearance in single-screw extruders.

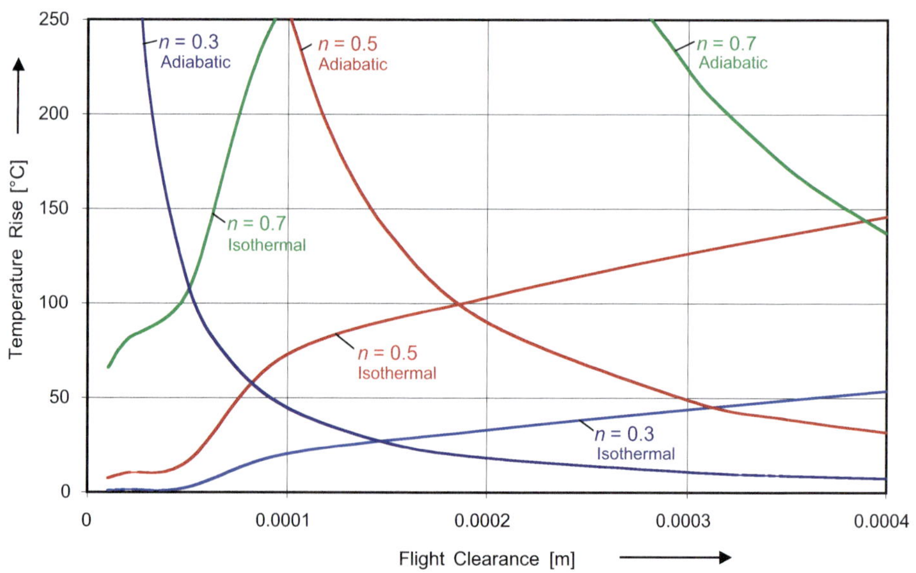

Figure 3.37 Adiabatic and isothermal melt temperature rise versus flight clearance

At higher values of the power law index, the crossover clearance value becomes larger. The isothermal melt temperature rise calculated for large values ($\delta > 0.001D$) of the flight clearance are probably unrealistic because the thermal development length will be greater than the width of the flight (see Equation 3.15). In this case, fully developed temperatures will not be reached in the flight clearance and Equation 3.17 and Equation 3.18 will not yield accurate values for the increase in melt temperature. Figure 3.37 shows that increases in melt temperature from 25 to over 100 °C can be expected in a typical flight clearance (δ = 0.001D) of a 114-mm extruder running at 100 rpm.

3.3.2.3 Temperature Distribution Numerical Calculations

Winter [41] performed numerical calculations of the developing temperature profile in the flight clearance for power law fluids assuming isothermal conditions at the barrel wall and adiabatic conditions at the screw flight surface. These assumptions are considerably more realistic than the purely adiabatic case or the purely isothermal case, although a better boundary condition would probably be a prescribed maximum heat flux. Winter calculated a typical maximum temperature increase of about 150 °C. This value is closer to the maximum temperature rise in the adiabatic case than to the maximum temperature rise in the isothermal case. These analyses indicate that the temperature increase in the flight clearance can be quite significant and can greatly influence the amount of degradation occurring in extruders.

Rauwendaal [55] developed a finite element method (FEM) program to determine temperature profiles in the melt-conveying zone of extruders. This program allows the calculation of three-dimensional velocities and temperatures at any point in the screw channel. The program is based on a 2.5-D analysis, meaning that the velocities are assumed to change little in the down-channel direction.

Results from Finite Element Analysis

The down-channel velocities for a 38-mm extruder running high density polyethylene (HDPE) at 100 rpm with a standard flight clearance are shown in Figure 3.38. The figure shows the velocities for an aspect ratio of 8:1 (the depth of the channel is magnified by a factor of 8). Since a screw channel usually has a small aspect ratio (approx. 0.1:1), displaying the results with a large aspect ratio shows the results more clearly than a normal 1:1 aspect ratio. The isovels (lines of constant velocity) are parallel to the barrel surface over approx. 70% of the channel width. Thus, the influence of the flight flanks on down-channel velocities extends over approximately 1.5 times the channel height at both the leading and trailing edge of the flight.

The cross-channel velocities (v_x) for the same 38-mm extruder are shown in Figure 3.39. The x-velocities are negative in the top one-third of the channel and positive in the bottom two-thirds. This brings about a recirculating flow in the screw channel. Leakage over the flight reduces the recirculating action. The cross-channel velocities are about one-third of the down-channel velocities when the typical square pitch flight helix angle is used. The influence of the flight flank extends over about 1.5 times the channel height.

Figure 3.38 Color contour plot of down-channel velocities

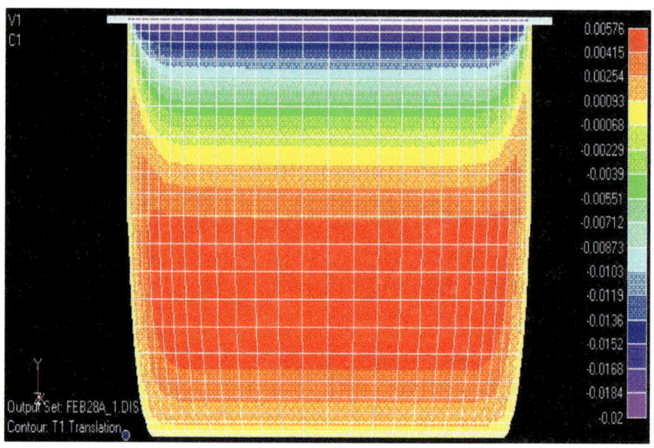

Figure 3.39 Color contour plot of cross-channel velocities

The normal velocities (v_y) are shown in Figure 3.40. Positive *y*-velocities occur at the trailing flight flank with a maximum velocity of about 0.02 m/s, approximately 10% of the maximum *z*-velocity. Negative velocities occur at the pushing or leading flight flank. The maximum and minimum normal velocities occur quite close to the flight flanks, about 0.2 *H* from the flight flank, where *H* is the channel height. In the mid region of the channel, the normal velocities are essentially zero. The downward flow rate at the pushing flight flank is the same as the upward flow rate at the trailing flight flank, and also the same as the cross-channel flow at the lower portion of the channel toward the trailing flight flank.

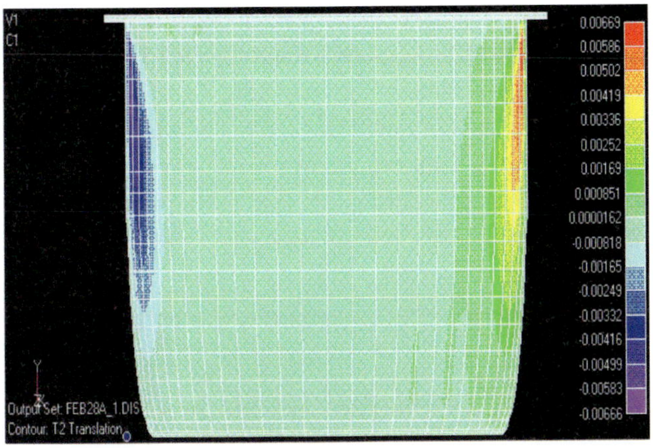

Figure 3.40 Color contour plot of normal velocities

The pressure field in the screw channel is shown in Figure 3.41. The isobars (lines of constant pressure) largely run perpendicular to the barrel, with only small deviations in the corners of the channel. These deviations indicate the normal pressure gradients responsible for the normal velocity components shown in Figure 3.40. The cross-channel pressure gradient is responsible for the cross-channel flow from the pushing to the trailing flight flank in the lower portion of the channel.

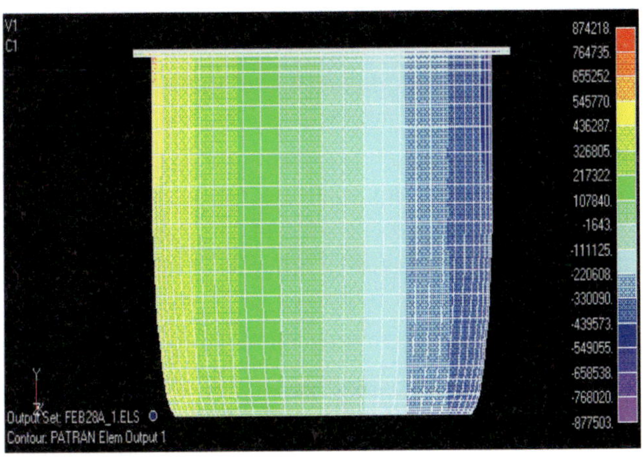

Figure 3.41 Color contour plot of pressures (in Pascal)

The temperature field is shown in Figure 3.42. The barrel surface temperature is set at 175 °C, and the screw surface is taken as adiabatic (zero heat flux). The melt temperatures at any point in the channel are considerably higher than the barrel temperature. The highest temperatures, approx. 31 °C above the barrel temperature, occur at about two-thirds of the channel height, where the cross-channel velocities are the lowest. This agrees well with experimental results [89]. The isotherms in the bottom half of the screw channel clearly show the influence of cross-channel circulation on the temperature field. In fact, the lower temperatures at the flight flanks and at the bottom of the screw channel are a direct result of the recirculation in the channel. Melt temperatures are low in this region because the recirculation causes low-temperature melt close to the barrel surface to flow along the screw surface.

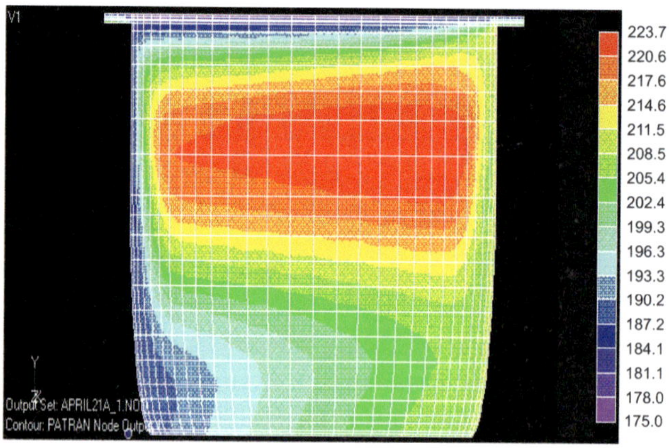

Figure 3.42 Melt temperature distribution in the screw channel

It is interesting to note that melt temperatures in the flight clearance are lower than in the screw channel. Since the shear rate in the flight clearance is higher than anywhere else, the highest temperatures might be expected to occur there as well. An explanation for the low temperatures in the clearance is that the clearance is quite thin. The high level of viscous heat generated in the clearance is efficiently conducted away to the barrel because of its close proximity. This effect was confirmed by the results of a numerical analysis by Pittman [90] and an analytical solution of non-isothermal drag flow by Rauwendaal and Ingen-Housz [91]. Further confirmation was provided by experimental work on leakage flow [92].

When the flight clearance is increased under the same process conditions, the major change that occurs is in the temperature field (Figure 3.43). The temperatures in the lower portion of the channel increase significantly with increasing clearance. This is due to a thicker, relatively stagnant layer of polymer melt at the barrel surface. This insulating, stagnant layer inhibits heat transfer between the material in the screw channel and the barrel. These results agree well with experimental work published elsewhere [89]. A larger flight clearance results in not only an increase in the maximum temperature in the channel, but an expansion of the size of the high-temperature region. This is due to a weaker recirculating flow which itself is the result of the large flow rate through the clearance. A large flight clearance reduces melt temperature control. High melt temperatures close to the screw surface can lead to degradation because the residence times are longest at the screw surface. The combination of high temperatures and long residence times at the screw surface with large flight clearance makes degradation more likely.

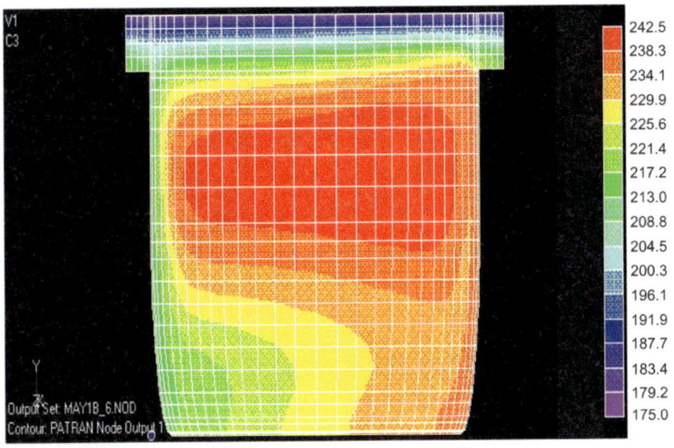

Figure 3.43 Melt temperature distribution with increased flight clearance

Conclusions from Finite Element Analysis

Finite element analysis can be very helpful in analyzing details of flow and heat transfer inside extruders that are very difficult to determine on operating extruders. FEA can predict three-dimensional velocity profiles in screw extruders as well as pressure and temperature fields.

When viscous heat generation is important, high melt temperature regions form in the center of the channel because of thermal convection caused by the recirculating flow pattern in the screw channel. Thus, the developing melt temperature non-uniformities are inherent to the flow in single-screw extruders.

When the flight clearance increases, melt temperatures inside the screw channel can rise considerably. Also, the high-temperature region expands toward the root of the screw. This can lead to degradation of the plastic because the residence times at the screw root are quite long.

The results indicate that it is very important that the flight clearance between screw and barrel is within reasonable limits. In practice, the radial flight clearance should be no more than three thousandths of the diameter of the screw. Also, the inherently non-uniform melt temperatures make distributive mixing sections almost indispensable.

The thermal development length increases with the screw diameter, as shown in Figure 3.44. For small diameters (no more than 60 mm), the thermal development time is approx. 10 to 20 s, and for larger diameters ($D > 100$ mm), the thermal development time is longer than 30 s. Considering that a typical residence time in the melt-conveying zone of an extruder is approx. 15 to 20 s, it is clear that fully developed temperature conditions will not always be achieved in larger extruders. Fully developed conditions may occur at a low screw speed, but at high screw speed it is unlikely that fully developed temperatures can be reached within the length of the extruder.

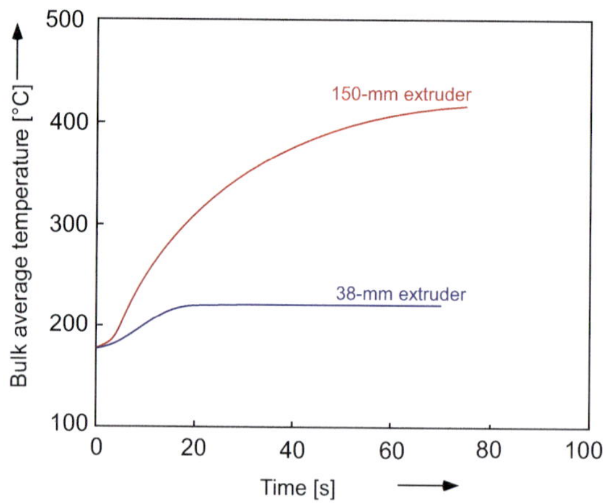

Figure 3.44 Evolution of bulk average temperature over time

The increases in melt temperature and temperature non-uniformities in large extruders are greater than in small extruders because the surface-to-volume ratio becomes less favorable with increasing screw diameter, and therefore it is more difficult to keep the melt temperatures low and uniform. This is well known in practice, and it is for this reason that large extruders generally run at lower screw speeds than small extruders. When the screw diameter is large and when the extruder is operated at high screw speed, it is also more important to incorporate good mixing along the extruder screw.

Figure 3.45 shows a flow chart used to analyze high melt temperature problems. Signs of high melt temperature are high melt temperature readings, smoke evolution at the die exit, discoloration of the plastic, degradation of the plastic, and low viscosity and low melt strength at the die exit. The next step is to check if the melt temperature is measured correctly. Melt temperature is frequently measured incorrectly; for instance, by using a combined pressure/temperature (P/T) transducer. P/T transducers cannot measure melt temperature accurately because the temperature sensor is not immersed in the melt stream.

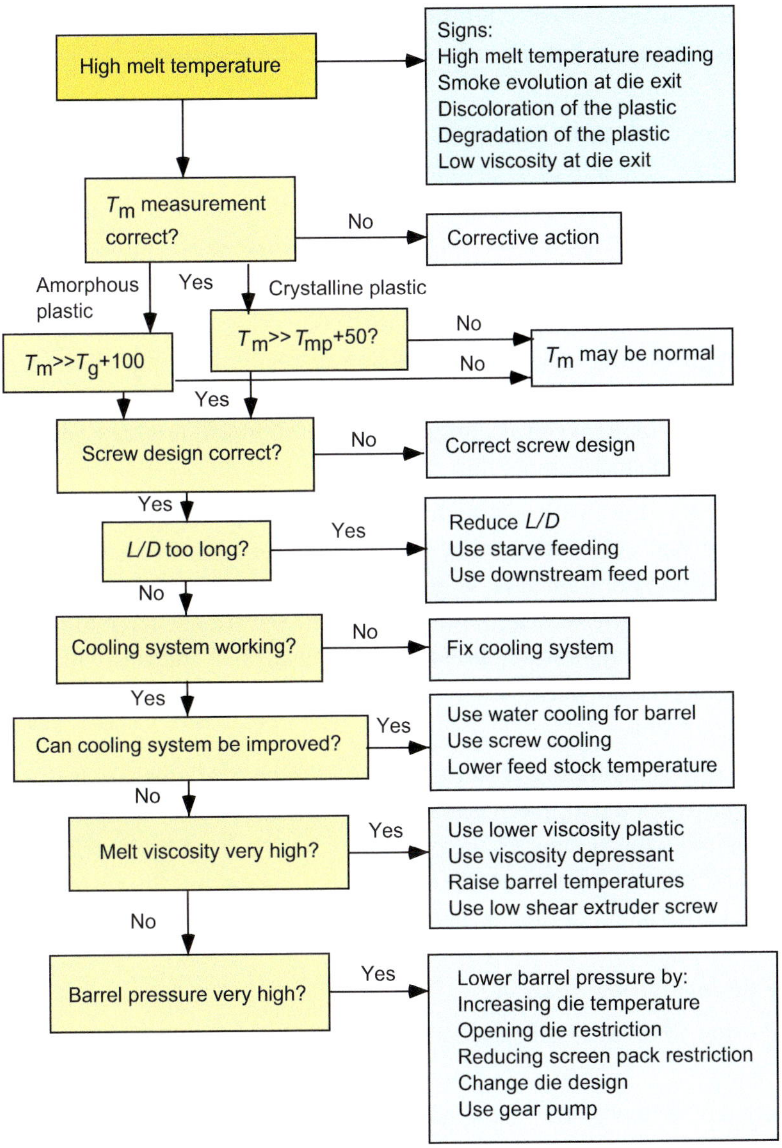

Figure 3.45 Flow chart high-melt temperature (CNV)

If the melt temperature measurement is correct, check whether the melt temperature is much above the glass transition temperature plus 100 °C for amorphous plastics or much above the melting point plus 50 °C for semi-crystalline plastics. If this is not the case, the melt temperature may well be normal. If that is the case, the screw design should be checked. Because most of the heating of the plastic occurs by frictional and shear heating by the screw, high melt temperatures are frequently caused by incorrect screw design. Usually the problem is that a screw for a medium or low viscosity plastic (i.e., a shallow-flighted screw) is used for a high viscosity plastic. High viscosity plastics require a deep-flighted screw to keep the viscous heating in check.

In some cases the length of the extruder may be longer than required for the plastic. In this case, the length of the machine can be reduced; however, this is not a simple modification. It is generally easier to use starve feeding to reduce the effective length of the extruder. If the extruder has multiple ports, it may also be possible to use a downstream feed port rather than the first feed port.

If the problem has not been solved at this point, the cooling system should be checked and fixed if necessary. If the cooling system is working but the melt temperatures are still high, it should be checked if the cooling can be improved. This may be done by using water cooling for the barrel, by using screw cooling, or by lowering the feed stock temperature.

If the problem is the result of a high viscosity of the plastic, explore the use of a lower viscosity plastic or a viscosity depressant. When high viscosity plastics are extruded at high screw speed, it is not uncommon that increasing the barrel temperatures can actually reduce the melt temperature. Higher barrel temperatures increase the melt temperature close to the barrel surface and this will reduce viscous dissipation in the plastic melt; this can lead to lower melt temperature at the end of the extruder.

If the high melt temperature is caused by very high barrel pressure (exceeding 30 MPa or approx. 5000 psi), there is a good chance that the melt temperature can be lowered by reducing the barrel pressure. This can be done by increasing the die temperature, by opening up the die restriction, by reducing the restriction of the screen pack, by changing the die design, or by using a gear pump.

3.3.2.4 Reducing Degradation

The following changes to the process can reduce the chance of degradation in the extruder:

- reduction of residence times and narrowing of the residence time distribution
- reduction of stock temperatures and elimination of high peak temperatures
- elimination of degradation-promoting substances

The residence time can generally be reduced by designing the screw for maximum throughput. Low stock temperatures and reduced peak temperatures can be achieved by designing the screw to minimize specific energy consumption and by avoiding stagnating regions if possible. Thus, the designs for both the screw and die must be as streamlined as possible.

In some special cases it may be beneficial to design an extruder screw for low output and small inventory. An example is the extrusion of a very small catheter for which production may occur at a screw speed of 2 rpm. At such a low screw speed, the average residence time in the extruder will range from 30 to 60 minutes, and degradation is likely to occur at such long residence times. In such a case, it is beneficial to design an extruder screw that has low output per revolution and a small total channel volume. This can be achieved by reducing the flight helix angle, reducing the channel depth, increasing the flight width, and increasing the number of flights. With such a low output screw the screw speed may be increased from 2 to 10 rpm and the average residence time reduced from 60 minutes to less than 10 minutes.

High stock temperatures are likely to be a problem in extrusion operations where the extruder is run at a high screw speed and where the polymer melt viscosity is high. The main screw design variable affecting viscous heating is the channel depth. Increasing the channel depth will reduce shear rate and thus viscous heating, but there are limits to how deep the screw can be cut. Among those limits are the physical strength of the screw, the solids-conveying capacity, and the melting capacity. It is not possible to increase the depth of the screw channel to the point where the melt-conveying rate exceeds the melting rate. This is a case in which a barrier type extruder screw can be useful.

One of the most demanding situations for controlling melt temperature is in foam extrusion. In the secondary extruder of a tandem foam extrusion line, the cooling screw is usually made with very deep channels to minimize viscous heating. Also, multiple (usually six) thin flights with a large helix angle are used to increase the heat-transfer capability of the screw.

If degradation occurs by a thermo-oxidative mechanism, air should be excluded from the extruder. This can be done by putting a nitrogen blanket on the feed hopper, vent port, or at the die, depending on where the air is being introduced. If degradation occurs by hydrolysis, moisture must be excluded from the process. If degradation occurs by a chemical reaction with the metal surfaces of screw and barrel, a non-reacting material of construction must be selected for the screw and barrel.

3.4 Extrusion Instabilities

Variation in extruder performance is perhaps the most frequent problem encountered in extrusion. One possible reason for the frequent occurrence of instabilities is the variety of possible causes, some of which are:

- Bulk-flow problems in the feed hopper
- Solids-conveying problems in the extruder
- Insufficient melting capacity

- Solid bed break-up
- Melt temperature non-uniformities in the die
- Barrel temperature fluctuations
- Screw temperature fluctuations
- Variations in the take-up device
- Melt fracture/shark skin
- Variations in screw speed
- Barrel wear/screw wear
- Insufficient mixing capacity
- Very low die-head pressure
- Insufficient pressure-generating capacity

As discussed in Chapter 1, proper instrumentation is vitally important to diagnose a problem quickly and accurately. Prerequisites for stable extrusion are a good extruder drive, a good temperature-control system, a good take-up device, and most importantly, a good screw design. More instabilities result from improper screw design than from any other cause, but a change in screw design is often only considered as the very last option.

The extruder drive should be able to hold the screw speed constant to about 0.1% or better; the same holds true for the take-up device. However, this is not always the case on actual extrusion lines. The extruder should be equipped with some type of proportioning temperature control, preferably a PID-type control or better. On-off temperature control is inappropriate for most extrusion operations.

3.4.1 Frequency of Instability

Various investigators [93, 94] have classified extrusion instabilities based on the time frame in which they occur. The frequency of the instability is often an indication of the cause of the problem. Most earlier investigations have distinguished only three or four types of instabilities based on the frequency. However, it is probably more appropriate to distinguish at least five types of instabilities:

1. High-frequency instabilities occurring faster than the frequency of screw rotation
2. Screw-frequency instabilities occurring at the frequency of screw rotation
3. Low-frequency instabilities occurring at about one-fifth to one tenth the frequency of screw rotation
4. Very slow fluctuations occurring at a frequency of at least several minutes
5. Random fluctuations

3.4.1.1 High-Frequency Instabilities

High-frequency instabilities are often associated with die-flow instabilities such as melt fracture, "shark skin", or draw resonance. They can also be caused by drive problems, melt temperature non-uniformities, or vibration.

Shark Skin

This defect manifests itself as a regular ridged surface distortion, in which the ridges run perpendicular to the extrusion direction. A less severe form of shark skin is the occurrence of a matted non-glossy surface. Shark skin is generally thought to be formed in the die land or at the exit. It is primarily dependent on the temperature and the linear extrusion speed.

Factors such as shear rates, die dimensions, approach angle, surface roughness, *L/D* ratio, and material of construction seem to have little or no influence on shark skin.

Shark skin is postulated to be caused by the rapid acceleration of the surface layers of the extrudate when the polymer leaves the die. If the stretching rate is too high, the surface layer of the polymer can actually fail and form the characteristic ridges of the shark skin surface [95]. High-viscosity polymers with narrow MWD seem to be most susceptible to shark-skin instability [96, 97].

The shark-skin problem can generally be diminished by reducing the extrusion velocity and increasing the die temperature, particularly at the land section. There is some evidence that running at very low temperatures can also reduce this problem [98] as can the selection of a polymer with a broad MWD. Using an external lubricant, either as an additive to the polymer or by coextruding a thin, low-viscosity outer layer, can also relieve a shark-skin problem.

Melt Fracture

Melt fracture is a severe distortion of the extrudate which can take many different forms: spiraling, bambooing, regular ripple, random fracture, among others (Figure 3.46).

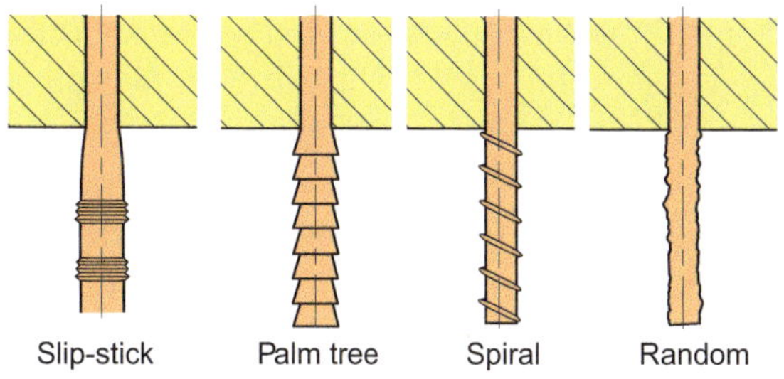

Figure 3.46 Various forms of melt fracture

Melt fracture is not a surface defect like shark skin. Instead, it is associated with the entire body of the molten extrudate. However, many workers do not distinguish between shark skin and melt fracture, but lump all these flow instabilities together under the term melt fracture. There is a large amount of literature on the subject of melt fracture (e.g., [99–111]). Despite the large number of studies on melt fracture, there is no clear agreement about its exact cause and mechanism. It is quite possible that the mechanisms are not the same for different polymers and/or different flow channel geometries [112]. Linear polymers tend to develop instabilities in the die land, while branched polymers tend to develop instabilities in the converging region of the die flow channel. However, there is relatively uniform agreement that melt fracture is triggered when a critical wall shear stress, on the order of 0.1 to 0.4 MPa (15 to 60 psi), is exceeded in the die.

A number of mechanisms have been proposed to explain melt fracture. Some of the more popular ones are:

- Critical elastic deformation in the entry zone
- Critical elastic strain
- Slip-stick flow in the die

The importance of entry zone design has been demonstrated by many workers. In general, the smaller the entry angle, the higher the deformation rate at which instability occurs. Gleissle [113] has proposed critical elastic strain as measured by recoverable strain as a criterion. Based on measurements with eleven fluids, he proposed the existence of a critical value of the ratio of the first normal stress difference to the shear stress. The average value was 4.63, with a standard deviation of about 5% for eleven widely different fluids. Much larger differences were found in the critical shear stress of the polymers; the average was 3.7×10^5 Pa with a standard deviation of about 55%. In 1961, Benbow, Charley, and Lamb [110, 114] introduced the slip-stick mechanism to explain flow instability and extrudate distortion. Above a certain critical stress, the polymer melt is believed to experience intermittent slipping due to lack of adhesion between the melt and the die wall. This slippage is thought to relieve excessive deformation energy absorbed as a result of flow through a die. A large number of workers have observed slippage by various techniques.

More recent work by Utracki and Gendron on pressure oscillations occurring during the extrusion of polyethylenes [115] led them to conclude that pressure oscillation does not seem to be related to elasticity or slip. They concluded that the parameter most closely related to pressure oscillations was the critical strain (Hencky) value ε_c of the melt. For linear low-density polyethylene (LLDPE), $\varepsilon_c < 3$, for high-density polyethylene (HDPE), $\varepsilon_c < 2$, while for low-density polyethylene (LDPE), $\varepsilon_c > 3.5$. The instability seems to be related to the inability of the polymer melt to sustain levels of strain larger than the critical strain.

Streamlining the flow channel geometry reduces the tendency toward melt fracture in branched polymers. Increased temperatures, particularly at the wall of the die land, increase the extrusion rates possible, while avoiding melt fracture. The critical wall-shear stress appears to be relatively independent of the die length, radius, and temperature. The critical stress seems to vary inversely with molecular weight, but seems to be independent of MWD. Certain polymers exhibit a superextrusion region, above the melt fracture range, where the extrudate is not distorted [116]. Superextension is particularly advantageous with polymers that melt fracture at relatively low rates, such as FEP. In superextrusion, the polymer melt is believed to slip relatively uniformly along the die wall. The occurrence of slip in extruder dies has been studied by a number of investigators [116, 117]. However, it is still not clear whether the slip is actual loss of contact between the polymer melt and the metal wall or failure of a thin polymeric layer very close to the metal surface.

The melt fracture problem can be reduced by streamlining the die, increasing the temperature at the die land, running at lower rates, reducing the molecular weight or the polymer melt viscosity, increasing the cross-sectional area of the exit flow channel, or by using an external lubricant. In some instances, the melt fracture problem can be solved by going to superextrusion; this process is often used in the wire coating industry where high line speeds are quite important for economic production.

Draw Resonance

Draw resonance occurs in processes where the extrudate is exposed to a free surface-stretching flow, such as blown film extrusion, fiber spinning, and blow molding. It manifests itself as a regular cyclic variation of the dimensions of the extrudate. An extensive review [112] and an analysis [119] of draw resonance were done by Petrie and Denn. Draw resonance occurs above a certain critical draw ratio, when the polymer is still in the molten state when it is taken up and rapidly quenched after take-up.

Draw resonance will occur when the resistance to extensional deformation decreases as the stress level increases. The total amount of mass between die and take-up may vary with time because the take-up velocity is constant but the extrudate dimensions may not be. If the extrudate dimensions decrease just before take-up, the extrudate dimensions above it must increase. As the larger extrudate section is taken up, a thin extrudate section can form above it. This can go back and forth as a cyclic variation of the extrudate dimensions. Draw resonance does not occur when the extrudate is solidified at the point of take-up, because the extrudate dimensions at the take-up are fixed [120, 121]. Isothermal draw resonance is found to be independent of the flow rate. The critical draw ratio for almost Newtonian fluids, such as nylon, polyester, polysiloxane, is approximately 20. The critical draw ratio for strongly non-Newtonian fluids, such as polyethylene, polypropylene, polystyrene, can be as low as 3 [122]. The amplitude of the dimensional variation increases with draw ratio and draw-down length.

Various investigators have performed theoretical studies of the draw resonance problem by linear stability analysis. Pearson and Shah [123, 124] studied inelastic fluids and predicted a critical draw ratio of 20.2 for Newtonian fluids. Fisher and Denn [125] confirmed the critical draw ratio for Newtonian fluids. Using a linearized stability analysis for fluids that follow a White-Metzner equation, they found that the critical draw ratio depends on the power law index n and a viscoelastic dimensionless number. The dimensionless number is a function of the die take-up distance, the tensile modulus, and the velocity at the die. Through their analysis, Fisher and Denn were able to determine stable and unstable operating regions. In some instances, draw resonance instability can be eliminated by increasing the draw ratio, but under most operating conditions, draw resonance is eliminated by reducing the draw ratio.

Ide and White [126–129] demonstrated experimentally and theoretically that polymers whose elongational viscosities increase with time or strain do not exhibit draw resonance, but undergo cohesive failure at high draw ratios. One polymer that behaves in this fashion is LDPE. Polymers whose elongational viscosity decreases with time or strain do exhibit draw resonance at low draw ratios and fail in a ductile fashion at high draw ratios. Examples of polymers that behave in such a fashion are HDPE and polypropylene (PP). Lenk [130] proposed a unified concept of melt flow instability. His main conclusions were that all flow instabilities originate at the die entrance and that melt fracture and draw resonance are not distinct and separate flow phenomena but are both caused by elastic effects that have their origin at the die entrance. Lenk's analysis, however, is purely qualitative and does not offer much help in the engineering design of extrusion equipment or in determining how to optimize process conditions to minimize instabilities.

3.4.1.2 Screw Frequency Instabilities

Screw frequency instabilities occur to a small extent in essentially every extrusion operation. They can be caused by variations in intake of a polymer from the feed hopper to the feed housing when the flow is interrupted every time the flight passes by the feed opening. This can cause a cyclic pressure change detectable if the extruder has accurate pressure readout. One way to reduce unsteady intake of polymer from the feed hopper is to use a double-flighted screw geometry in the feed section. Generally, a better solution is to change the shape of the feed opening. According to Wheeler [131], the screw frequency instability is more likely to occur at very low die-head pressures.

Screw frequency variations can also be caused by the pressure difference across the screw flight. This pressure difference is responsible for the recirculating flow in the cross-channel direction. When pressure is measured along the screw or at the very end of the screw the pressure pattern will have a saw-tooth shape (Figure 3.47).

In most cases, the major cause of screw frequency instabilities is the pressure difference between the leading and trailing edges of the flight in the pump section. This pressure fluctuation is often called "screw beat". This pressure difference is inherent to the conveying process and occurs even if no pressure is developed in the pump section because the pressure difference is drag-induced.

If the flight clearance can be neglected, the pressure difference ΔP across the screw flight is:

$$\Delta P = \frac{6\mu v_{bX} W}{pH^2} \cong \frac{6\mu\pi^2 D^2 N \sin^2\varphi}{pH^2} \tag{3.19}$$

where μ is the viscosity, D the screw diameter, N the rotational speed, φ the screw flight helix angle, p the number of parallel flights, and H the channel depth.

This expression is valid for a Newtonian fluid. The pressure difference increases with viscosity, diameter, screw speed, and helix angle; and it decreases with channel depth. When the helix angle increases from 17.5 to 25.0°, the pressure difference will double. Thus, large-helix angle screws will exhibit more screw frequency pressure fluctuations than small-helix angle screws. The screw-frequency pressure fluctuation can be reduced by placing a multi-flighted screw section, such as a Saxton [132] or CRD [133–139] mixing section, at the end of the screw (Figure 3.48). A multi-flighted mixing section at the end of the screw will reduce the amplitude of a pressure fluctuation but will increase its frequency.

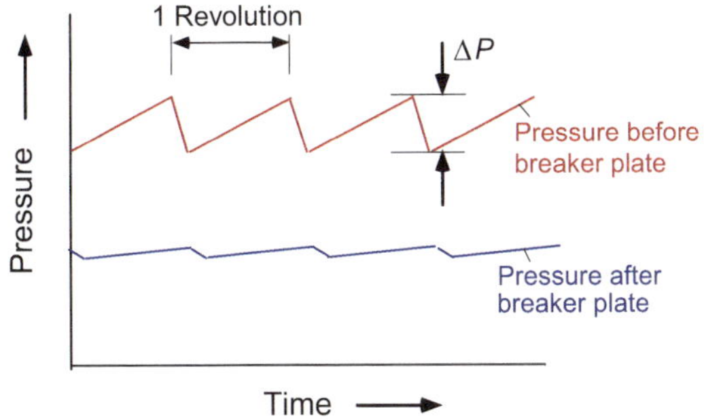

Figure 3.47 Pressure variation over time ("screw beat")

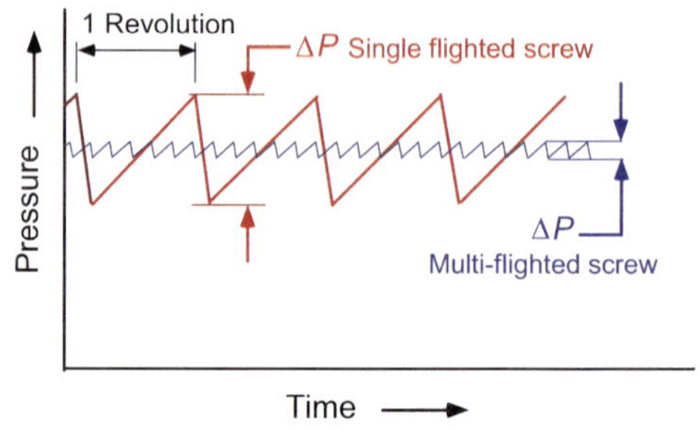

Figure 3.48 Pressure fluctuation with single and multi-flighted screw

These pressure fluctuations will be most severe at the very end of the screw, and will decrease with increasing distance from the screw because the polymer melt is slightly compressible. Thus, the actual pressure fluctuation will be very much dependent on the location of the pressure transducer. When a breaker plate is used, the pressure fluctuation will diminish significantly because the breaker plate will largely break up the flight-induced pressure fluctuation (see Figure 3.47).

Obviously, screw frequency pressure fluctuation will be problematic when the value of ΔP is large relative to the actual die-head pressure. This will occur when the die-head pressure is low (as pointed out by Wheeler [93]), when the polymer melt viscosity is high, the screw diameter large, the screw speed high, the helix angle or pitch large, or when the channel depth is shallow.

3.4.1.3 Low-Frequency Instabilities

Low-frequency instabilities have been associated with solid bed breakup [94, 140]. Fenner et al. [94] attempted to theoretically predict solid bed breakup. They proposed that it occurred because of acceleration of the solid bed in the plasticating zone of the extruder. They also claimed that no solid bed acceleration occurs in the absence of a melt film between the solid bed and the screw. In practice, formation of a melt film can be avoided by cooling the screw [141].

Earlier, Maddock [142] found that screw cooling helped in reducing surging. The most likely reason that screw cooling reduces surging is that it reduces the throughput rate by a substantial amount, about 20% in the experiments of Edmondson and Fenner [141]. Therefore, the melting process will be completed over a shorter axial distance, reducing the stresses acting on the solid bed.

Solid bed breakup is also more likely to occur on screws with a high compression ratio. Fenner, Cox, and Isherwood [69] found solid bed breakup with screws having high compression ratios (3:1 and 4:1), but not with screws having low compression ratios (2.25:1). A low compression ratio screw would seem a better solution than a high compression ratio screw with screw cooling. The chance of formation of a melt film on the screw surface can be reduced by using a barrier screw geometry.

Fluctuations occurring over approx. 10 to 30 s can be caused by temperature fluctuations along the extruder barrel. Such temperature fluctuations may not be noticeable from the temperature readouts, because of the slow response of many temperature sensors, and because the sensors are often located a considerable distance from the polymer/metal interface. However, if the actual temperature at the interface fluctuates, there will be a corresponding fluctuation in the flow rate. In time-proportioning temperature control systems, power is added or removed at relatively short intervals, typically approx. 15 to 20 s, and these bursts of heating or cooling energy will cause short-term changes in the polymer/metal interface temperature with corresponding variations in throughput rate. The variation in throughput can be as much as 5 to 10%. Therefore, from a stability point of view, the true proportioning temperature control is significantly better than the time-proportioning temperature control.

Throughput variations caused by wall temperature changes have been described by Gitschner and Lutterbeck [143]. They were able to show a very clear correlation between the on and off cycling of barrel cooling and die-head pressure fluctuations. They also found that the pressure fluctuations correlated very closely with the resulting throughput fluctuations. Clearly, these temperature-induced throughput fluctuations can be considerably more severe in the case of on–off temperature control.

3.4.1.4 Very Slow Fluctuations

Very slow fluctuations are often associated with poor temperature control, changes in ambient conditions (room temperature, relative humidity), plant voltage variation, and similar causes. A steady, slow reduction in output is often caused by build-up of contaminants on the screen pack or plate-out on the screw.

3.4.1.5 Random Fluctuations

Random fluctuations are often associated with irregular feeding. Maddock [142] discussed a case in which the extruder performance was very sensitive to the level of fill in the feed hopper. Random fluctuations can also result from a combination of cyclic fluctuations. Figure 3.49 shows the pattern of regular sinusoidal variation, and Figure 3.50 shows a combination of three sinusoidal variations with different amplitudes and frequencies. This variation appears to be random, but it is actually made up of three different components, each one a very regular sinusoidal variation. This situation occurs frequently in extrusion, because in many cases, the process is affected by variations from different sources which will typically have different frequencies and amplitudes. Obviously, the more sources of variation acting on the process, the more complicated and random the information from the process will tend to be.

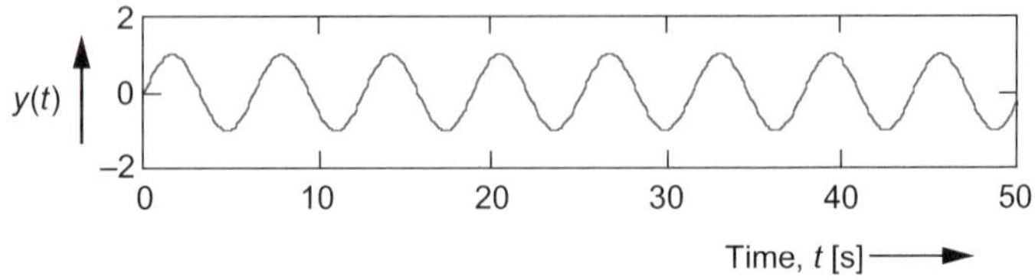

Figure 3.49 Regular sinusoidal variation

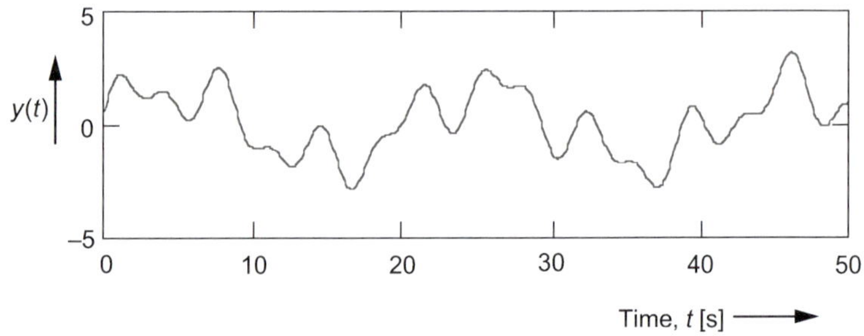

Figure 3.50 Combination of three sinusoidal variations

The pattern of variation shown in Figure 3.50 is often detected in measurements of melt pressure. Obviously, the response time of the measuring instrument must be short enough to capture high-frequency variations occurring in the process. Fast Fourier Transform (FFT) analysis can be used to analyze a complex signal and decompose it into the base frequencies. As a result, FFT is a powerful tool in troubleshooting complex extrusion problems. Becker et al. [144, 145] used FFT analysis of melt pressure signals to decipher extrusion instabilities. Reinhard et al. [146] described the application of spectral analysis to problems of surging in extrusion.

3.4.2 Functional Instabilities

Another method of classifying instabilities is by the functional zone in which the instability originates. Thus, the following instabilities can be distinguished:

- Solids-conveying instabilities
- Plasticating instabilities
- Melt-conveying instabilities
- Devolatilization instabilities
- Mixing instabilities
- Die-forming instabilities

Each of these will be discussed in detail.

3.4.2.1 Solids-Conveying Instabilities

There are three major causes for solids-conveying instabilities: flow problems in the feed hopper, internal deformation of the solid bed in the screw channel, and insufficient friction against the barrel surface. Flow problems in the feed hopper can be detected by observing the flow from the feed hopper when it is disconnected from the extruder. Solids-conveying problems in the extruder itself are difficult to diagnose. One method for diagnosing such a problem is to Teflon-coat the screw. Though the coating may not last long, the retarding force acting on the solid bed will be substantially reduced, and solids conveying will be improved. If coating the screw eliminates the instability, this is a strong indication of a solids-conveying problem. A more permanent solution can be provided by a grooved barrel section or a nickel-plated screw impregnated with a fluorocarbon polymer.

The stability of solids conveying is strongly related to the uniformity of the feed stock. The best stability is achieved with uniform pellet size and shape. Large variations in particle size and shape invariably lead to variations in extruder performance. This is often observed when regrind is added to the virgin feed stock. Since the particle size and shape of regrind is usually non-uniform, increasing the amounts of regrind in the feed stock will reduce the stability of the extrusion process. In many cases, this reduction in stability will put an upper limit on the amount of regrind that can be added to the feed stock for the extruder.

3.4.2.2 Plasticating Instabilities

Plasticating problems are likely to occur on screws with a high compression ratio and a short compression section length and when the overall length of the extruder is short. A short extruder will run into melting-related instabilities at lower outputs than a longer extruder. This is one of the reasons behind the trend in the extrusion industry toward longer extruders. In the 1950s and 1960s, most single-screw extruders were about 20D long. In the 1970s and 1980s, most extruders were about 25D long. In the 1990s and 2000s, most single-screw extruders were about 30D long.

An instability caused by insufficient melting capacity can be diagnosed by preheating the feed stock. If preheating reduces or eliminates the instability, then the problem is most likely insufficient melting capacity. Melting can be improved by changing the processing conditions or by changing the screw geometry. The following processing conditions can improve melting:

- Preheating the feed stock
- Increasing the barrel temperature when the extruder is running at low screw speed
- Reducing the barrel temperature when the extruder is running at high screw speed

- Reducing the screw speed
- Increasing the barrel pressure

Several changes in screw design can improve the melting capability; they include:

- Increased flight helix angle in the transition section
- Reduced flight clearance in the transition section
- Increased length of the transition section
- Multi-flighted design of the transition section
- The use of a fluted mixing section at the end of the transition section

3.4.2.3 Melt-Conveying Instabilities

Most melt-conveying or pumping problems are caused by improper design of the metering section. The most common problem is excessive channel depth; the second most common problem is insufficient length of the metering section. If the channel depth is too large, the metering end of the screw can be cooled to reduce the effective depth. Conversely, if the channel depth is too shallow, the metering end of the screw can be heated to increase the effective depth of the channel and to increase the melt-conveying capability. However, if the metering depth is incorrect, it is better to switch to a screw design with proper dimensions of the metering section.

In some cases, the melt-conveying capability cannot be improved sufficiently by a simple change in screw design. One example is a two-stage extruder screw that must operate at high discharge pressure. If the second stage of the screw is not long enough, it may not be possible for the screw to generate the required pressure. Such a condition will result in vent flow; that is, molten polymer flowing out of the vent port. One possible solution to such a problem is to place a gear pump between the extruder and the die. In this setup, the gear pump can generate the required die-head pressure, and it is only necessary for the screw to generate enough pressure to feed the gear pump. Other possible solutions to vent flow problems are:

- Reduce die-head pressure
- Increase the length of the extruder
- Use internal screw heating in the second stage of the screw
- Reduce the barrel temperature in the second stage

3.4.2.4 Devolatilization Instabilities

Devolatilization instabilities can be caused by plugging of the vent port, variation in the vacuum level, or by variations of the volatile level in the feed stock. The efficiency of devolatilization can be improved by:

- Increasing the barrel temperatures in the first stage
- Increasing the vacuum level at the vent port

- Preheating the feed stock
- Use of a stripping agent

Screw design can also affect the devolatilization efficiency a great deal. A multi-flighted extraction section can improve degassing. Further, it is important that the polymer is fully melted in the first stage of the screw. A fluted mixing section placed at the end of the first stage of the screw will help to ensure complete melting.

3.4.2.5 Mixing Related Instabilities

Extrusion instabilities are often related to a screw with insufficient mixing capacity. Mixing can be slightly improved by increasing the die-head pressure, but this is a relatively ineffective method, and it also increases the chance of degradation. The mixing capacity of a screw can be improved significantly by adding one or more mixing sections to the design of the screw. The polymer melt must be well mixed when it leaves the extruder screw and enters into the die. For this reason it is beneficial to have an efficient distributive mixing device at the very end of the screw.

Mixing Sections

There are certain characteristics that are desirable for mixing sections in general:

1. The pressure drop in the mixing section should be minimal. This can be achieved by including forward pumping capability.
2. Flow within the mixing section should be streamlined; dead spots should be avoided.
3. The mixing section should wipe the barrel surface completely; thus, circumferential grooves should be avoided.
4. The mixing section should be operator-friendly; it should be easy to assemble, install, run, clean, and disassemble.
5. The mixing section should be easy to manufacture and reasonably priced.

3.4.2.6 Distributive Mixing Sections

In addition to the general characteristics desirable for mixing sections, there are some specific characteristics important for distributive mixing: 1) The plastic melt should be subjected to significant shear strain, and 2) the flow should be split frequently with reorientation of the fluid elements. Splitting and reorientation can improve mixing exponentially and, therefore, is critical in achieving efficient distributive mixing. Distributive mixers can be divided into several main groups: cavity mixers, pin mixers, slotted-flight mixers, variable-channel-depth mixers, and variable-channel-width mixers.

Cavity Mixers

One well-known cavity mixer is the cavity transfer mixer (CTM). It consists of a screw section and a barrel section, both containing hemispherical cavities (Figure 3.51). As the plastic melt travels through the mixer, the material is frequently cut and reoriented. This action gives the mixer good mixing capability. There are a few practical issues, however, that make the CTM a less than ideal mixer. Because it has no forward pumping capability, the CTM is a pressure-consuming mixing device that will reduce the extruder output and increase the temperature buildup in the plastic melt. Because of the hemispherical cavities, the streamlining is not ideal. This can be a problem in product changeover as well as with thermally sensitive materials. Other drawbacks of the CTM are high cost, difficult installation and cleaning, and the fact that the barrel is not completely wiped during processing.

Another cavity mixer is the Twente Mixing Ring (TMR) developed by Semmekrot at Twente University in Enschede, the Netherlands. It is somewhat similar to the CTM, but it has a freely moving mixing ring with circular holes instead of a barrel with cavities (Figure 3.52). This eliminates a number of the disadvantages of the CTM; the TMR is easier to install, easier to clean, and the barrel is completely wiped.

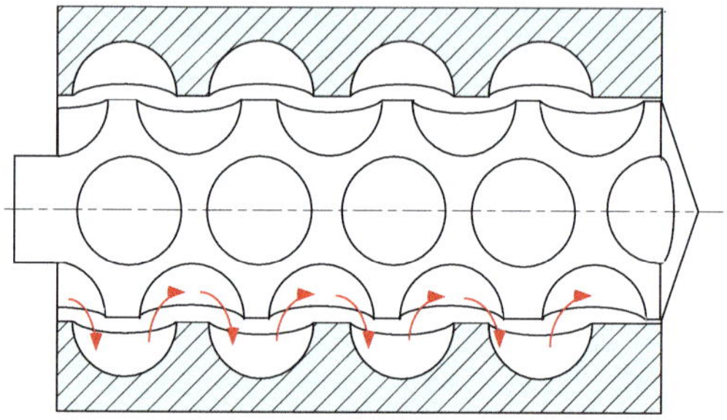

Figure 3.51 Cavity transfer mixer

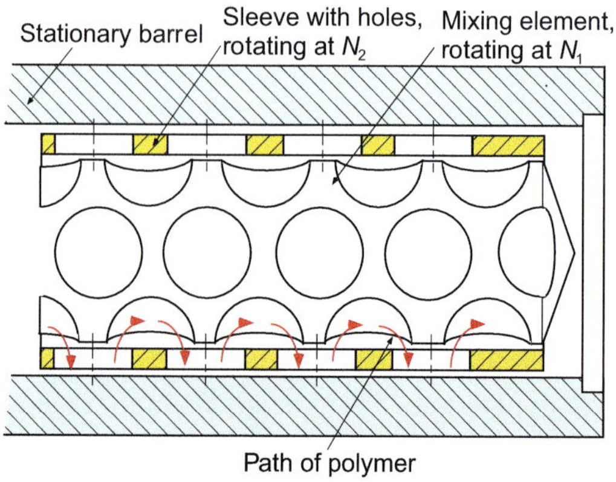

Figure 3.52 Twente Mixing Ring, $N_1 > N_2$

Pin Mixers

Pin mixers have been in use for many years and come in many shapes and sizes. Often the pins are circular, but other shapes can be used as well; for example, square, rectangular, or diamond-shaped. One example of a pin mixing section is shown in Figure 3.53.

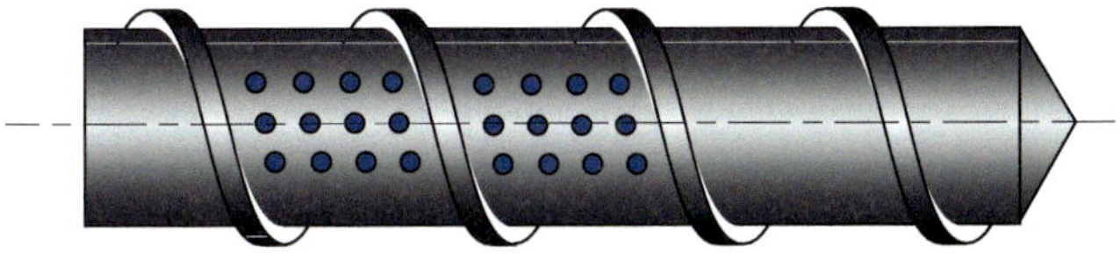

Figure 3.53 Examples of pin mixing sections

Pin mixing sections achieve a moderate level of splitting and reorientation and, as a result, give a moderate improvement in mixing. However, the pins create a restriction to flow and, thus, reduce extruder output. The greater the number of pins, the more the output is reduced. Another drawback of pins is that they tend to create regions of stagnation, particularly at the corner of the pin and the root of the screw.

A special type of extruder with pins is the pin barrel extruder. The pins protrude all the way into the screw channel, and the screw flight is slotted at the various pin locations (Figure 3.54). The advantages of this extruder are good mixing capability and low energy consumption. These characteristics have made the pin barrel extruder quite popular in rubber extrusion. In many extruders, the pins are adjustable, making it easier to install and pull the screw and clean the extruder.

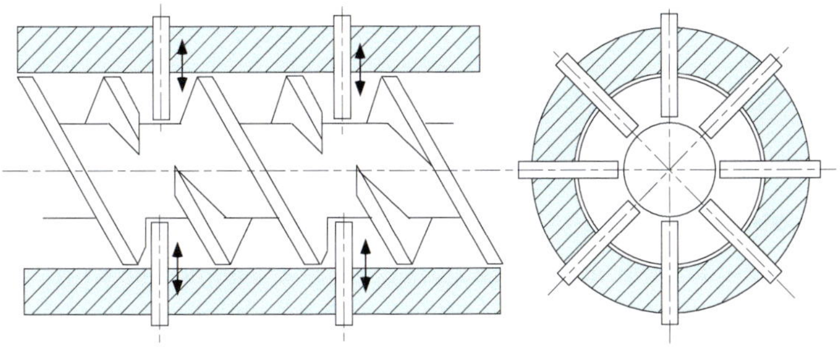

Figure 3.54 The pin barrel extruder

Slotted Flight Mixers

There are a great number of slotted-flight mixers. A simple form is the Axon mixing section (Figure 3.55). Better slotted mixers are the Dulmage (Figure 3.56) and the Saxton mixing sections (Figure 3.57). Both these mixers provide frequent splitting and reorientation, resulting in effective mixing action, and, because of the forward orientation of the flights, some forward pumping capability. Thus, they combine good mixing with high output capability. The drawback of the Dulmage mixer is that the barrel is not completely wiped by the mixing section because of the circumferential slots.

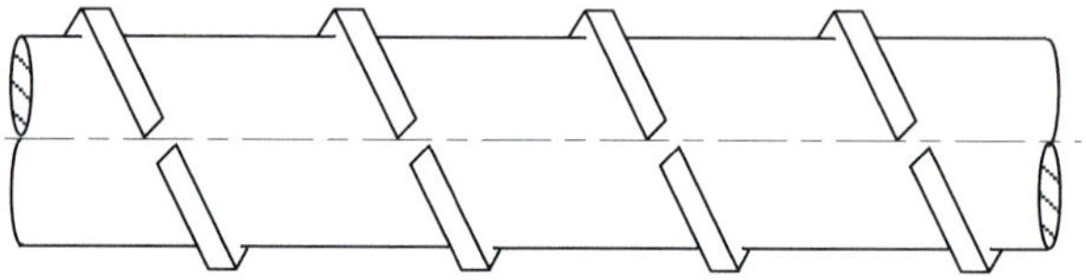

Figure 3.55 The Axon mixing section

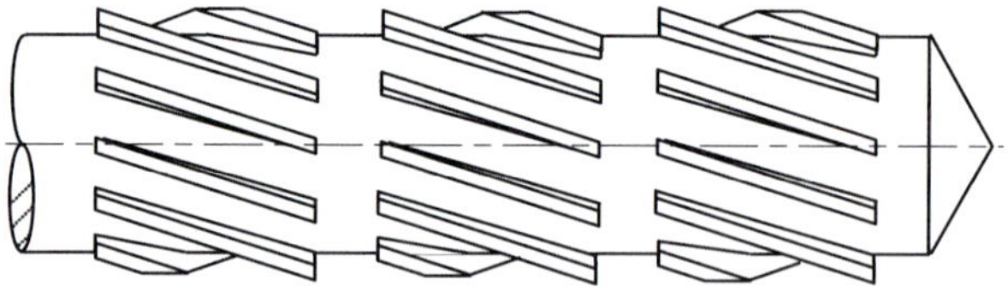

Figure 3.56 The Dulmage mixing section

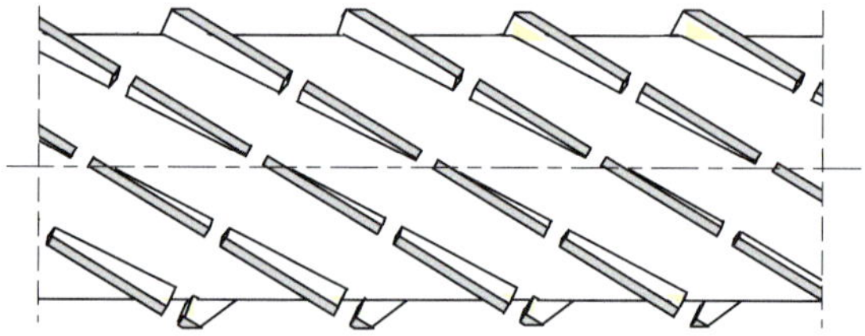

Figure 3.57 The Saxton mixing section

Variable-Depth Mixers

The channel depth of variable-depth mixers such as the double-wave screw (Figure 3.58) is varied to obtain improved mixing. In the double-wave screw mixer the channel depth varies periodically in each channel in such a way that when one channel decreases in depth, the other increases and vice versa. However, there is no strong mechanism for flow splitting and reorientation, and therefore the mixing capability of this mixer is moderate. Other variable-depth mixers are the Pulsar and Strata-Blend mixing sections (Figure 3.59). These mixers do not achieve efficient flow splitting and reorientation.

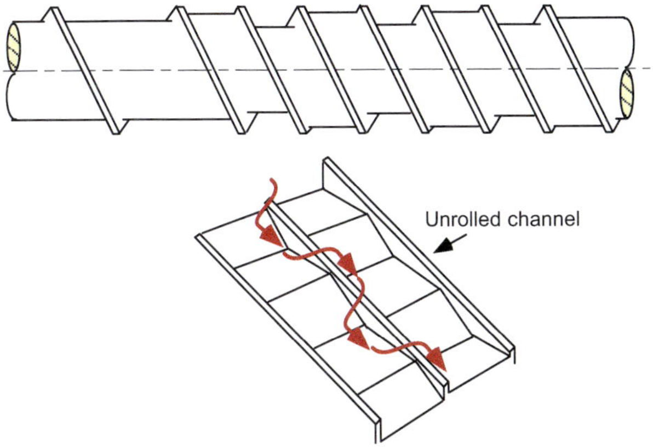

Figure 3.58 Double-wave screw

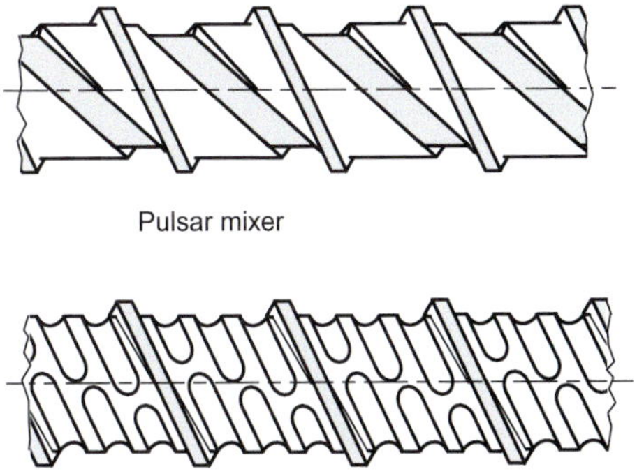

Figure 3.59 The Pulsar mixer and Strata-Blend mixer

Summary of Distributive Mixers

Table 3.9 is a compilation of the important characteristics of the various distributive mixers, listing how they perform with respect to different criteria. The last column in Table 3.9 is the most important when it comes to distributive mixing effectiveness. The Dulmage, Saxton, CTM, TMR, and CRD all do very well in this category. Both the Saxton and the CRD mixer combine good mixing with low cost, good streamlining, ease of use, and low pressure drop. The advantage of the CRD mixer is that it also provides dispersive mixing capability. The geometry of the CRD mixer is shown in Figure 3.67 and Figure 3.68.

Table 3.9 Comparison of Various Distributive Mixers

Mixers	Pressure drop	Dead spots	Barrel wiped	Operator-friendly	Machining cost	Splitting reorientation
Pins	Medium	Yes	Partial	Good	Low	Medium
Dulmage	Low	No	Partial	Good	Medium	Good
Saxton	Low	No	Fully	Good	Medium	Good
CTM	High	Yes	Partial	Poor	Very high	Good
TMR	High	Few	Fully	Medium	High	Good
CRD	Low	No	Fully	Good	Medium	Good
Axon	Low	No	Fully	Good	Low	Medium
Double-wave	Low	No	Fully	Good	High	Low
Pulsar	Low	No	Fully	Good	Medium	Low
Stratablend	Low	Few	Fully	Good	Medium	Low

3.4.2.7 Dispersive Mixing Sections

The following characteristics are desirable for dispersive mixing:

- The mixing section should have regions where the material is subjected to high stresses, preferably elongational stresses.
- The high stress regions should be designed in such a way that the exposure to high shear stresses occurs only for a short time, while exposure to elongational stresses is maximized.
- All fluid elements should pass through the high stress regions many times to achieve efficient dispersive mixing action.
- All fluid elements should pass through the high stress regions the same number of times for uniform mixing.

There are several types of dispersive mixing sections: blister rings, fluted mixing sections, and planetary gear extruders.

Blister Ring

The blister ring is simply a circumferential shoulder on the screw with a small clearance between the ring and the barrel (Figure 3.60). All material must flow through this small gap where it is exposed to high shear stresses. Since no forward drag flow occurs in the blister ring, relatively high pressure drops occur across the blister ring. The stress level in the gap is not uniform; therefore, the mixing action is not uniform.

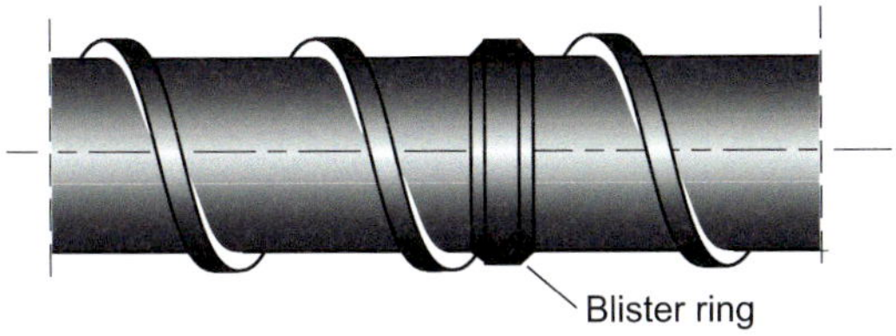

Figure 3.60 Blister ring

Fluted Mixing Sections

These mixers have inlet and outlet flutes separated by barrier flights. The material must pass through the narrow gap of the barrier flights to exit from the mixer, and this is where the mixing action takes place. One of the earliest fluted mixers was the Egan mixing section developed by Gregory and Street in which the flutes have a helical orientation (Figure 3.61).

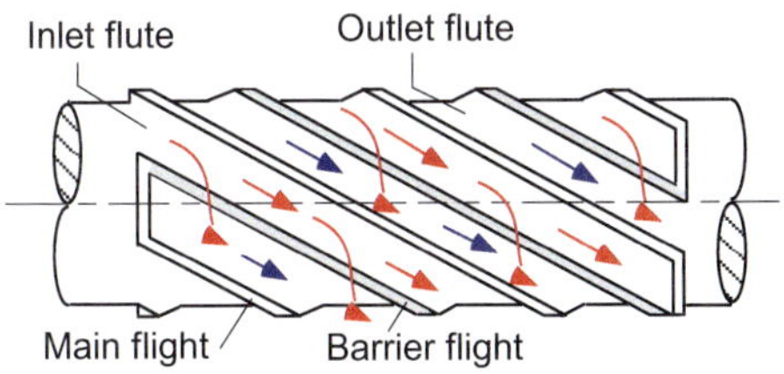

Figure 3.61 Egan mixing section

Another fluted mixer is the Union Carbide mixer (UC mixer) developed by LeRoy and popularized by Maddock, which has straight flutes (Figure 3.62). Because of the straight flutes, the LeRoy mixer has no forward pumping capability and thus, tends to have a high pressure drop. It is typically machined with a ball mill, and as a result, the flutes have a semicircular cross section. This tends to result in inefficient streamlining at the entry and exit of the flutes. Despite these shortcomings, the LeRoy mixer is probably the most commonly used mixer in single-screw extruders.

It is important to design mixing sections with a low pressure drop. This is particularly true for dispersive mixers. A high pressure drop reduces output, increases melt temperatures, increases residence time, and increases the chance of degradation. Higher melt temperatures reduce melt viscosity and the stresses in the melt in the mixing section, and therefore reduce dispersive mixing. Because a high pressure drop causes high temperatures, a high pressure drop should be avoided.

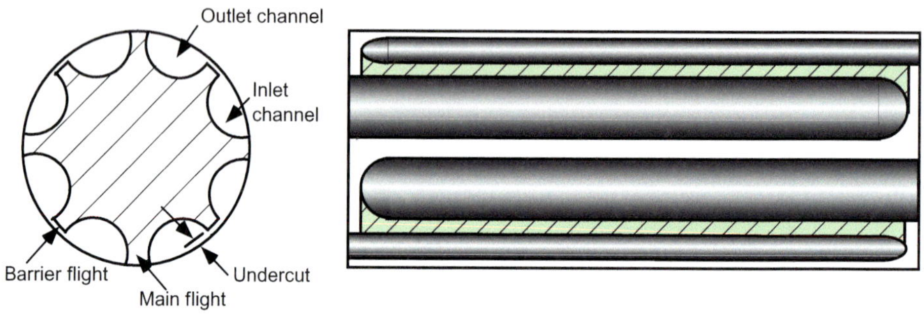

Figure 3.62 LeRoy mixing section

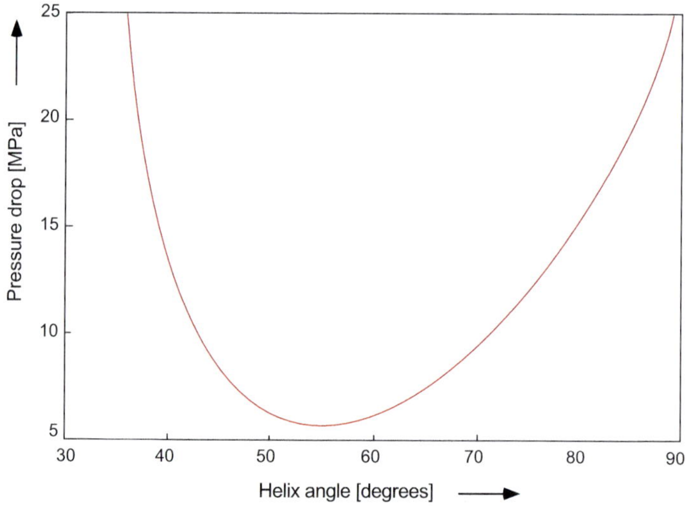

Figure 3.63 Pressure drop versus helix angle for a Newtonian fluid

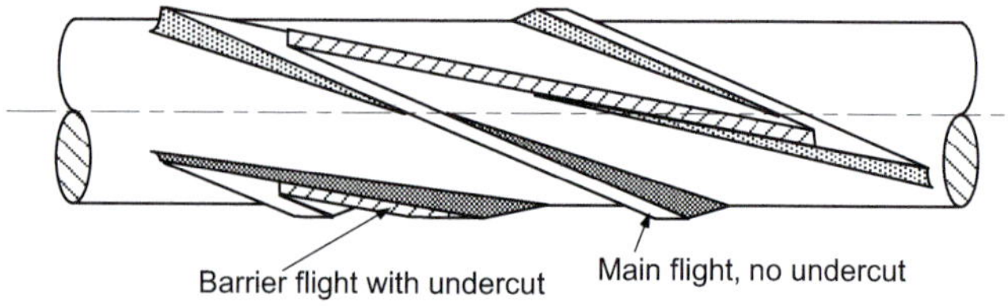

Figure 3.64 Zorro mixing section

The strong influence of helix angle on pressure drop is illustrated graphically in Figure 3.63. Clearly, a 90° helix angle (as in the LeRoy or Maddock mixer) is not a good choice in terms of pressure drop. Likewise, a 30° helix angle (as in the Egan mixer) is not a good choice either. The optimum helix angle is approx. 50°, and less than 50° for non-Newtonian fluids. The optimum helix angle depends on the degree of non-Newtonian behavior, and for typical plastics, the optimum angle is approx. 45°.

To promote good streamlining, the helix angle of the barrier flight can be made larger than the main flight, as shown in Figure 3.64. This makes the entry channel wide at the entrance and the exit channel wide at the exit. To minimize hangup, the channels should taper to zero depth at the end of the entry channels and at the beginning of the exit channels. The geometry of this Zorro mixing section is shown in Figure 3.64.

Planetary Gear Mixers

Planetary gear mixers have six or more planetary screws that revolve around the circumference of the main screw. The planetary barrel section must have helical grooves corresponding to the helical flights on the planetary screws. The planetary barrel section is generally separate, with a flange-type connection to the other barrel section (Figure 3.65). These machines are commonly used in Europe, but not commonly used in the United States. Some of the benefits of planetary gear mixers are:

- Good homogeneity of the melt at low temperature level
- Uniform shear exposure
- High output per screw revolution
- Low production cost per unit throughput
- Self-cleaning action for easy material change
- Good dispersive and distributive mixing of various additives

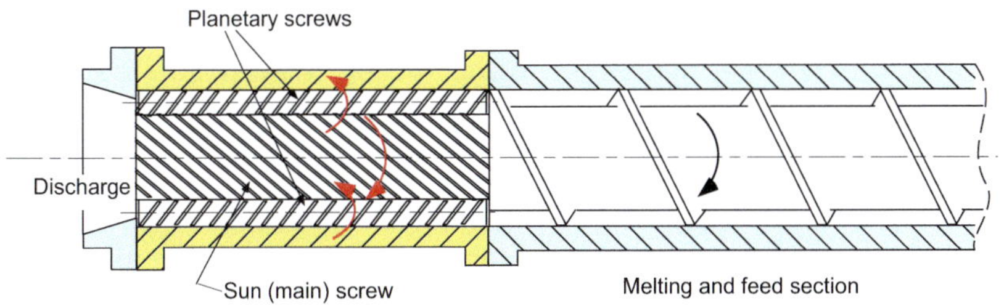

Figure 3.65 Schematic of planetary gear mixer

These characteristics make the planetary gear extruders well suited for foam extrusion and processing of heat-sensitive materials, such as rigid and flexible PVC. They are also used to process blends (e. g., PVC and ABS), powder coatings, epoxy, polyester, acrylic, polyurethane, and chlorinated polyethylene.

The Chris Rauwendaal Dispersive (CRD) Mixer

Current dispersive mixers have two important drawbacks. First, they rely mostly on shear stresses to disperse materials rather than on elongational stresses; and second, the material passes over the high stress region only once. New mixing technology developed by Rauwendaal eliminates the disadvantages of existing dispersive mixers [133–139]. The CRD mixer uses a slanted pushing flight flank to create a wedge-shaped lobal region (Figure 3.66).

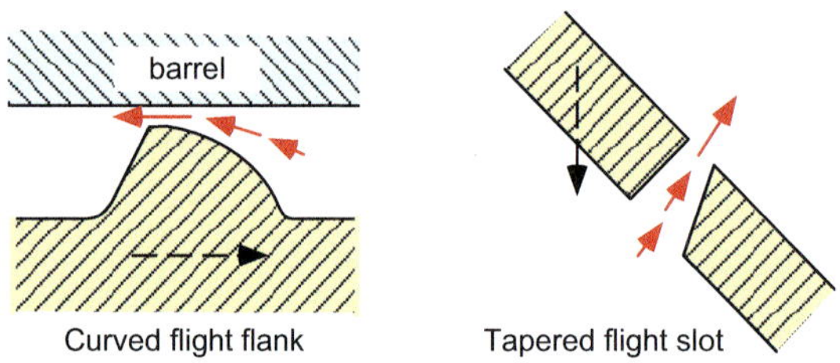

Figure 3.66 Wedge-shaped regions in CRD mixer

The flights in the CRD mixer are slotted to improve the distributive mixing capability; the slots are not straight but tapered. As a result, the fluid accelerates as it passes through the slots and thus is exposed to elongational flow. Therefore, the fluid is exposed to elongational stress as it flows over the mixing flights, and again when it passes through the slots in the flights. It is important to incorporate good distributive mixing in a dispersive mixing element to randomize the fluid. This ensures that all fluid elements are exposed to the mixing action several times.

The wedge shape creates strong elongational flow. The CRD mixer uses multiple mixing flights with a relatively large flight clearance to ensure that all fluid elements pass through the high stress region several times. Figure 3.67 shows a CRD5 mixer which has four flights with tapered slots. One out of four flight segments is a wiping flight section. The wiping flight sections are staggered in such a way that the mixer completely wipes the barrel.

Because of the large clearances of the mixing flights, wiping flights are used to avoid a stagnant film at the barrel surface and improve pumping. The wiping flights can be continuous flights, as in the CRD8 mixer, or wiping flight segments, as in the CRD5. The CRD8 mixer (Figure 3.68) has eight flights, six mixing flights and two wiping flights. All the flights are slotted to provide the best possible mixing action. Mixers with separate wiping flights are easier to manufacture than mixers that have wiping segments along the mixing flights. Good wiping action is important in maintaining good heat transfer characteristics in the mixing section.

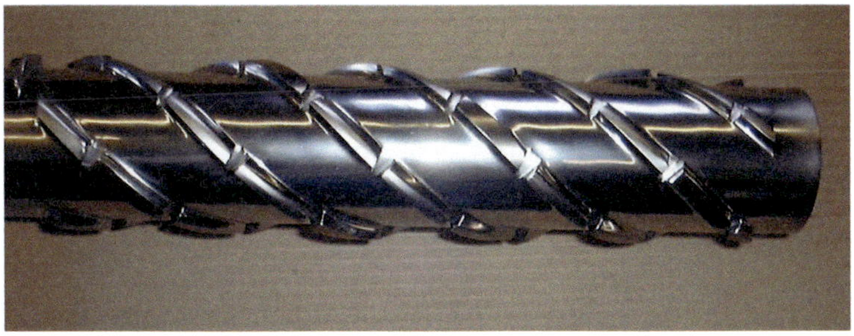

Figure 3.67 CRD5 mixer

Figure 3.68 CRD8 mixer

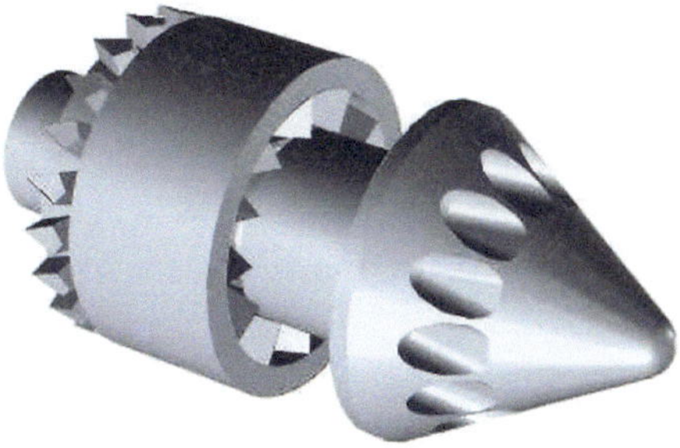

Figure 3.69 CRD non-return valve for injection molding applications

The wedge-shaped mixing-flight geometry can also be used in fluted mixers and barrier-type extruder screws. As a result, a wide range of CRD fluted mixers and barrier screws are available. The CRD mixing technology has also been applied to injection molding screws. A special slide-ring non-return valve (NRV) has been developed that incorporates CRD mixing action. A solid model of this CRD-NRV is shown in Figure 3.69.

At the beginning of the CRD-NRV, there is a mixing ring with tapered pins called elongational mixing pins (EMP). The slide ring has internal EMPs for additional mixing. Finally, the conical tip of the CRD-NRV has many conical holes machined into it and therefore acts as a third mixing ring. An important advantage of the CRD-NRV is that it fits in the same space as a conventional NR, making installation quick and easy. Since many processors are adding masterbatches right at the extruder or injection molding machine, there is increased demand for more-efficient mixing devices.

The CRD mixer has the advantages of being easy to manufacture and is relatively inexpensive, and it allows single-screw extruders to achieve a level of dispersive mixing equal to that of twin-screw extruders. There are a number of benefits associated with the elongational flow generated in the mixer. Dispersive mixing is more efficient in elongational flow than in shear flow. Gels cannot be dispersed in shear flow, but they can be dispersed in elongational flow. There is also less viscous dissipation in elongational flow than in shear flow. As a result, CRD mixers achieve lower melt temperatures than shear based mixers.

Summary of Dispersive Mixers

Just as we compared distributive mixers in Table 3.9, we can now compare dispersive mixers (Table 3.10). Most existing dispersive mixers are based on shear flow and expose the polymer melt to only one high-stress mixing event. As a result, these mixers tend to have limited dispersive mixing capability.

Table 3.10 Comparison of Dispersive Mixers for Single-Screw Extruders

Mixer	Pressure drop	Dead spots	Barrel wiped	Cost	Number passes	Type of flow
Blister	High	Fair	Fair	Low	1	Shear
Egan	Medium	Good	Yes	Medium	1	Shear
LeRoy/Maddock	Medium	Fair	Yes	Medium	1	Shear
Zorro	Low	Good	Yes	High	1	Shear
Helical LeRoy	Low	Good	Yes	Medium	1	Shear
Planetary gear	Medium	Good	Yes	Highest	Multiple	Shear
CRD mixer	Low	Good	Yes	Medium	Multiple	Elongation

3.4.2.8 Solving Mixing Problems

In addition to having efficient mixing devices along the screw, several other issues are important in achieving good mixing. The method of feeding plays an important role in the mixing action of the extruder.

Flood Feeding versus Starve Feeding

Most single-screw extruders are flood fed, but flood feeding is often detrimental to achieving good mixing in the extruder. Flood feeding generates high pressures in the solids-conveying and plasticating zones of the extruder. Such high pressures tend to agglomerate ingredients that must then be dispersed and distributed [147–149]. Obviously, this can be highly counterproductive.

In starve-feeding, a feeder meters the material into the extruder. As a result, there is no accumulation of material at the feed opening. The first several turns of the screw are partially filled with material, so there is no development of pressure in this part of the extruder. The screw channel does not become completely filled with material until some distance from the feed opening. At this point in the extruder, the pressure will begin to build up. In effect, starve feeding reduces the effective length of the extruder. Therefore, the pressures along the extruder are lower than in flood feeding, and there is less chance of agglomeration, resulting in improved mixing action in the extruder.

A number of researchers have analyzed the effect of starve feeding on the mixing capability of extruders and injection molding machines [150–154]. Without exception, these investigators found major improvement in the mixing quality with starve feeding compared to flood feeding. Starve feeding has become the standard mode of operation for twin-screw extruders used in compounding. However, the benefits of starve feeding are not limited to twin-screw extruders. Mixing in single-screw extruders can also be improved by starve feeding, in some cases quite significantly.

Initial Scale of Segregation

The mixing action required in an extruder is determined by both the initial and final scale of segregation. The required final scale of segregation is usually in the vicinity of 1 micron. If the initial scale of segregation is the size of a pellet (about 3000 microns), the mixing action must produce a 3000-fold reduction in the scale of segregation. In simple shear mixing, this requires a total strain rate of 3000 units [155]. If the average shear rate is 50 s^{-1}, it will be necessary to expose the polymer melt to this shear rate for 60 s to produce a total shear strain of 3000. A typical residence time in the melt-conveying zone is 15 to 20 s. Clearly, the mixing action taking place during this time will be insufficient to reduce the scale of segregation to the micron level.

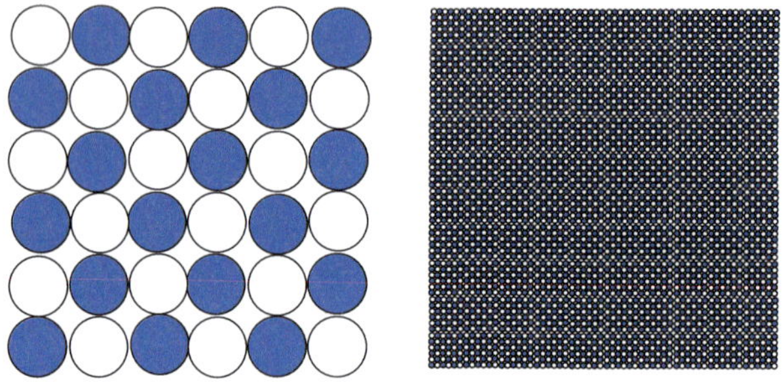

Figure 3.70 A coarse scale of segregation (left) and a fine scale of segregation (right)

This situation can be improved by running the mixing process with ingredients in powder form rather than in pellet form. If the size of the powder particles is 100 microns, and the powder is well mixed before extrusion, the initial scale of segregation will be 100 microns. In this case, to achieve a final scale of 1 micron, the mixing action must produce a 100-fold reduction in the scale of segregation. At a shear rate of 50 s^{-1}, this will take a shear time of 2 s. Clearly, a single-screw extruder will easily be able to accomplish this mixing task. The difference between a coarse and fine scale of segregation is illustrated in Figure 3.70.

One of the most difficult mixing tasks is to mix a low percentage of a color concentrate (CC) in pellet form with natural pellets. Consider a case in which the pellet size is 3 mm (3000 microns) and 1% CC is added to the natural material. For every CC pellet, there will be one hundred natural pellets. The initial scale of segregation in this case will be about 30 mm (30,000 microns). Since the CC pellets will not all be the same distance from one another, the actual scale of segregation may be as high as 100 mm (100,000 microns). In order to achieve a final scale of segregation of 1 micron, the mixing action must accomplish a 100,000-fold reduction in scale. If the average shear rate is 50 s^{-1}, the polymer melt will have to be exposed to this shear rate for 200 s (more than 3 min) to produce a total shear strain of 100,000. It is very unlikely that this can be accomplished in a simple conveying screw in a single-screw extruder. To accomplish this mixing task successfully, it will be necessary to use a very efficient mixing screw.

3.4.2.9 Melt Temperature Variation

The variation of the melt temperature distribution in the cross-sectional area of pipes has a significant impact on the cooling process and the final product quality [208, 209]. The melt temperature variation in circumferential direction can lead to ovality and non-uniform pipe wall thickness distribution. Time-dependent melt temperature variation causes geometrical variations in extrusion direction.

These effects are well known in the plastic pipe production industry. We will introduce an analytical method to quantify the mentioned quality problems under the assumption of a known melt temperature distribution during entering the pipe vacuum calibration unit. The cooling simulation software chillWARE® is used to realize the relevant simulations [209].

The analysis is employed for a PE100 pipe with an outer diameter of 250 mm and a wall thickness of 24 mm (SDR 10.4 is ratio of diameter to wall thickness). The pipe material is a standard PE100 material.

Causes of Wall Thickness Variation

It should be noted that non-uniform melt temperatures is only one of several possible causes of wall thickness variation. Overeijnder lists [183] the following possible causes.

1. Variations in elastic stresses in the melt caused by the folding of the polymer melt in the channels of a twin-screw extruder – this is typical for RPVC.
2. Incorrect design of the flow channel of the extrusion die can destabilize the melt flow, creating waves in the pipe.
3. Incorrect temperatures of oven and screw can create temperature differences that influence the discharge of the melt from the screw channels. In this case, wall thickness variation is related to screw speed.
4. In multi-layer pipe extrusion viscosity differences between the layers can cause flow variation. This is referred to as wave instability or interfacial instability.
5. In some operations unmelted polymer particles reach the discharge end of the extruder. These can create waviness in the pipe.
6. It is possible that the frictional force of the melt against the calibration sleeve is greater than the melt strength – this causes velocity variation. This problem is more likely to occur in a thin-walled pipe; it is sometimes referred to as slip-stick in the calibration sleeve. This problem can also occur with improper alignment between the extrusion head and calibrator.
7. In some situations the cleats of the puller do not have enough traction on the pipe, causing slipping. This can create ring-shaped waves in the pipe.
8. In the extrusion of foam pipes the inner wall can become irregular when the cell sizes become too large (> 0.15 mm). This is sometimes referred to as the orange peel effect.

The wave amplitude typically varies from 0.1 to 3% of the wall thickness. The wavelength varies significantly. The wavelength is smaller than the wall thickness for items 2, 5, and 8. The wavelength is about equal to the wall thickness for items 1, 4, and 6. The wavelength is greater than the wall thickness for items 3 and 7.

Melt Temperatures in Extruders

Melt temperatures in extruders are the result of viscous dissipation and heat transfer. Conductive heat transfer is slow because polymers have very low thermal conductivity – about one hundred times lower than the thermal conductivity of steel. If we consider the metering section of a single-screw extruder filled with molten polymer, the heat transfer in the melt close to the barrel surface is relatively efficient. As a result, the thin melt layer at the barrel surface can cool down quickly [184]. However, heat transfer in the melt further away from the barrel surface is inefficient because of the low thermal conductivity.

The resulting melt temperature distribution is inherently non-uniform as shown in Figure 3.71. This figure shows the melt temperature distribution in a 63.5-mm extruder running a fractional melt index HDPE (0.2 g/10 min) at a screw speed of 100 rpm.

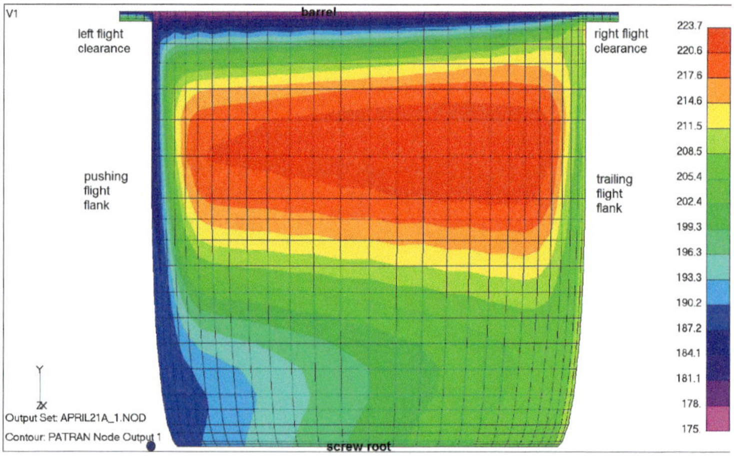

Figure 3.71 Melt temperature distribution in 63.5-mm extruder

Figure 3.71 is a color contour plot of the melt temperature distribution in a cross section of the screw channel in the metering section of the screw where the channel is completely filled with molten polymer. The channel depth has been magnified by a factor of eight to show the temperature distribution more clearly. It is clear that the melt temperature differences can be quite large. Unfortunately, these melt temperature differences are inherent to the polymer extrusion process.

Non-uniform melt temperatures in the screw channel result in temperature variation in the melt flowing through the die and in the melt emerging from the die. One of the results of these non-uniform melt temperatures is dimensional variation in the extrudate. This can happen in any extrusion operation. Here we will focus on pipe extrusion. In pipe extrusion the dimensional variation resulting from non-uniform melt temperatures is typically a waviness of the internal diameter. The outside diameter generally does not vary significantly because calibration results in a smooth OD of the pipe.

Melt temperature variation has been studied in detail at the University of Bradford, England [185–187]. Figure 3.72 shows melt temperatures measured with a wall thermocouple, a thermocouple mesh, and infrared.

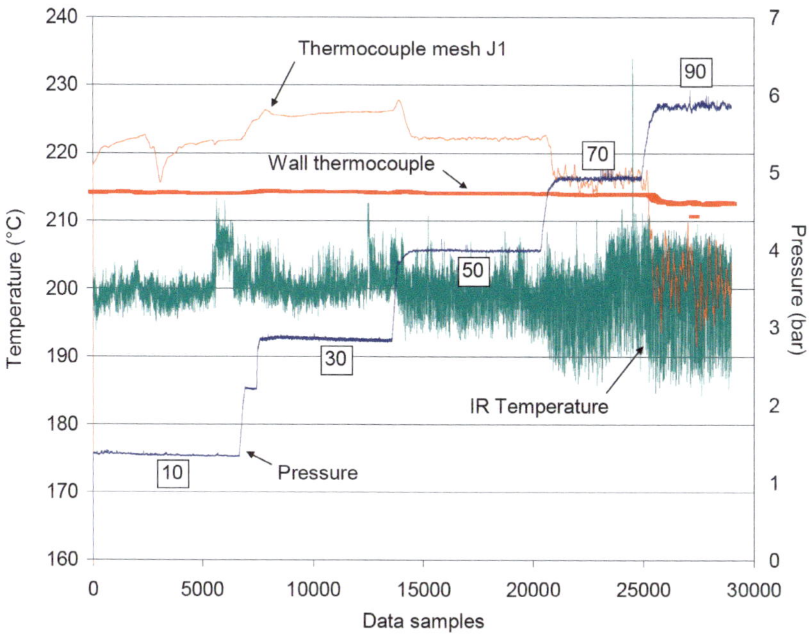

Figure 3.72 Melt temperature vs. time comparing thermocouple mesh to infrared, courtesy University of Bradford

Figure 3.72 shows that the wall thermocouple does not react significantly to melt temperature variation. The thermocouple mesh shows significant variation (10–20 °C) at screw speeds of 70 rpm and higher. The melt temperature variation increases with screw speed. The infrared measurement shows significant variation at all screw speeds. At 90 rpm the IR temperature variation ranges from 20–25 °C.

chillWARE® Cooling Simulation

The chillWARE® cooling simulation software enables the possibility to simulate the cooling process of extruded pipes, sheets, films, and profiles. The cooling situation can be analyzed during the whole cooling process at arbitrary process positions. In addition to the pure temperature distributions in the pipe cross-sectional area, residual stresses and thermal shrinkage are simulated as well. Exemplary applications of the software are described in the following references [189–191, 209].

The outer diameter and the wall thickness change during the cooling process. This behavior is caused by the change in the material density as exemplarily shown in Figure 3.73 for a reference PE100 material. The thermal density change generates a volume shrinkage of the material during the cooling process. The measurements in Figure 3.73 show a volume shrinkage of 23% during the cooling process. The volume shrinkage is separated in the radial, axial, and tangential directions, whereas the distribution depends on the material, the SDR-class, and the processing velocity. Of course, the absolute values for the change in geometry (outer diameter and wall thickness) depend on the melt temperature. Further information concerning the temperature-dependent material properties can be obtained from [188].

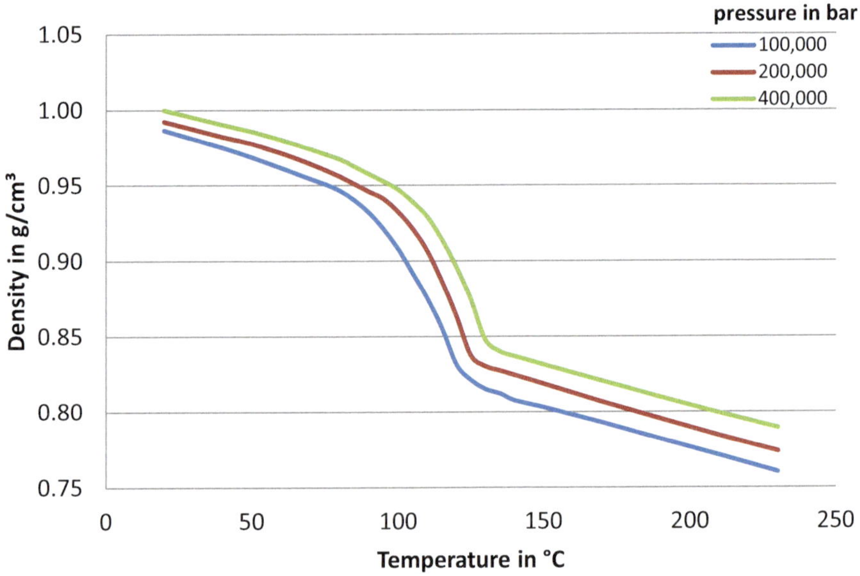

Figure 3.73 Variation of the material density with changing temperature for different pressures for a PE100 material. The crystallization process can be observed in the region around 120 °C

The cooling simulation chillWARE® is based on the finite difference and finite elements method and calculates the temperature distribution in radial pipe layers within the entire cooling section. The material properties of the polymers, such as the temperature-dependent thermal conductivity or the crystallization behavior, have a major importance for the modelling. By adjusting a variety of parameters (e.g., coolant temperature, production rate, or details of the cooling section) the optimal operating point of the cooling section can be determined. In the software the user-defined target internal temperature of the plastic products is used as a target value at a defined position in the cooling section.

Influence of Melt Temperature on Final Product Geometry

The pipe geometry is mainly defined by the melt temperature, the mass throughput, the extrusion die geometry, the outer vacuum calibration, and the vacuum and temperature settings in the downstream cooling sections. The major impact on the final outer pipe diameter is realized by the vacuum calibration unit (example shown in Figure 3.74). A perfect vacuum calibration realizes a full contact between the pipe outer surface and the inner calibration surface, which leads to an optimal heat transfer and a freezing of the outer pipe surface. Therefore, the pipe outer diameter is restraint-guided until the pipe leaves the calibration. The calibration time depends on the calibration length and the processing velocity and is usually not sufficient to realize the required final cooling. Therefore, additional cooling sections (e.g., (vacuum) spray cooling tanks, immersion baths) are following behind the calibration. The temperature levels and distribution inside the pipe wall after leaving the calibration have a significant effect on the final product geometry. Usually, it is not possible to compensate occurring melt temperature variations in the calibration, so that a non-uniform melt temperature distribution at the extrusion die leads to a non-uniform pipe temperature distribution after the calibration.

Figure 3.74 Example for an adjustable and conical vacuum calibration sleeve (source: CCA GmbH)

Exemplary Simulation of Geometric Variations Caused by Non-uniform Melt Temperatures

The effect of melt temperature variations on the product geometry is analyzed exemplarily with the simulation software chillWARE®. The reference production parameters are shown in Table 3.11.

Table 3.11 Process Parameters Applied for the Simulation of the Reference Process

Parameter	Unit	Value
Material	-	PE100
Final outer diameter	mm	250
Final wall thickness	mm	24
Mass throughput	kg/h	350
Melt temperature	°C	218
Processing velocity	m/min	0.35
Outer diameter of calibration	mm	256.6
Wall thickness at extrusion die	mm	31.04

The cooling section contains a vacuum calibration with a length of 600 mm, two vacuum spray cooling tanks with a length of 9000 mm each, and two spray cooling tanks with a length of 9000 mm each. The cooling process as a result of the chillWARE® simulation is shown in Figure 3.75.

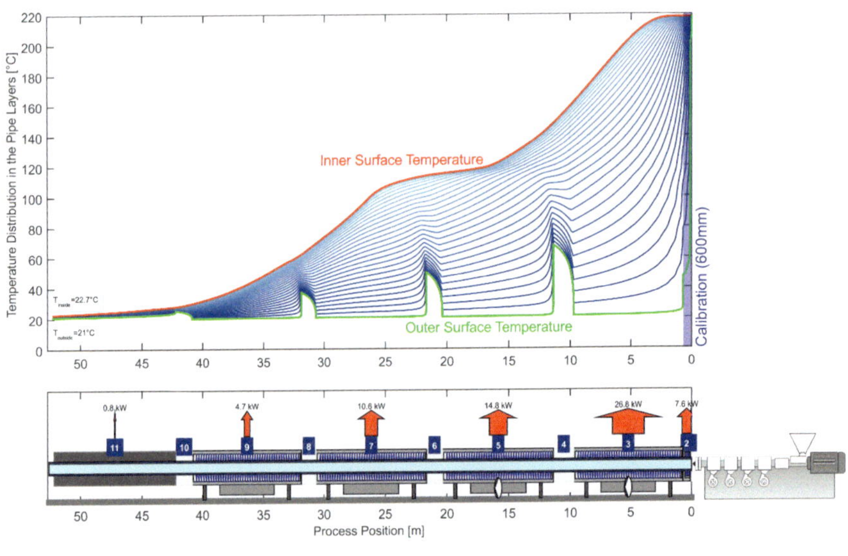

Figure 3.75 Cooling process for the reference production process. 7.6 kW thermal power is dissipated in the vacuum calibration unit. The outer surface temperature is 47.4 °C; the inner temperature is still at 218 °C melt temperature. The average cross-sectional temperature is 194 °C

The cross-sectional temperature distribution at the end of the calibration is shown in Figure 3.76.The outer diameter of the pipe experiences a thermal shrinkage of 3.5% from the extrusion die until the final product is finished at the end of the production line. The amount of thermal shrinkage depends on the melt temperature. The melt temperature is virtually decreased from 218 °C down to 188 °C in order to simulate variations in melt temperature. It is assumed that the polymer volume flow rate remains unchanged.

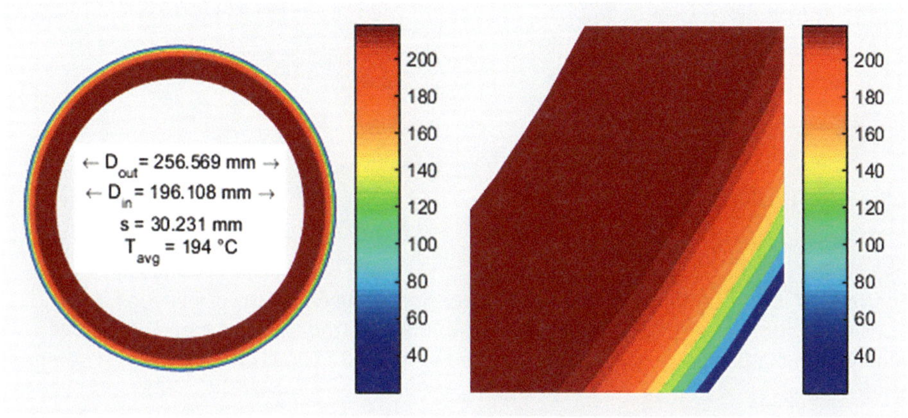

Figure 3.76 Cross-sectional temperature distribution at the end of the calibration for the reference process (homogenous melt temperature of 218 °C at the extrusion die). The temperatures of the inner pipe layers are still near the melt temperature

The corresponding simulation results directly after the vacuum calibration are presented in Figure 3.77. The average temperature is decreased from 194 °C to 167.7 °C. The outer diameter is fixed in the calibration; the thermal shrinkage of the outer diameter is reduced because of the lower average temperature level and the lower outer surface temperature. The dissipated thermal power is reduced from 7.6 kW to 6.7 kW due to the lower overall temperature gradients. Most interesting is the simulation of the resulting change in the outer diameter. The reference cooling process (cooling units and cooling fluid temperature levels) is unmodified and optimized for a target outer diameter of 250 mm and a wall thickness of 24 mm. The simulation results show that the final product outer diameter is enlarged from 250.00 mm to 250.99 mm (+1004%) and the pipe wall thickness is scaled up to 24.54 mm (+1022%). The geometric changes can be compensated by adapting the vacuum level in the first spray cooling tank or modifications in the calibration, if an adjustable calibration is available. This is only possible for a continuous/stable change in the melt temperature. This is usually not the case, if the changes in the melt temperature are a result from phenomena in the screw extruder.

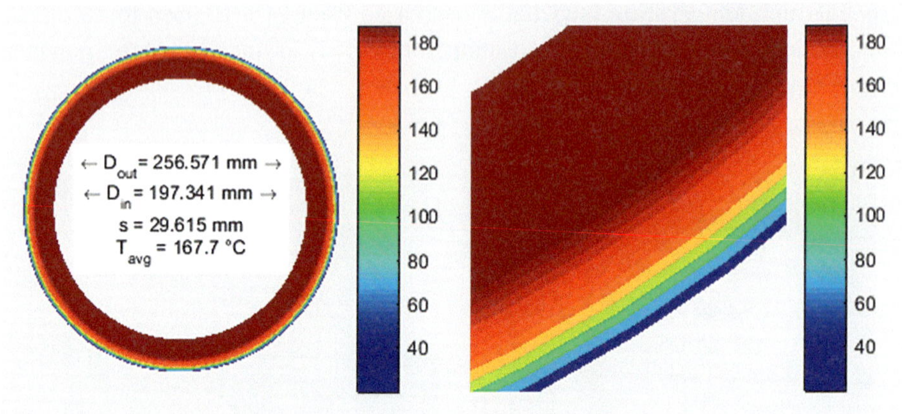

Figure 3.77 Cross-sectional temperature distribution at the end of the calibration for the modified process (melt temperature decreased to 188 °C at the extrusion die). The average temperature of the pipe cross section is reduced from 194.0 °C to 167.7 °C

An example for a resulting waviness in the product surface (change of the outer pipe diameter) is shown in Figure 3.78. The introduced reference process and the modified variant are applied; the melt temperature variation occurs with a frequency of f = 0.833 Hz, which corresponds to an extruder screw speed of 50 rpm. The resulting time-dependent outer diameter $D_a(t)$ is calculated with the reference outer diameter $D_{a,ref}$ = 250 mm and the enlarged diameter $D_{a,mod}$ = 250.99 mm as:

$$D_a(t) = 0.5 \cdot \left(\left(D_{a,\text{mod}} - D_{a,ref} \right) \cdot \sin(2\pi \cdot f \cdot t) + D_{a,\text{mod}} + D_{a,ref} \right)$$

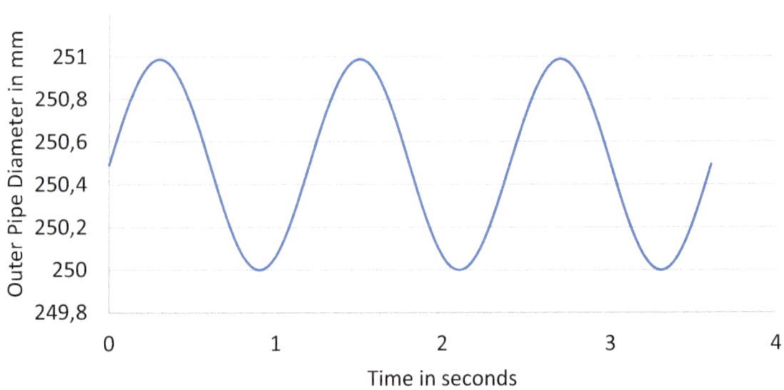

Figure 3.78 Resulting sinusoidal variation in the pipe surface (outer diameter) caused by a non-uniform melt temperature

Conclusions

The effect of non-uniform melt temperatures on the geometry of a pipe with outer vacuum calibration is analyzed in this section. The cooling simulation software system chillWARE® is employed to quantify the resulting variations of the outer pipe diameter and the wall thickness for a PE100-pipe with an outer diameter of 250 mm and a wall thickness of 24 mm (SDR 10.4). Melt temperature variation of 30 °C (188 °C–218 °C) is applied as initial conditions for the simulation. Most important for the final product geometry is the cooling process in the outer vacuum calibration and the subsequent thermal shrinkage caused by the temperature-dependent density of the thermoplastic material. The simulation results show that the outer diameter is increased by 1.004% and the pipe wall thickness enlarged by 1.022%. Furthermore, the generation of a wavelike product surface for a screw extruder driven with 50 rpm is calculated based on the simulation results. This contribution shows the importance of a stable thermal control of the melt temperature in order to guarantee an adequate product quality. Non-uniform melt temperatures can cause essential problems regarding the final product geometry and surface quality.

3.4.3 Solving Extrusion Instabilities

There are many different causes of extrusion instabilities. Even though the mechanism of the instability is not always clear, the following actions can often reduce extrusion instabilities:

- Reduce the screw speed
- Reduce the screw temperature
- Reduce the barrel temperature at the delivery end
- Reduce the channel depth in the metering section
- Increase the length of the compression section
- Increase the rear barrel temperatures
- Increase the die-head pressure

The first approach to solving the problem is generally adjustment of the barrel temperature profile or other process conditions. If temperature adjustment does not solve the problem, the hardware (thermocouples, controllers, screw, barrel, drive, and so on) should be checked. If the problem is not associated with hardware, it must be a functional problem, and the next step is to determine which functional zone is causing the problem. The troubleshooting flow chart in Figure 3.79 can help to systematically explore the possible causes of the instability.

If the problem cannot be solved by changing the processing conditions (which is, of course, the first choice), the problem can generally be solved by making a material change or by making a change in the screw or barrel design. In most cases, material changes are not possible. In that case, the problem usually must be solved by a new screw design. Another option is to add a gear pump at the end of the extruder.

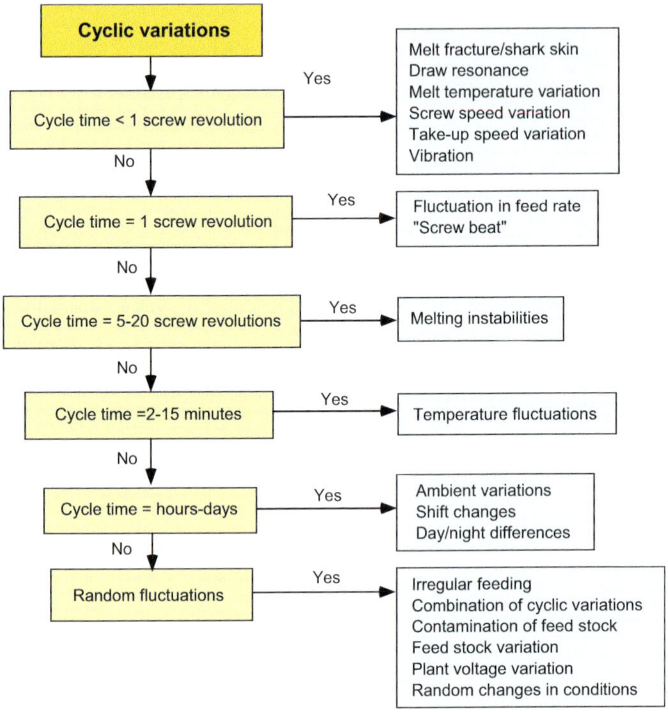

Figure 3.79 Troubleshooting flow chart for fluctuation in extruder performance

Figure 3.80 illustrates some of the important interactions that take place in the extrusion process. Because of these complicated interactions, it is often difficult to predict the effect of a change in process conditions or in screw design. For instance, increasing the barrel temperature will generally increase the polymer melt temperature at the discharge end of the extruder. However, it is also possible for the melt temperature to reduce with increasing barrel temperature. This is possible because, when the barrel temperature is increased, the local melt viscosity can reduce, which will reduce the local viscous heating. This effect can result in lower melt temperature at the discharge end of the extruder.

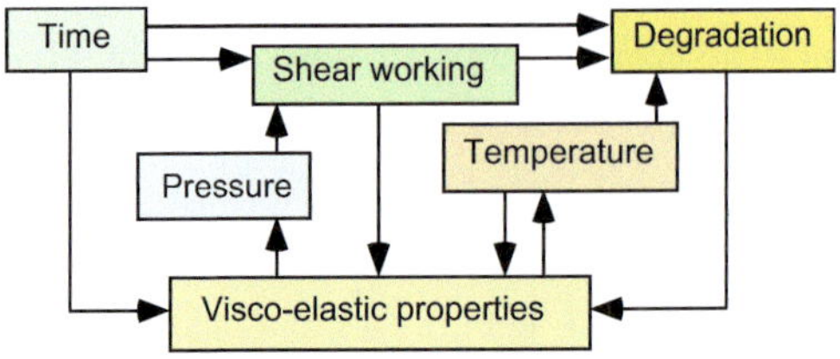

Figure 3.80 Troubleshooting flow chart for fluctuation in extruder performance

3.5 Air Entrapment

Air entrapment is a rather common problem in extrusion. It is caused by air being dragged in with particulate material from the feed hopper. Under normal conditions, the compression of the solid particulate material in the feed section will force the air out of the solid bed. However, under some circumstances, the air cannot escape back to the feed hopper and travels with the polymer until it exits from the die. As the air pockets exit the extruder, the sudden exposure to a much lower ambient pressure may cause the compressed air bubbles to burst in an explosive manner. Even if the air bubbles do not burst, the extrudate is generally rendered unacceptable as a result of the air inclusions.

There are a number of possible solutions to air entrapment. The first approach should be to change the temperature in the solids-conveying zone to achieve a more positive compacting of the solid bed. Often, an increase in the temperature of the first barrel section will reduce air entrapment, but in some cases, a decrease in the temperature will cause an improvement. In any case, the temperatures in the solids-conveying zone are important parameters in the air entrapment process. Both the barrel and screw temperatures are important. Thus, if screw temperature adjustment capability is available, it should definitely be used to reduce the air entrapment problem.

The next step is an increase in the die-head pressure to alter the pressure profile along the extruder to achieve a more rapid compacting of the solid bed. The die-head pressure can be increased by simply adding screens in front of the breaker plate. Another possible solution is to starve feed the extruder. However, this may reduce extruder output and requires additional hardware, such as an accurate feeding device.

The aforementioned recommended solutions can be implemented rather easily. However, if these measures do not solve the problem, more drastic steps must be taken. One possibility that must be considered is a change in particle size or shape. If this is a reasonable option, it will most likely solve the problem. A rather safe solution is to utilize a vacuum feed hopper system, but these systems are rather complex and expensive. Another possible solution is to use a grooved barrel section. Pressure development in a grooved barrel section is much more rapid than in a smooth barrel. This causes a rapid compacting of the solid bed and, therefore, decreases the chance of air entrapment. Instead of grooving the barrel, one can opt for reducing the friction on the screw, which would have a similar effect. Luker [45] described a coating that might be used for this purpose. Air entrapment is also often successfully eliminated by vented extrusion using a multistage extruder screw. Increasing the compression ratio of the screw is also likely to reduce air entrapment.

It should be noted that bubbles in the extrudate are not only a sign of air entrapment, but may also indicate moisture, surface agents, or volatile species in the polymer itself, or degradation as shown in the fishbone diagram in Figure 3.81. Thus, before concentrating on solving an apparent air entrapment problem, one should make sure that the problem is indeed caused by air entrapment.

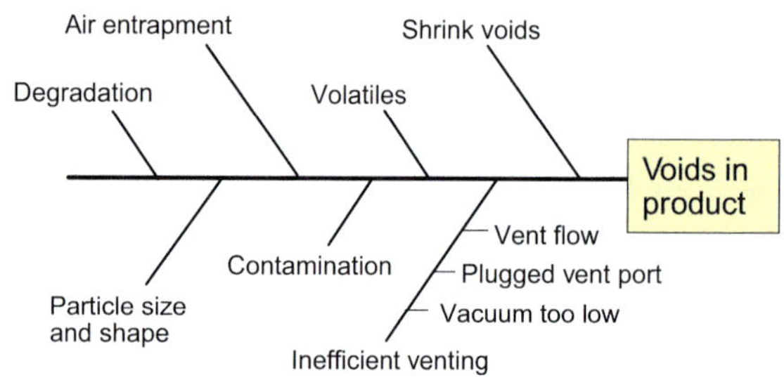

Figure 3.81 Fishbone chart for voids in extruded product

In some cases, the pellets themselves contain small air bubbles. In this case, one of the few possible solutions is vented extrusion because most of the other recommended solutions will not work in this situation. The fishbone diagram in Figure 3.81 helps in systematic troubleshooting. The diagram can be used to put together a troubleshooting flow chart for this kind of problem. An example of such a flow chart, developed for troubleshooting voids in the extruded product, is shown in Figure 3.82.

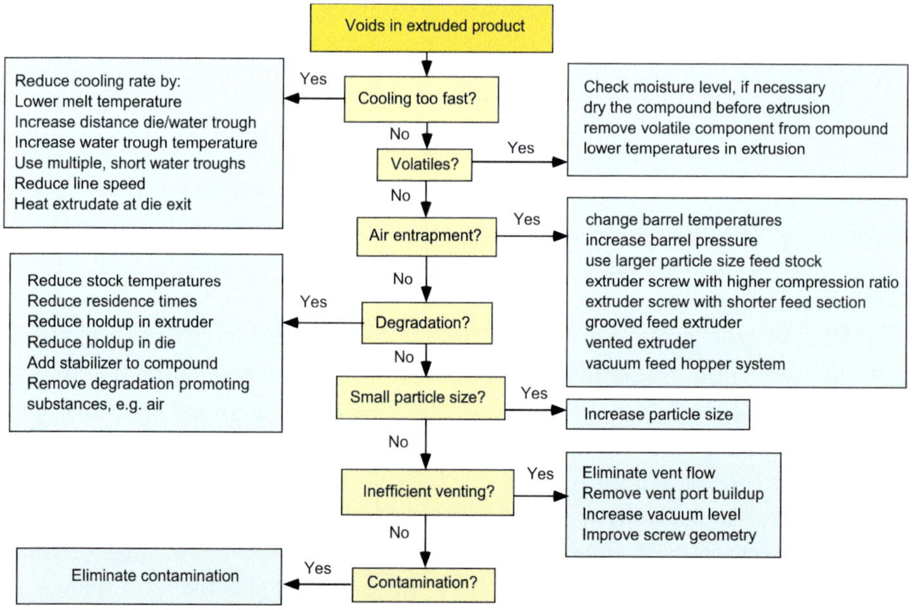

Figure 3.82 Troubleshooting flow chart for voids in extruded product

3.6 Gel Problems

The first problem with gels is that different people mean different things by the word "gel". Basically, a gel is a visual defect caused by differences in refractive index within a plastic product. As a result of the flow processes in extruders, gels usually take the shape of elongated ellipses, often called "fish eyes". Gels create problems in thin-walled products such as film, tubing, and fiber because they can form visible defects. In thick-walled products, gels are usually not visible and, therefore, not considered to be a problem.

Gels can come from several sources including high molecular weight (HMW) materials, cross-linking, degradation, and contamination. HMW material and cross-linked particles generally form elongated ellipses without a dot in the center of the fish eye. Dirt and foreign matter (such as silicas used in catalysts), inorganic material, additives, or fines from regrind generally form gels with a dot in the center. These most resemble fish eyes.

Gels constitute one of the common appearance problems in thin-gage extrusion (film, tubing, etc.). Gels usually consist of HMW droplets of cross-linked material in the plastic. When gels form in the polymerization process at the resin producer, they are called P-gels. These gels tend to form as a result of stagnant regions in the polymerization reactor. Gels formed in the extrusion process are called E-gels. High melt temperatures and long residence times are the main causes of gel formation in extrusion.

3.6.1 Measuring Gels

When there is a problem with gels in an extruded product, it is important to determine whether the gels are in the incoming raw material or are created during the extrusion process. Clearly, the resin supplier may be in disagreement with the processor concerning this issue! To determine whether the incoming material contains gels, the material must be tested. One method to test for P-gels is to press a thin plaque, using the polymer pellets as supplied by the resin producer, and visually examine the plaque for gels. The plaque must be prepared in such a way as to minimize the exposure to high temperatures to be sure that gels are not created in the sample preparation process. The number of gels per unit area can be counted using an overhead projector and polarized film to project an image on a screen. The number of gels per unit area is a measure of the amount of gels in the material as supplied. Obviously, the conditions used to press the plaque and the thickness of the plaque have to be standardized for the measurements to be meaningful.

For extruded film, there is an existing standard for manual gel counting that does not appear to be used much today [6]. A number of companies use automatic gel counting methods based on laser or CCD camera technology. However, these methods are not standardized, and each company tends to use its own procedures. A proposed ASTM procedure has been submitted for these automatic gel detection methods, but many issues remain to be resolved (for example, a standardized method to report the results, what size gels to count, and how to report the distribution of gel sizes).

Various end-use applications have different requirements with regard to gels. As a result, it is difficult to develop one standardized test method that satisfies all requirements. In fiber extrusion, in particular PP extrusion, a screen-buildup test is used in some cases. This test reflects other factors as well as the gel level in the polymer, because other materials can be trapped in the screen as well. Also, not all gel particles may be captured in the screen. For best gel capture capability, a 3D fiber filter with a specified rating should be used. The buildup can be quantified by monitoring the increase in pressure drop over time. The time to reach a specific pressure drop is a measure of the percentage of gels greater than a certain size removed by the filtration process. A regular wire mesh screen pack is not suitable for this test because wire mesh screens have very limited gel-capturing capability.

3.6.2 Gels Created in the Extrusion Process

To avoid E-gels, it is important to avoid dead spots in the extruder. This can be accomplished by making sure that both the screw and the die have a streamlined design. Mixing sections with stagnant regions, such as the Maddock mixing section, should be avoided. It is also important that the screw, barrel, and die surfaces are smooth without grooves, scratches, or gouges that might collect melted plastic and cause degradation. Another method for reducing gel formation in the extruder is to start up the extruder with a highly stabilized version of the plastic, or even a different plastic, to coat the critical surfaces with a degradation-resistant layer of plastic. This can reduce the chance of degradation and gel formation.

It is also important to check the resin feed tubes, blenders, feeders, hoppers, and other bulk handling hardware components for fines, streamers, or contamination from another plastic. To avoid fines, streamers, and contamination, the bulk handling equipment should be completely blown down and cleaned when a material change is made.

3.6.3 Removing Gels Produced in Polymerization

Unfortunately, it is quite difficult to remove P-gels in a regular extrusion process. Adding dispersive mixing elements to the extruder screw is usually not sufficient to achieve the amount of mixing necessary to eliminate P-gels. Also, the screen pack typically used before the breaker plate does not have enough gel-capturing capability to significantly reduce the gel level. New dispersive mixers capable of generating elongational flow such as the CRD mixer (see Section 3.4.2.7) can disperse gels [133–139] and provide a valuable tool in reducing gel problems.

One of the best tools available to remove gels is a depth filtration medium, such as sintered metal or a random 3D fiber. These depth filters have been in use for decades and have proven themselves in high quality film and fiber applications in which gels must be kept to the lowest possible level. Several companies sell large area depth filtration devices that are well-suited for gel removal. Unfortunately, such filters are expensive, maintenance intensive, and require replacement on a regular basis. However, they are effective in removing gels from the plastic melt. Figure 3.83 shows a flow chart for troubleshooting gel problems.

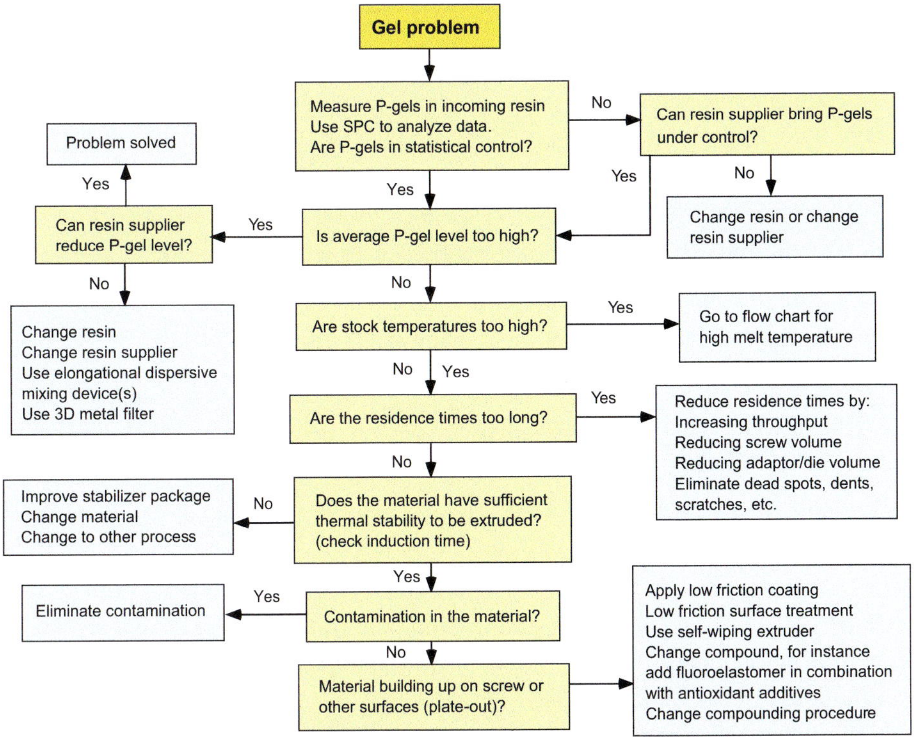

Figure 3.83 Flow chart to troubleshoot gel problems

3.7 Die-Flow Problems

Die-flow problems typically result in appearance problems. These can be related to melt fracture, die-lip buildup, gels, V- or W-patterns, specks, color variation, lines, and optical properties (for example, transparency, mattness, gloss, haze, etc.).

3.7.1 Melt Fracture

Melt fracture has been discussed in Section 3.4.1.1. It manifests itself as extrudate surface roughness, “sharkskin”, “orange peel”, and other distortions. Melt fracture can be reduced or eliminated by:

- Streamlining the die-flow channel
- Reducing the shear stress in the land region (operating below the critical shear stress for melt fracture)
- Using a processing aid (for example, a fluoroelastomer for polyethylene)
- Using “super extrusion” (operating above the critical shear stress for melt fracture)
- Ultrasonic vibration

Streamlining the die flow channel is always a good idea, but it will increase the cost of a die. For a high-volume product, it generally makes sense to design and manufacture a fully streamlined die. For a small volume product, this may not make economic sense.

Reducing the shear stress in the land region can be accomplished by:

- Increasing the die land temperature
- Opening up the die land region (increasing the die gap)
- Reducing the extrusion rate
- Using a process aid (for example, an external lubricant or a viscosity depressant)
- Increasing the melt temperature
- Reducing the polymer melt viscosity
- Using a more shear-thinning plastic

Several polymer processing aids (PPA) are available to eliminate or reduce melt fracture. An effective method for eliminating melt fracture in HMW polyolefins is to add a small amount of fluoroelastomer [157], approx. 500 to 1000 ppm (parts per million). After a fluoroelastomer PPA is added to a polyolefin, it usually takes a certain amount of time for a critical coating of fluoropolymer to form on the die. This conditioning time can vary from 5 minutes to more than one hour [158].

Silicon-based polymers such as polydimethyl siloxanene (PDMS) have been used as PPAs for many years. Dow Corning has developed ultrahigh-molecular-weight PDMS additives that work as process aids for polyethylene and polypropylene. Because these materials solidify with the polymer, they reportedly do not affect printability or paint adhesion. HMW PDMS has been used to reduce the surface roughness of extruded LLDPE tape [159].

Super extrusion is a technique in which the shear stress in the die land is above the critical shear rate for melt fracture. This is possible with polymers that exhibit a second stable region above the melt fracture region. Linear polymers such as HDPE, FEP, and PFA exhibit super extrusion behavior. The melt fracture behavior of a polymer can be determined using a capillary rheometer and running a polymer melt at different shear rates and observing the corresponding condition of the extrudate. A typical flow curve for a linear polymer looks like the one shown in Figure 3.84.

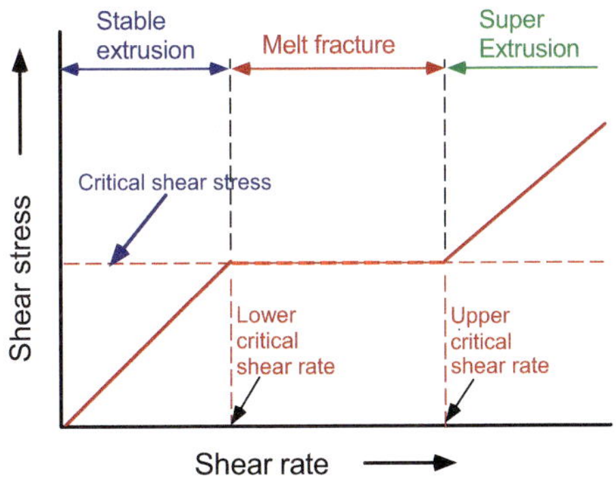

Figure 3.84 Flow curve of a linear polymer showing melt fracture and super extrusion regions

Melt fracture can be avoided by running under conditions in which the shear stress in the die is below the critical level for melt fracture or above the critical shear stress level for melt fracture. This requires operating at a shear rate below the lower critical shear rate for melt fracture or above the upper critical shear rate for melt fracture.

Ultrasonic vibration of the die can be accomplished by mounting external transducers that deliver ultrasonic energy in the kHz range to the die. Little quantitative information is available on this technique, but it is known that it has been successfully practiced in the extrusion industry. The principle behind using ultrasonic vibration in extrusion is related to the shear-thinning characteristics of polymers. The melt viscosity of polymers is reduced by orders of magnitude when the rate of deformation is increased. This applies not only to steady deformation but also to cyclic deformation. When a polymer melt is exposed to a high-frequency vibration, its viscosity will decrease by a large amount depending on the degree of shear-thinning.

When ultrasonic die vibration is used, the polymer melt layer at the die wall is most exposed to the high-frequency deformation. This results in a great reduction in melt viscosity at the die wall with several beneficial effects, including:

- Reduction of die-head pressure
- Reduction of extrudate swell
- Reduction of melt fracture
- Reduction of die-lip buildup (die drool)

3.7.2 Die-Lip Buildup ("Die Drool")

Die-lip buildup is a common problem in the extrusion industry, in which material accumulates right at the die exit as illustrated in Figure 3.85.

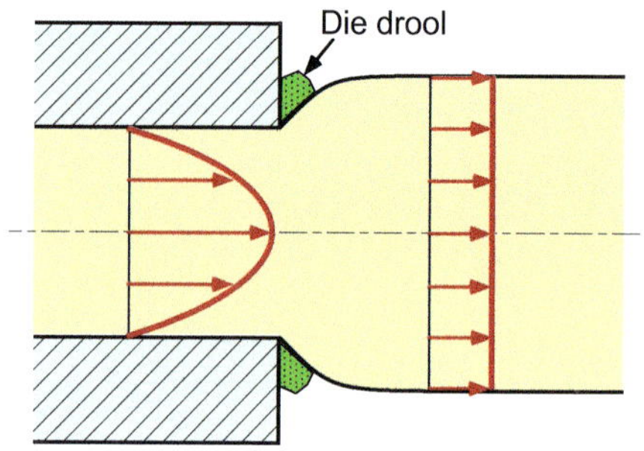

Figure 3.85 Illustration of die-lip buildup

Material buildup at the die exit can cause lines in the extruded product. This problem is often referred to as "die drool". It typically results from incompatible components in the compound, though it can also happen in non-compounded plastics. Die drool can be caused by gas or moisture in the molten plastic, by degradation, or by poor dispersion of fillers or additives. Die drool can be reduced by changing the material, the process, or the die design.

To reduce die drool by changing the material:

- Remove the incompatible component
- Add a fluoroelastomer
- Add a compatibilizer
- Change the compounding procedure

To reduce die drool by changing the process:

- Adjust the die temperature (usually lower)
- Blow air at the die exit
- Use a scraper at the die exit
- Use ultrasonic vibration

To reduce die drool by changing the die design:

- Use a low-friction coating in the die
- Use another die material, for instance ceramic
- Use a longer land length
- Use a small taper in the land region of the die

It is also important to inspect the condition of the internal die surfaces for scratches, bad plating, a pitted surface, or general poor surface quality. If any of these conditions are found, they must be corrected to minimize die-lip buildup. Interestingly, several of the measures that reduce die drool also reduce melt fracture. Processing aids that reduce melt fracture often reduce die drool as well [160, 161].

A number of studies on die drool were made by Professor Martin Zatloukal and coworkers at Thomas Bata University in Zlin, Czech Republic [179–182]. Zatloukal [182] published a very comprehensive review of studies on die drool with 179 references. In a study on die drool with HDPE [181] it was found that thermally induced degradation can lead to HDPE melt viscosity/elasticity enhancement. This promotes material accumulation at the end of the extrusion die.

3.7.3 V- or W-Patterns

V- or W-patterns are often related to line tension, more specifically, uneven line tension. In advanced cast film lines (biaxially oriented polypropylene (BOPP) or biaxially oriented polyethylene terephthalate (BOPET)) there are tension control systems that allow tension adjustment in specific regions of the film. A gear pump is often necessary to minimize output variation in these lines. The die-head pressure in such extrusion operations can be very high (up to 600 bar).

V- or W-patterns can also be related to uneven flow out of the die. The die design should be checked to be sure that the melt temperature is uniform in the direction of extrusion as well as across the die. In some cases, a static mixer can be placed inside the die, just before its manifold region, to provide mixing just before the polymer melt enters the manifold region.

Figure 3.86 shows a micrograph taken at 100× magnification under transmitted light of a V- pattern caused by uneven flow of PE out of the extrusion die. The process was pipe extrusion of PE for water conveying applications.

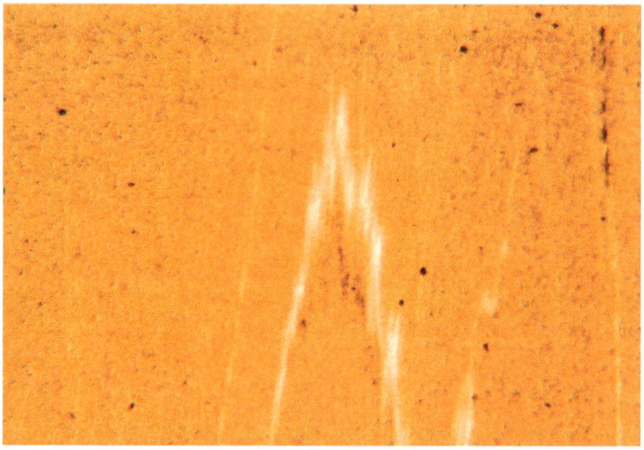

Figure 3.86 Optical micrograph of V-pattern in PE

3.7.4 Specks and Discoloration

Specks are a common problem in extruded products, particularly in thin or transparent products. The specks can be black, brown, yellow, or almost any other color different from the matrix material. Specks are usually caused by contamination, degradation, or wear. Degradation can manifest itself as discoloration, specks, pinholes, loss of volatiles (smoking), or changes in the physical properties in the extruded product. To find the cause of specks and discoloration:

- Check for contamination
- Check for high temperatures
- Check for stagnation
- Check the thermal stability of the plastic (and perhaps improve stabilizer package)
- Check for foreign particles (or wear)

Degradation can generally be reduced by:

- Reducing stock temperatures in the extruder
- Reducing the residence time in the extruder
- Eliminating the presence of degradation-promoting substances, for example, oxygen
- Adding a thermal stabilizer or improving the stabilizer package

Poor color uniformity can be caused by mixing problems in the extruder, by variation in the color additive or color concentrate, by variation in the addition of the color additive or concentrate, and by compatibility problems between the virgin and masterbatch (see the Case Study 4.12 in Chapter 4). Mixing a small amount of color concentrate into a natural polymer evenly is actually quite difficult because of the large initial scale of segregation. Mixing can be enhanced by improving the mixing capability of the extruder screw. Another way to improve mixing is to reduce the initial scale of segregation of the mixture. This can be accomplished by:

- Reducing the particles size of the color concentrate and virgin material
- Using liquid colorants

Solids-conveying problems in extrusion caused by liquid colorants can be avoided by using a porous carrier resin. Several resin suppliers (for example, DSM, Akzo, Montell) can now supply porous carriers for use with liquid additives. These carriers can be for colorants, but also for other additives such as antioxidants, peroxides, or silanes. Mixing can be further improved by starve-feeding the extruder as discussed in Section 3.4.2.8.

Figure 3.87 shows an optical micrograph taken at 500× magnification under transmitted and polarized light of black specks caused by agglomeration of pigment particles in LDPE. The black specks were observed as fish eyes. The process was film extrusion of LDPE for packaging applications.

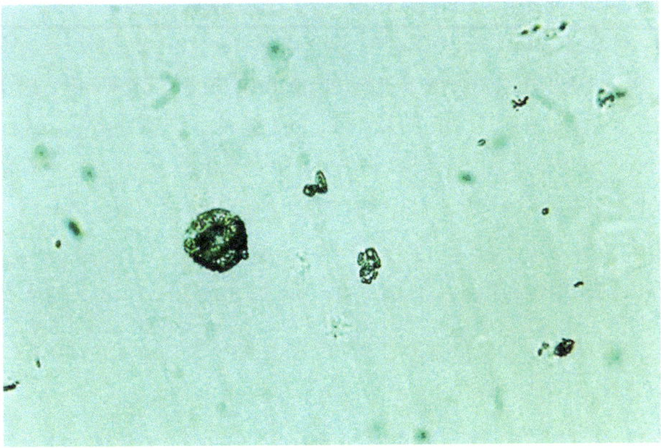

Figure 3.87 Optical micrograph of black specks in LDPE

3.7.5 Lines in Extruded Product

Lines can be caused by the die, the breaker plate, or the extruder screw, by die-lip buildup, and by downstream equipment such as calibrators, cooling baths, or catapullers (haul off). Visible lines right at the die exit must be formed at the die exit (e.g., by "die drool"), in the die (by poor internal surface conditions in the die or by buildup in the die), or upstream (at the breaker plate, screen pack, or screw). A single line is often formed in a crosshead die. When lines appear with in-line dies, the number of lines often will match the number of spider supports in the die. The breaker plate can cause a large number of lines in the product.

In blown-film extrusion, the number of lines in the film frequently corresponds to the number of ports in the die and, as a result, such lines are often referred to as port lines. These port lines are obviously related to the die design, but they are also very dependent on the melt flow characteristics of the polymer. High molecular weight polymers have very long relaxation times and are more likely to exhibit port lines or other types of die lines.

Lines can also originate downstream of the die, for instance, in the calibrator. Lines can result from an object touching the extrudate or by local cold or hot sections. These problems can generally be diagnosed easily by observing at what point in the extrusion process the lines originate.

3.7.5.1 Weld Lines

Lines in the extruded product can result from weld lines. These form when the polymer melt is split and recombined in the die or even before the die. Weld lines are also called knit lines; they can form in tubing and pipe dies where a mandrel is held in place by spider supports. The polymer melt is split at the start of the spider leg and flows together again behind (downstream) the spider support. Because of the limited mobility of long polymer molecules it takes a certain amount of time for the molecules to re-entangle. This re-entanglement process is also called a "healing" process. Longer molecules take longer to re-entangle. As a result, high molecular weight (high viscosity) polymers are more susceptible to weld lines than low molecular weight (low viscosity) polymers.

The severity of the weld line problem will be determined by:

- the length of time from the point where the melt streams recombine to the exit of the die (residence time), and
- the healing time of the polymer melt.

If the residence time is longer than the healing time, the weld line will disappear inside the die and not cause a problem in the extruded product. However, if the residence time is shorter than the healing time, the weld line will not disappear inside the die and the weld line will cause a problem in the extruded product. The weld line problem can be reduced or eliminated by increasing the residence time in the die or by reducing the healing time of the polymer melt.

The residence time in the die can be increased by reducing the flow rate (extruder throughput) or by changing the die geometry. The flow splitter has to be located as far away from the die exit as possible. Some die geometries reduce weld line problems. For instance, spiral mandrel dies for pipe, tubing, and blown film spread out the weld line as the melt flows through the spiral mandrel section; this largely eliminates problems with weld lines. Tubing and pipe dies with a rotating mandrel and/or die can also effectively spread out the weld lines and eliminate weld line problems. Some dies incorporate relaxation zones to enhance the healing process. Relaxation zones are basically local regions in the die flow channel where the cross-sectional area of the channel is increased.

The healing time depends on the molecular characteristics of the polymer and the melt temperature. Reducing the molecular weight of the polymer will speed up the re-entanglement process. Also, higher melt temperatures will increase the mobility of the polymer molecules and reduce the healing time. The molecular architecture also plays an important role. The molecules of linear polymers tend to align more readily and, as a result, entangle more slowly when separate flow streams meet. This is a particular problem in liquid crystalline polymers (LCPs) that have rod-like molecules. As a result, LCPs are highly susceptible to weld lines. Another method to promote re-entanglement is to subject the material in the die to a high-frequency vibration. Some processors use ultrasonic vibration of the die by mounting external transducers that deliver ultrasonic energy in the kilohertz range. Since polymers are not only shear-thinning but also frequency thinning, the effective viscosity of the polymer melt is reduced by high-frequency vibration. Other benefits of high-frequency vibration are reduced extrudate swell, reduced extrudate distortion at the die exit, reduced melt fracture, and reduced die-lip buildup (die drool).

3.7.6 Optical and Appearance Properties

Optical properties such as transparency, mattness, gloss, and haze are strongly determined by the cooling conditions of the extruded product. In crystalline polymers, the crystal growth is very temperature- and stress-dependent. As a result, the morphology of the extruded product will depend on how rapidly the polymer melt is cooled as it leaves the die. Slow cooling generally promotes crystal growth. Rapid cooling reduces crystallization; in fact, in some semi-crystalline polymers, the crystallinity can be suppressed completely with rapid cooling.

In sheet extrusion, in which the molten sheet of polymer is forced through a set of rolls, the surface conditions are strongly determined by the surface texture of the rolls. If the rolls are polished, they will impart a polished surface onto the polymer sheet. This is why these rolls are often called polishing rolls. Obviously, a variety of different textures can be machined into the rolls and, consequently, a number of different textures can be imparted to the extruded sheet.

Figure 3.88 shows an optical micrograph taken at 500 × magnification under transmitted and polarized light of grown spherulites caused by crystallization of PP. The process was film extrusion of PP for packaging applications.

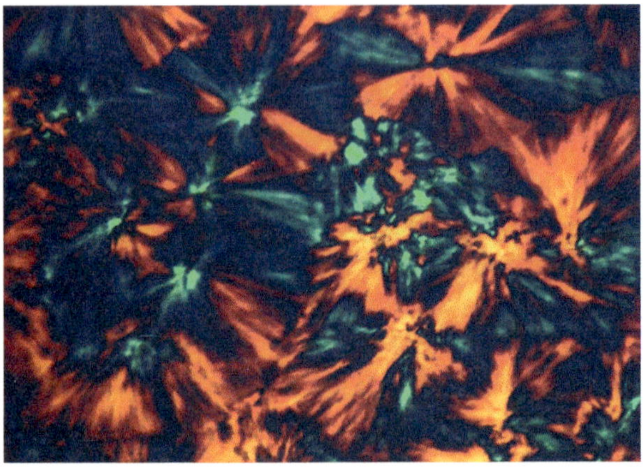

Figure 3.88 Optical micrograph of grown spherulites in PP

4 Case Studies

In this chapter, several actual case studies will be discussed. They describe a variety of problems, systematically analyze each problem, and offer a solution.

4.1 Film Coextrusion – Degradation in the Middle Layer

4.1.1 Description of the Problem

This case involved a three-layer coextruded film with an A–B–A configuration. The A-layers were made of a random copolymer PP (natural), each 3-micron thick, while the B-layer was a 9-micron homopolymer PP (natural). The middle layer was extruded on a 200-mm single-screw extruder; the outer layers were extruded on a 120-mm single-screw extruder. The polymer streams went into a feed block and from there to a cast film line. The extrusion line was fully instrumented. The film was biaxially oriented subsequent to extrusion.

The problem in this film was an appearance problem caused by gels. The defects looked like fish eyes. In general, defects in extruded film with biaxial orientation become magnified during the orientation process, which in some cases can lead to film breakage. The first objective was to determine the location of the introduction of the defect. To accomplish this, the region around the gel was examined using an optical microscope.

4.1.2 Analysis of the Problem

The film was cut through the gel to allow examination of the film cross section. Then the sample was embedded in epoxy resin and polished after curing of the epoxy. A micrograph (see Figure 4.1) was taken at 200 × magnification, illuminated with transmitted and polarized light. The microscope was a Leica optical microscope, Laborlux 12 Pol S, equipped with polishing and microtome capability. The illumination could be transmitted, reflected, and polarized. The micrograph shows that the gel was located in the middle layer. The discoloration of the material in the middle layer indicated degradation. At this point, the plant was visited and the operation of the extruders was checked, particularly the 200-mm extruder with $L/D = 24$.

The extruder operating conditions are shown in Table 4.1.

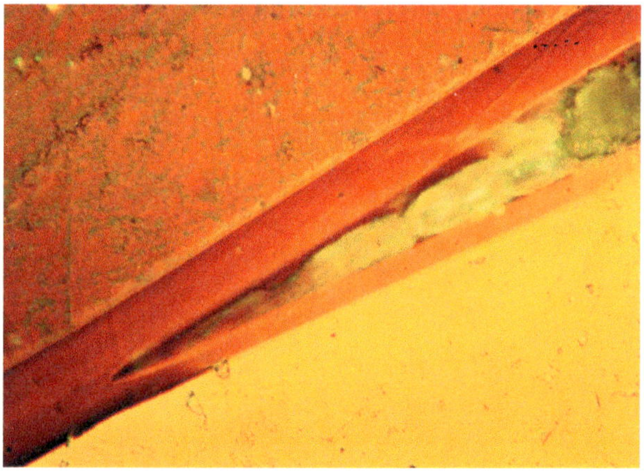

Figure 4.1 Optical micrograph of three-layer film

Table 4.1 Operating Conditions of 200-mm Extruder

Screw rotation speed, rpm	255
Extruder back pressure, bar	65
Barrel temperature profile, °C	190 (= feed zone), 200, 210, 220, 230, 240, 250 (= screw tip)
Temperature of polymer granules, °C	25
Melt temperature, °C	240
Output, kg/h	190

The extrusion line had a melt temperature probe at the discharge end of the extruder barrel, just before the screen pack. Thus, the melt temperature had been taken at only one location, where the temperature was found to be 240 °C. However, the melt temperature at the die exit was determined to be higher, close to 280 °C. This temperature was measured using an infrared temperature probe (Raytek, model Raynger PM). Though it was not possible to measure the melt temperature in the individual layers, it could be assumed that the high overall melt temperature represented mostly the temperature of the middle layer because it was the thickest part of the structure.

The screw design for the 200-mm extruder was also checked. The geometry is summarized in Table 4.2. The screw geometry was typical for the processing of polyolefins. The screw compression ratio of 2.7 was appropriate for polypropylene.

Table 4.2 Summary of Geometry of Existing Screw

Diameter, mm	200
Total length, *L/D*	24
Feed zone:	
Channel depth, mm	28.5
L/D	9
Compression zone:	
L/D	6
Metering zone:	
Channel depth, mm	10.5
L/D	9

4.1.3 Solution

The solution strategy was visualized based on the fishbone diagram in Figure 4.2, customized to the problem. Following this diagram, the causes were checked one by one.

The recommended melt temperature for this process is approx. 260 °C. Therefore, the barrel temperatures of the 200-mm extruder were lowered to achieve a melt temperature at the die exit of 260 °C. It was possible in this case to lower the melt temperature by adjusting the barrel temperatures, but a melt temperature variation of about 10 °C across the width of the extrudate was observed. When the melt temperature was lowered to 260 °C, the gel problem disappeared.

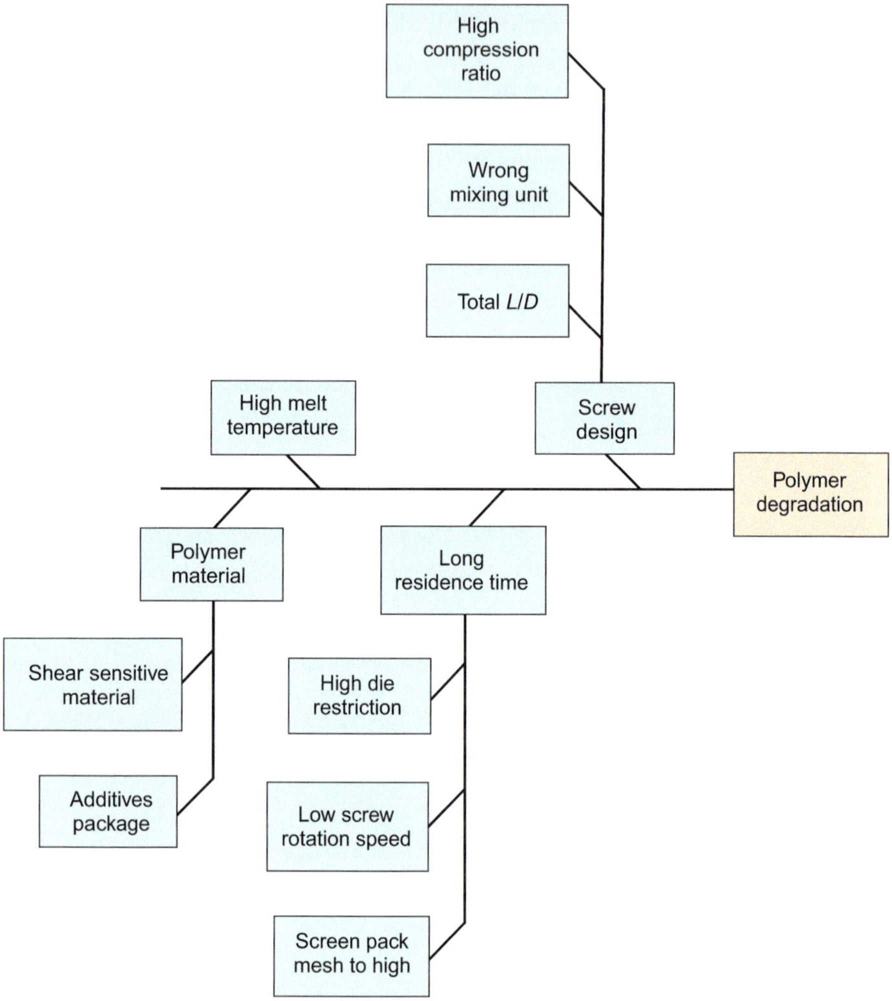

Figure 4.2 Fishbone diagram of degradation problem

The most effective method to reduce the melt temperature variation was to improve the mixing capability of the screw. A distributive mixing element was added to the metering zone. The final screw design is summarized in Table 4.3. Once a screw with the modified screw geometry was running, the melt temperature variation decreased significantly.

Table 4.3 Screw Geometry of New Screw

Feed zone:	
Channel depth, mm	28.5
L/D	9
Compression zone:	
L/D	6
Metering zone:	
Channel depth, mm	10.5
L/D	6
Mixing zone:	
L/D	3

4.2 Film Coextrusion with Interfacial Problems

4.2.1 Description of the Problem

This problem occurred in coextruded cast film. The film was a two-layer film in which one layer was an ionomer and the other layer polyolefin. The total film thickness was 60 microns. The customer described the problem as delamination in large sections of the film.

4.2.2 Analysis of the Problem

In order to observe the different layers and the conditions at the interface clearly, a crosscut of the film was made and the sample was embedded in epoxy resin and polished after the epoxy was cured. The magnification was 500×, and reflected light was used because TiO_2 was present in both polyolefin layers. A photomicrograph is shown in Figure 4.3. The ionomer layer in the micrograph was about 20 microns thick, and the polyolefin layer was about 40 microns thick. The micrograph showed that the delamination was occurring at the interface between the ionomer and polyolefin layers. The delamination was observed as voids at the interface.

Figure 4.3 Photo micrograph of coextruded film

4.2.3 Solution

The viscosities of the different materials were analyzed. The melt index of the polyolefin was 2 [g/10 min], while the melt index of the ionomer was 4.3 [g/10 min]. The higher melt index of the ionomer indicated that the melt viscosity of this material was lower than that of the polyolefin. Large differences in melt viscosity can create shear stress differences at the interface and these can cause delamination. To reduce the viscosity difference, a different ionomer was selected with a melt index closer to the melt index of the polyolefin. After this material change was made, the delamination problem disappeared.

4.3 Lines in the Extruded Film

4.3.1 Description of the Problem

This problem occurred in a mono-layer blown film line operating with a 60-mm extruder and a 200-mm diameter die with a die gap of 1 mm. The material extruded was an LDPE with a melt index of 2 [g/10 min]. The company experienced a problem with a marked line in the extruded film in the direction of the machine. In some cases, the line was so severe as to cause tearing of the film. The company tried several LDPE resins with different melt index values in attempts to solve the problem; however, the problem could not be solved by resin changes.

The extruder operating conditions are shown in Table 4.4. These operating conditions were appropriate for LDPE extrusion in bag manufacturing.

Table 4.4 Operating Conditions of Extruder

Screw rotation speed, rpm	80
Barrel temperature profile, °C	130 (= feed zone), 140, 150, 150, 160 (= screw tip)
Temperature of polymer granules, °C	25
Melt temperature, °C	200
Output, kg/h	55

4.3.2 Analysis of the Problem

The solution was visualized based on a fishbone diagram customized to the problem (Figure 4.4), and each one of the possible causes was checked.

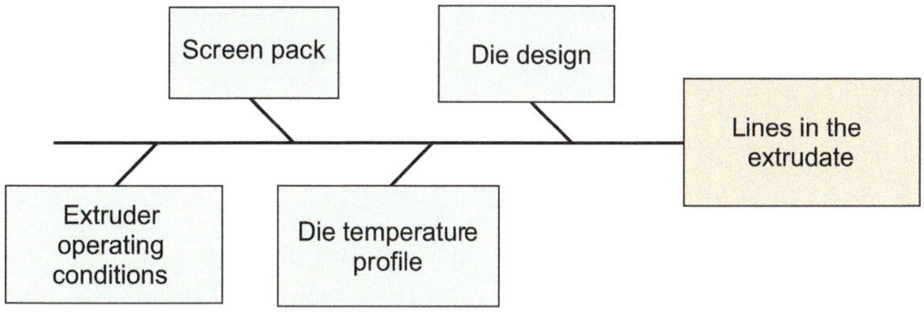

Figure 4.4 Fishbone diagram of problem with lines in film

At this point, the tooling was disassembled to analyze the geometry of the die; the tooling had been designed in-house. It was discovered that there were sharp corners in the die as shown in Figure 4.5. Sharp corners were found in both the mandrel and the housing. The screen pack and its support were also verified, but they were not the origin of the problem because there was just one marked line in the extrudate.

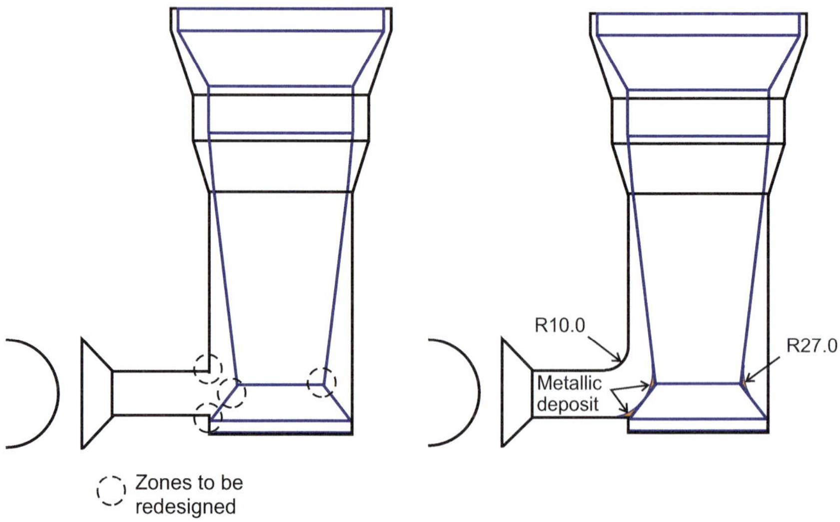

Figure 4.5 Die design before (left) and after (right) design modification

4.3.3 Solution

The mandrel and housing were modified to eliminate the sharp corners, thereby providing streamlined flow channels. Figure 4.5 shows the die design before and after modification. The sharp corner in the housing was eliminated by grinding away material. Metal was applied to the mandrel corner, and the region was smoothed out, eliminating the sharp edge of the mandrel. When the tooling was reinstalled in the plant, the problem disappeared completely.

■ 4.4 Color Variation in Polypropylene Carpet Fiber

4.4.1 Description of the Problem

The problem showed up as color variation in carpet made from the homopolymer PP fiber. The PP had a melt index of 18 [g/10 min]. The color was added to the PP by adding a small amount of color masterbatch (5%) to the natural PP, then the mix was processed in a 60-mm single-screw extruder. The color concentrate (CC) contained 50% color by weight, and the base polymer of the CC was the same as the natural PP.

The extruder operating conditions are shown in Table 4.5. These operating conditions were reasonable for PP extrusion in this application.

Table 4.5 Operating Conditions of Extruder

Screw rotation speed, rpm	132
Extruder back pressure, bar	60
Barrel temperature profile, °C	195 (= feed zone), 200, 205, 215, 235, 235 (= screw tip)
Temperature of polymer granules, °C	25
Melt temperature, °C	240
Output, kg/h	76

4.4.2 Analysis of the Problem

A sample of the polymer melt was taken at the die exit, and this material was pressed into a thin plaque. The pressed film samples were analyzed for color variation using optical microscopy with transmitted light. The magnification was 50×. Figure 4.6 shows an example of a cross section of the film. From the micrograph it is clear that there are non-uniformities in the color distribution within the sample, showing up as darker and lighter regions and as striations.

Figure 4.6 Micrograph of film sample (crosscut)

4.4.3 Solution

The solution strategy was visualized based on the fishbone diagram in Figure 4.7, customized for the problem. Each possible cause was carefully checked. The first step in the troubleshooting process was to check the rheology of the virgin PP and the color masterbatch. It turned out that the viscosities of the CC and the natural PP were quite close. Therefore, it was concluded that the mixing problem was not associated with differences in viscosity between the CC and the natural PP.

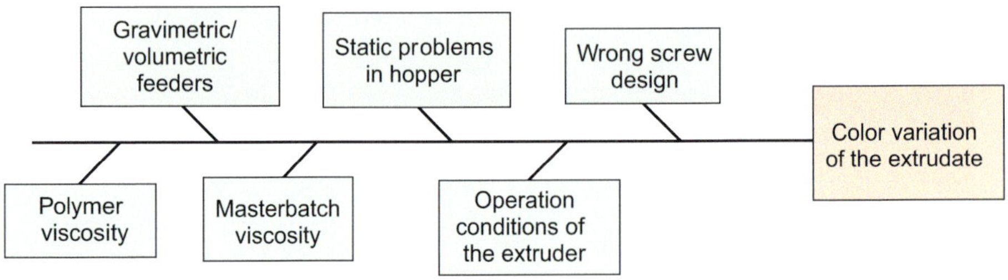

Figure 4.7 Fishbone diagram for color variation

The second step was to check the geometry of the screw used to extrude the material. The screw geometry in use at the time of the problem is summarized in Table 4.6. The screw had a conventional three-zone design with a compression ratio of 3.1 and without any mixing elements. Since simple conveying screws have poor mixing capability, a new mixing screw was designed and manufactured.

Table 4.6 Important Data of Current Screw Design

Diameter, mm	60
Total length, *L/D*	30
Feed zone: Channel depth, mm *L/D*	 11 10
Compression zone: *L/D*	 7
Metering zone: Channel depth, mm *L/D*	 3.5 13

The new mixing screw geometry is summarized in Table 4.7. The new screw was designed with a barrier geometry in the melting section of the screw, followed by two dispersive mixing elements (2*D* each), followed by one 2*D* distributive mixing element. With this screw, the color could be processed without any color variation in the extrudate or in the final product. In this case, two dispersive mixers could be used without problems because of the low melt viscosity (high melt index) of the polymer. Low melt viscosity raises little concern regarding an increase in temperature due to viscous dissipation.

Table 4.7 Data of New Mixing Screw

Diameter, mm	60
Total length, *L/D*	30
Feed zone:	
Channel depth, mm	12
L/D	10
Barrier melting zone:	
L/D	14
Dispersive mixing zone:	
Channel depth, mm	8
L/D	4
Distributive mixing zone:	
Channel depth, mm	8
L/D	2

4.5 Plastic Film with Poor Transparency

4.5.1 Description of the Problem

This case involved a PP film manufactured in a cast film line. The mono-layer was made of homopolymer PP (natural). The film was extruded on a 120-mm single-screw extruder with $L/D = 28$. The polymer stream flowed into a feed block, and from there to a cast film die. The extrusion line was fully instrumented.

The film exhibited a serious appearance problem, in this case poor transparency. The film looked opaque throughout the entire roll. The extruder operating conditions are shown in Table 4.8. The extrusion line had a melt temperature probe at the discharge end of the extruder barrel, just before the screen pack. Melt temperature, measured only at this one location, was 225 °C. The screw design was also checked and its geometry is shown in Table 4.9. The screw geometry indicates a screw design intended for processing polyolefins. The screw compression ratio of 4 is relatively high for polypropylene.

Table 4.8 Operating Conditions of 120-mm Extruder

Screw rotation speed, rpm	170
Extruder back pressure, bar	200
Barrel temperature profile, °C	225 (= feed zone), 225, 215, 215, 215, 215, 215 (= screw tip)
Temperature of polymer granules, °C	30
Melt temperature, °C	225
Output, kg/h	650

Table 4.9 Screw Geometry

Diameter, mm	120
Total length, *L/D*	28
Feed zone: Channel depth, mm *L/D*	 21.3 7
Compression zone: *L/D*	 8
Metering zone: Channel depth, mm *L/D*	 5.3 11
Mixing zone: Channel depth, mm *L/D*	 20 2

The first issue was to take into account all the variables related to film transparency in a semi-crystalline material such as PP.

4.5.2 Analysis of the Problem

The problem was visualized based on the fishbone diagram in Figure 4.8, and the variables were checked one by one. Crystallization is a major concern in this case. Therefore, optical microscopy was used to examine the type of crystals occurring in the film under different operating conditions for the extruder and with different cooling rates.

Film manufactured under one specific set of operating conditions was cut to allow examination of the film cross section. A micrograph was taken at 500 × magnification with transmitted and polarized light as illumination (see Figure 4.9). A Leica optical microscope, Laborlux 12 Pol S, equipped with polishing and microtome capability was used for the examination. Transmitted, reflected, and polarized light can be used with this microscope.

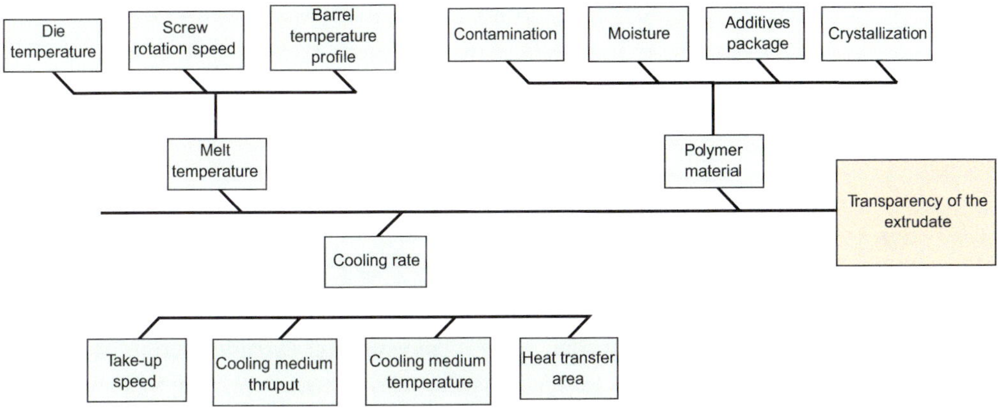

Figure 4.8 Fishbone chart of transparency problem

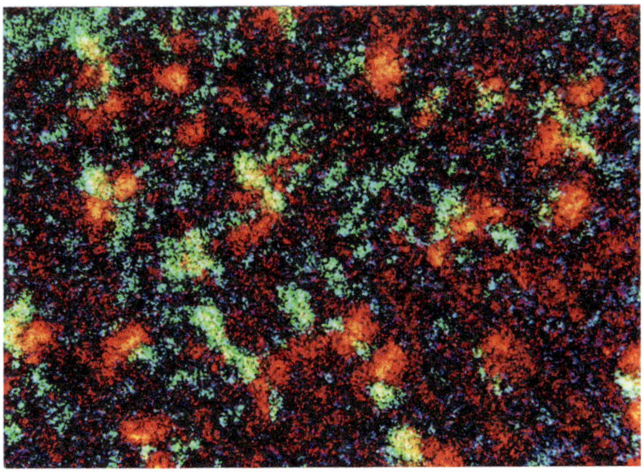

Figure 4.9 Optical micrograph of spherulites in a polypropylene film

The micrograph shows the type of crystals or spherulites in the film. It is possible to observe both amorphous and crystalline regions. Several micrographs of film manufactured at different operating conditions were obtained.

4.5.3 Solution

Crystallization was confirmed as the key variable affecting film transparency based on the micrographs at different line operating conditions. Transparency was better at higher cooling rates and lower chill roll temperatures. The lower limit for the chill roll temperature was taken as the temperature at which condensation started (about 12 °C). Cooling medium conditions and take-up speed were modified to increase the cooling rate. The chill roll temperature was lowered to 12 °C. Under these conditions the transparency of the film met requirements.

4.6 Wear Problem in Film Extrusion

4.6.1 Description of the Problem

This problem was observed in a PP film which had dark inclusions. The PP was extruded on a 120-mm extruder using a conventional extruder screw. The inclusions were quite noticeable because the film was transparent, and, in some cases, the inclusions were so large that problems with film breakage occurred.

A micrograph of the film with a magnification of 200× was prepared with transmitted and polarized light. The micrograph is shown in Figure 4.10. The inclusion, clearly a foreign object, was identified by a magnet as a metal particle. The metal particles were as large as 200 microns (0.2 mm).

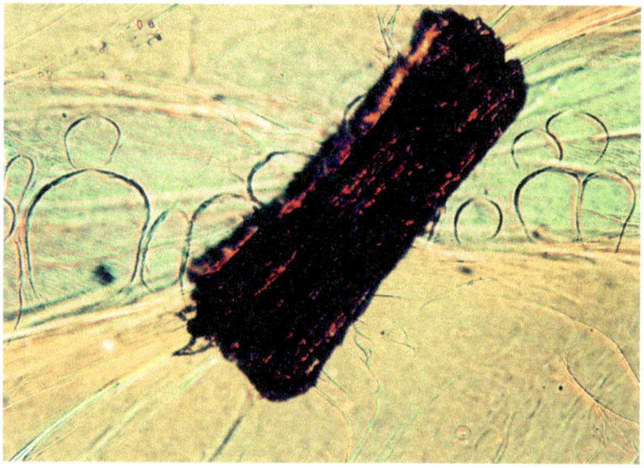

Figure 4.10 Micrograph of foreign object

4.6.2 Solution

Both the screw and the barrel were examined for wear. The screw did not show appreciable wear; however, the barrel did. The company ordered a new barrel, and when the new barrel was installed there were no more inclusions observed in the film.

In this case, it had been a long time since the extruder had been checked for wear. It was recommended that the extruder screw and barrel be checked once a year to determine the extent of wear. As a good preventive maintenance program, it is necessary to examine the screw and barrel at least once a year.

■ 4.7 Multilayer Film – Several Appearance Problems

4.7.1 Description of the Problem

This problem concerned a six-layer coextruded blown film with appearance problems. The film was of natural color and the structure of the film was nylon–EVOH–tie layer–PET–tie layer–LLDPE. The total film thickness was 160 microns. In some products, the middle PET layer was replaced with nylon. Three defects were discovered in the film:

- Opaque film
- "Milky" appearance on one side of the film
- Film curling

4.7.2 Analysis of the Problem

The film was microtomed and the cross section of the film was examined in detail by optical microscopy (see Figure 4.11). A micrograph was taken at 200 × magnification under illumination with transmitted and polarized light, using a Leica optical microscope (Laborlux 12 Pol S) equipped with polishing and microtome capability.

Figure 4.11 Micrograph of coextruded six-layer film, 200 ×

Since all the polymer layers in this structure were transparent, it was not possible to distinguish the layers. Colorimetry was used to distinguish the layers: iodine was used to color the PA6 red, the EVOH brown, and the tie layers gray; the iodine did not affect the color of PET or LLDPE. A micrograph of the colored film (see Figure 4.11) clearly indicated that the actual film structure was nylon-EVOH-tie layer-PET-tie layer-LLDPE based on the developed colors. The "milky" appearance problem and the film-curling effect were both located on the nylon film side.

The die diameter was 250 mm with a die gap of 1.3 mm. The die temperature profile was 215, 220, and 220 °C. This is a low temperature profile, considering the presence of nylon (PA6) and PET in the film structure. Based on the raw material data sheets, the melting points were around 220 °C for both PA6 and PET. These are typical values for film grade resins.

4.7.3 Solution

The solution was analyzed based on previous information and the fishbone diagram shown in Figure 4.12, customized to the appearance problems listed in the problem description. Efforts toward a solution were focused on the operating conditions of the nylon extruder and blown film die. The barrel temperature profile was increased to 220, 250, 260, and 260 °C, and the die temperature profile to a constant 260 °C in order to improve the melt homogeneity of the nylon. The problems of "milky" appearance on one side of the film and opacity disappeared.

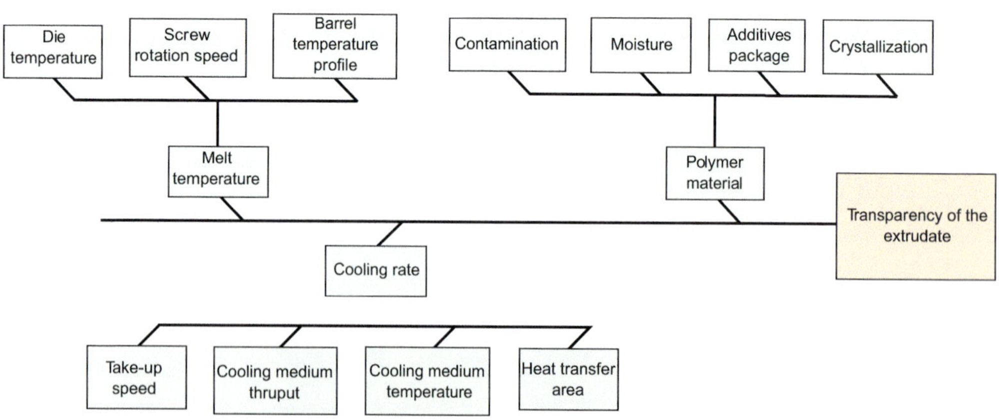

Figure 4.12 Fishbone chart of appearance problem

The curling effect on the polyamide side of the film was related to the nature of the asymmetric film structure and the film crystallization or cooling process. A structure with the same materials in the outside layers, an A-B-C-A, is less likely to exhibit curling. The curling problem can be minimized by passing the film through a water bath and drying system located between the nip rolls and the winder, or by changing to a nylon resin with a lower degree of crystallinity. In this case, a water bath with a drying blower was put in place and the curling effect vanished.

4.8 Dispersion Problem in a High-Density Polyethylene Bottle

4.8.1 Description of the Problem

The problem was exhibited as gloss, luster, and color variations in a plastic bottle made from HDPE to be used for a cosmetic application. The HDPE had a melt index of 1 [g/10 min]. The color was imparted by adding a small amount of color masterbatch (5%) to the natural HDPE, and the mix was processed in a 40-mm single-screw extruder. The CC contained 50% color by weight, and the base polymer of the CC was an LDPE. The color was silver/white with a pearl and luster effect.

The extruder operating conditions are shown in Table 4.10. These operating conditions were reasonable for HDPE extrusion and blow molding. The masterbatch included a pigment critical for the extrusion process because of its shear sensitivity. At very high screw-rotation speed, the plastic bottle lost its gloss, pearl, and luster effect almost entirely. This was not the desired result, and the final product was considered out of specifications.

Table 4.10 Operating Conditions of 40-mm Extruder

Screw rotation speed, rpm	90
Extruder back pressure, bar	200
Barrel temperature profile, °C	100 (= feed zone), 150, 160, 160 (= screw tip)
Temperature of polymer granules, °C	25
Melt temperature, °C	180
Output, kg/h	30

4.8.2 Analysis of the Problem

A sample cut of the plastic bottle was taken and examined using optical microscopy with transmitted light. A Leica optical microscope (Laborlux 12 Pol S) equipped with polishing and microtome capability was used. The magnification was 200 ×. Figure 4.13 shows an example of a micrograph of a cross section of the sample. From the micrograph it was clear that there were non-uniformities in the sample, appearing as pearl pigment agglomerates looking like tiny plates. Obviously, the problem was poor dispersion.

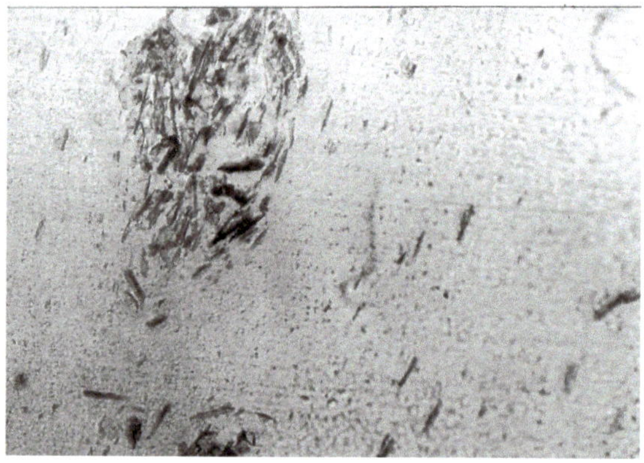

Figure 4.13 Micrograph of sample (crosscut)

4.8.3 Solution

The solution strategy was visualized based on the fishbone diagram shown in Figure 4.14, customized to the problem, and each one of the causes was checked. The first step in the troubleshooting process was to check the rheology of both the HDPE and the color masterbatch. The viscosities of the CC and the HDPE were discovered to be similar. Therefore, it was concluded that the dispersion problem was not associated with differences in viscosity between the CC and the natural HDPE.

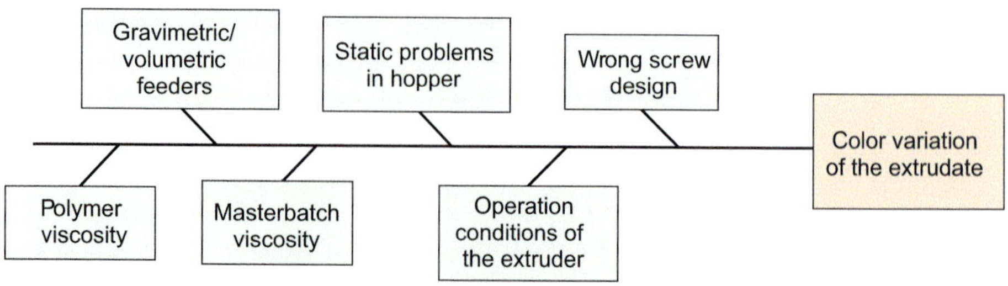

Figure 4.14 Fishbone diagram for color variation

The second step was to check the geometry of the screw used to extrude the material. The screw geometry with square pitch is summarized in Table 4.11. The screw had a conventional three-zone design with a compression ratio of 2.77 and no mixing elements. Since simple conveying screws have poor mixing capability, it was decided to design and manufacture a new screw.

Table 4.11 Geometry of Original Screw

Diameter, mm	40
Total length, *L/D*	20
Feed zone:	
Channel depth, mm	6.25
L/D	8
Compression zone:	
L/D	8
Metering zone:	
Channel depth, mm	2.25
L/D	11

The new screw geometry is summarized in Table 4.12. The new screw concept incorporated a melting zone with variable screw pitch section, followed by one $2D$ dispersive mixing element. In the variable pitch section, the pitch began at $0.8D$ and gradually increased to $1.2D$. This design considered the shear sensitivity of the pigment. One dispersive mixer was enough for the application.

Table 4.12 Geometry of New Mixing Screw

Diameter, mm	40
Total length, *L/D*	20
Feed zone:	
Channel depth, mm	8
L/D (length/diameter)	8
S/D (pitch/diameter)	0.8
Melting zone:	
L/D	5
S/D	0.8–1.2
Metering zone:	
Channel depth, mm	2.5
L/D	5
S/D	1.4
Dispersive mixing zone:	
L/D	2

With this screw, the critical silver-white pigment could be processed without any color variation in the final product, which also exhibited the required gloss, pearl, and luster effects.

4.9 Polymer Degradation

4.9.1 Description of the Problem

This case involved a mono-layer cast film made of homopolymer PP (natural). The film was extruded on a 200-mm single-screw extruder. The polymer stream flowed into a feed block and from there to a cast film line. The extrusion line was fully instrumented. The film was oriented in the machine direction (MD). The film product had serious appearance problems, including dull surface and poor quality of the rolls of film. The defects looked like gels, and the film rolls were out of specifications.

The first issue was to determine the nature of the defect. Therefore, the region around the gel was examined by optical microscopy.

4.9.2 Analysis of the Problem

The film was cut through the gel to allow examination of the film cross section; then the sample was embedded in epoxy resin and polished after curing of the epoxy. A micrograph was taken at 200 × magnification, illuminated with transmitted and polarized light. The microscope used was a Leica optical microscope (Laborlux 12 Pol S) equipped with polishing and microtome capability for which transmitted, reflected, and polarized light could be used. The micrograph shown in Figure 4.15 shows the nature of the gel. The discoloration of the material with a black speck in the center indicated degradation.

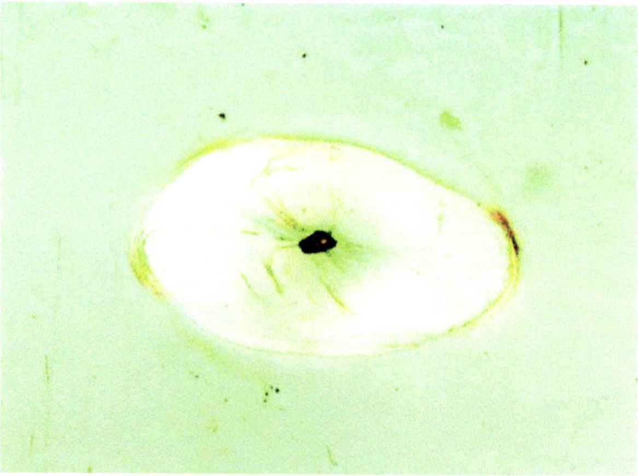

Figure 4.15 Optical micrograph of film cross section

At this point, the troubleshooter visited the plant and checked the operation of the extruder. A computer simulation based on the extruder operating conditions was done to obtain the melt temperature, T_m, because the melt temperature transducer was not operating properly. The melt temperature, T_m, is plotted against the extruder length, *L*/*D*, in Figure 4.16.

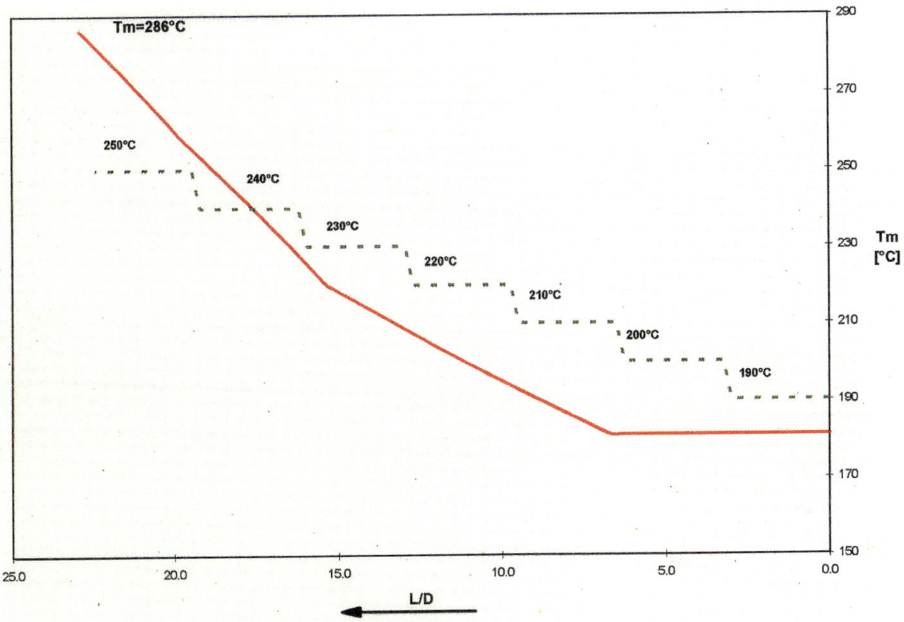

Figure 4.16 Extruder barrel temperatures and melt temperature

Based on the barrel temperatures in Figure 4.16, the melt temperature was calculated to be 286 °C. This is a very high value for a homopolymer PP melt temperature in this application. The graph showed that the melt temperature increased rapidly in the last 5*D* of the screw, corresponding to the metering/mixing zone. In general, the screw design, including a 4*D* dispersive mixer with a barrier, was considered appropriate for polypropylene.

4.9.3 Solution

The solution strategy was visualized based on the fishbone diagram in Figure 4.17, customized to the problem, and each one of the causes was checked. The recommended melt temperature for this process is approx. 260 °C. Therefore, the barrel temperatures of the 200-mm extruder were lowered in an attempt to achieve a melt temperature at the die exit of 260 °C. This was not entirely possible here, because of the resulting high back-pressure. The minimum obtained melt temperature was 270 °C, and at this temperature, the problem was less noticeable but still present.

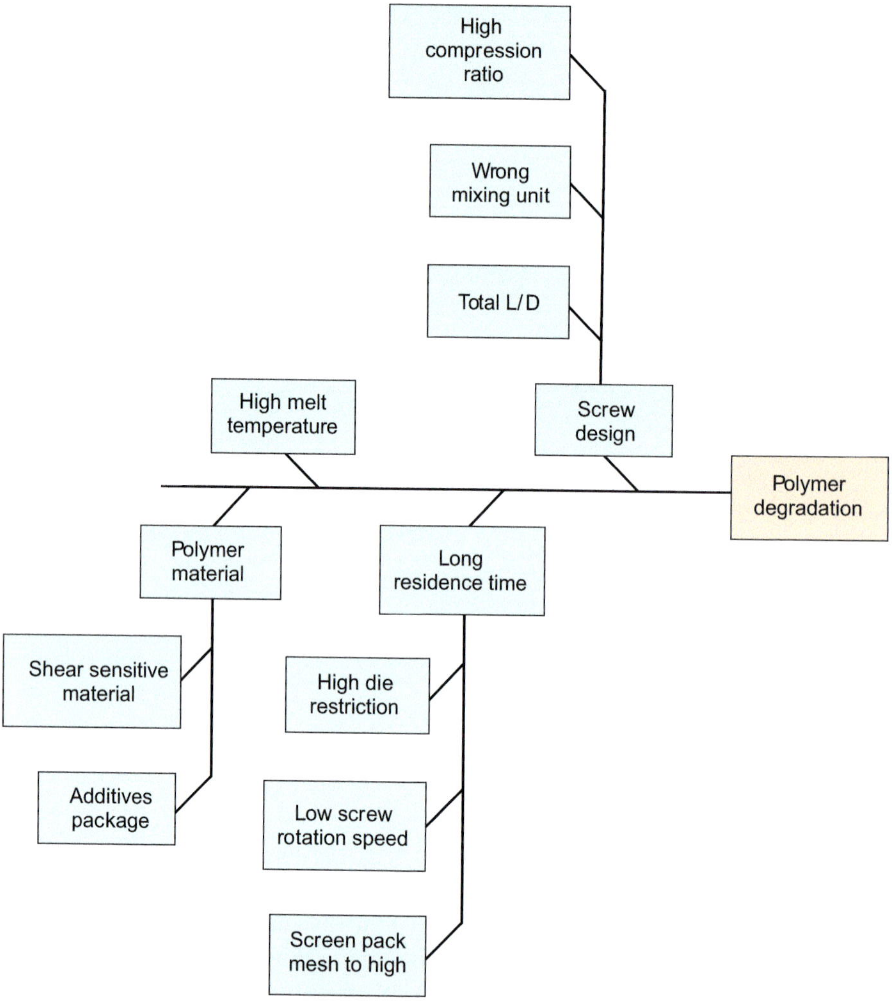

Figure 4.17 Fishbone diagram of polymer degradation

The polymer material and the additives package were investigated, since problems with the machine setup were ruled out. First, a polymer sample without the additives package was sent out for a thermogravimetric analysis (TGA) (see Section 2.5.2). The recorded TGA is shown in Figure 4.18. The obtained TGA registered the first loss of weight (1%) at 363 °C. This ruled out degradation of this PP type at this specific temperature.

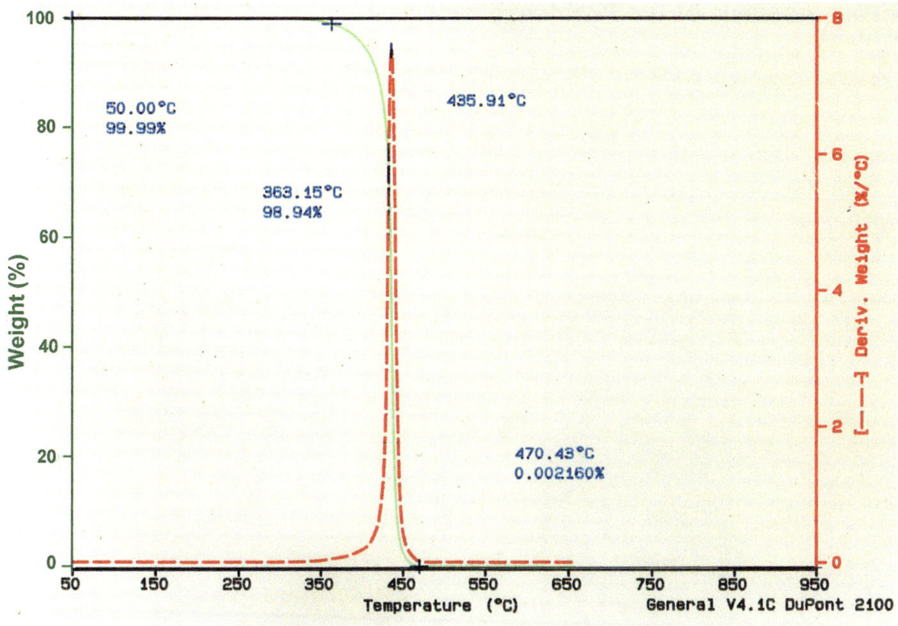

Figure 4.18 TGA for the homopolymer PP

Finally, it was clear that the problem was related to the additives package, a combination of slip and anti-block agents. The additive supplier was contacted, and it was decided to change the composition of the package to less thermal- and shear-sensitive materials. When the change was implemented, the problem disappeared completely.

■ 4.10 Heat-Sealing Problems in a Coextruded Film

4.10.1 Description of the Problem

This case concerned a coextruded barrier film with heat-sealing problems. Heat sealing has been the standard technique used for packaging of liquid and solid products. The natural color film exhibited excellent barrier properties, but a very poor sealing performance. The final user of this film was not informed about the exact film structure. Good film-sealing properties were required in the high speed packaging equipment.

4.10.2 Analysis of the Problem

The film was microtomed, and the cross section was examined in detail by optical microscopy as shown in Figure 4.19. A micrograph was taken at 200× magnification, illuminated with transmitted and polarized light. The microscope was a Leica optical microscope (Laborlux 12 Pol S) equipped with polishing and microtome capability.

Figure 4.19 Micrograph of coextruded barrier film, 200×

Since the polymer layers in this structure were transparent, it was not possible to observe differences between layers. Therefore, colorimetry was used to distinguish the different layers: iodine was used to color polymers such as nylon or EVOH, but it did not affect the color of polyolefins (PO) such as PE or PP. The micrograph confirmed that the actual film structure was PO-barrier-PO, based on the developed color. The distribution of layer thicknesses, as measured by image analysis, was 75/15/80 μm, and the total film thickness was 170 microns.

First, a film sample was sent for differential scanning calorimetry (DSC) to determine the polymer melting points for each layer. The DSC instrument was a heat-flow type calorimeter with a standard cell (TA Instruments, 2910). The recorded DSC shown in Figure 4.20 exhibited three melting points at 110.65, 120.32, and 179.06 °C, corresponding to LDPE, LLDPE, and EVOH, respectively. Due to their good cost/performance balance, polyethylenes are widely used as a sealing layer in coextrusion for packaging of diverse products. Finally, the actual film structure was revealed.

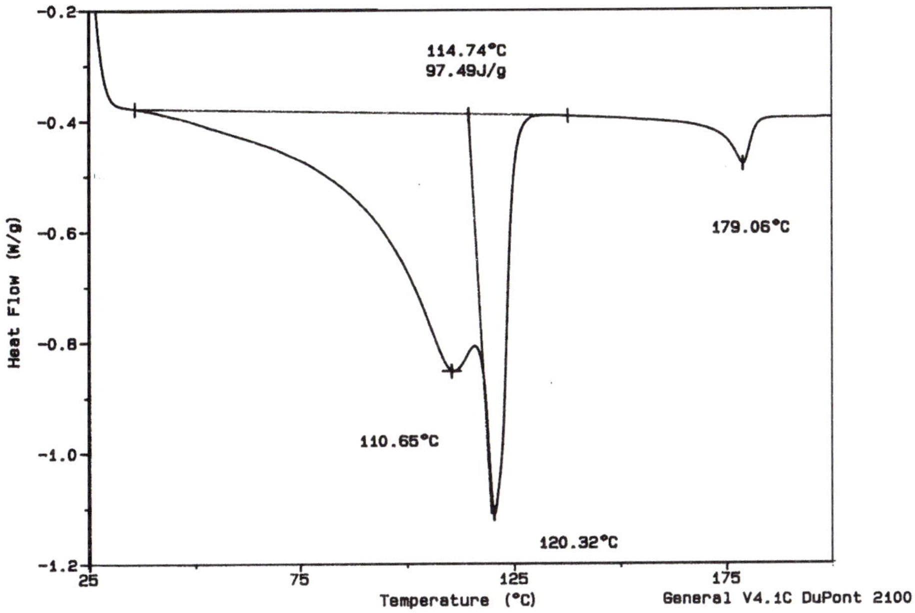

Figure 4.20 Differential scanning calorimetry of coextruded barrier film

4.10.3 Solution

The solution was analyzed based on previous information and the fishbone diagram customized to heat sealing problems shown in Figure 4.21.

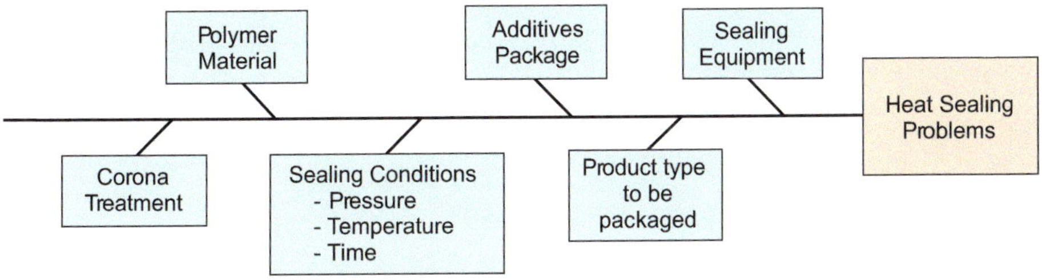

Figure 4.21 Fishbone diagram for a heat sealing problem

The packaging and sealing machinery was state-of-the-art technology VFFS (vertical fall, fill, and sealing) and HFFS (horizontal fall, fill, and sealing) equipment. The additives package being used contained slip and anti-block agents at 700 and 1200 ppm, respectively. Therefore, the solution investigation was focused on the operating conditions of the sealing machines and on the polyethylene types of the external layers in the film structure.

At this point, the troubleshooter visited the converting plant and checked the operation of the packaging and sealing machines. The setpoint temperatures, clamp pressure, and programmed cycle for the sealing machines were appropriate for the film structure. Therefore, a different film structure from the same film supplier was tried: a symmetric structure of LDPE (+15% mPE)-EVOH-LDPE (+15% mPE). A blend of LDPE with 15% mPE was chosen, because it exhibits better heat sealing properties than pure LDPE. It is characteristic of the melting behavior of polymers that low melting temperature and low heat of fusion enhances sealability. The new structure was checked in the converting plant with excellent results: it exhibited good barrier and heat sealing properties.

4.11 Output Problem in a Blown Film Line

4.11.1 Description of the Problem

This problem involved a monolayer blown film line operating with a 60-mm extruder, L/D = 25, a 250-mm diameter die, and a die gap of 0.8 mm. For several years, the material extruded was an LDPE with a melt index of 2 [g/10 min]. The company switched the blown film line to the production of LLDPE film with a melt index of 1 [g/10 min], assuming that the line would run successfully under similar conditions to those used with LDPE. The company experienced a low-output problem and melt fracture. The company tried several LLDPE resins with different melt index values, but changing the resin did not solve the problem.

The extruder operating conditions for LLDPE processing are shown in Table 4.13. The specific mass output, Q/N = 0.4 kg/h/rpm, was very low for LLDPE processing.

Table 4.13 Extruder Operating Conditions

Screw rotation speed, N, rpm	200
Temperature of polymer granules, °C	25
Melt temperature, °C	Not available
Back pressure, bar	Not available
Output, Q, kg/h	80

4.11.2 Analysis of the Problem

In this case, we were dealing with two complex problems. The low output was very closely related to screw design, because the specific mass output was very low, and the melt fracture was an appearance problem related to the fishbone diagram shown in Figure 4.22. The first step was to check the geometry of the screw in use. The screw geometry with square pitch is shown in Table 4.14.

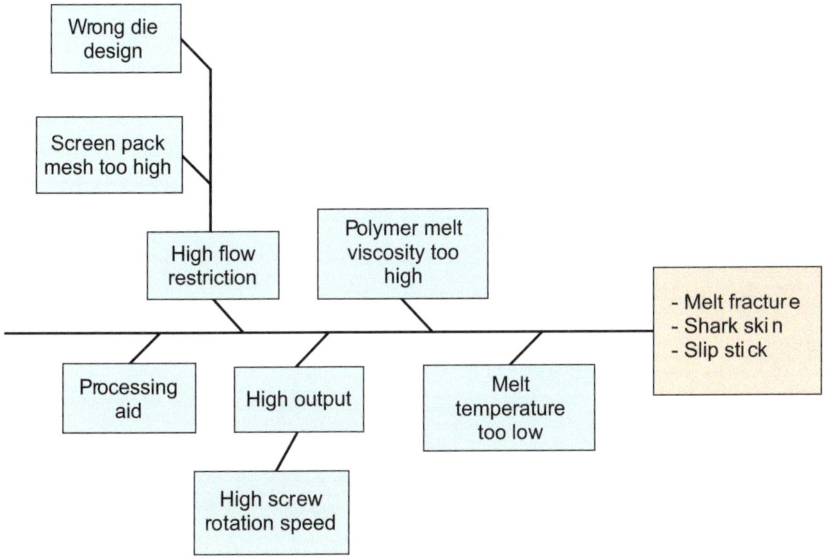

Figure 4.22 Fishbone diagram for melt fracture

Table 4.14 Screw Geometry

Diameter, mm	60
Total length, *L/D*	25
Feed zone:	
Channel depth, mm	9
L/D	8
Compression zone:	
L/D	4.7
Metering zone:	
Channel depth, mm	3
L/D	9.5
Mixing zone:	
L/D	2.8

The screw had a compression ratio of 3, a very long metering zone, and a distributive mixer. Especially because of its short compression and long metering zones, this screw was not adequate for LLDPE processing.

At this point, the tooling was disassembled to analyze the geometry of the blown film die. The tooling had a spiral mandrel die design, and the die gap (0.8 mm) was determined to be too small for LLDPE, creating a high flow restriction.

4.11.3 Solution

A new screw with a higher mass output was designed and manufactured. The new screw design delivered a specific mass output of 0.6 kg/h/rpm. The new screw concept incorporated a barrier melting zone with a 2*D* distributive mixing element for melt-temperature homogeneity. A dispersive mixing element was not necessary because the final product requirement asked for natural color. Some of the important screw design parameters are shown in Table 4.15.

Table 4.15 Screw Design Parameters of New Screw

Diameter, mm	60
Total length, *L/D*	25
Feed zone:	
Channel depth, mm	10
L/D	10
S/D	1
Barrier melting zone:	
Channel depth, mm	10 → 3
L/D	13
S/D	1.0 → 1.6
Mixing zone:	
Channel depth, mm	11
L/D	2

The mandrel of the blown film die was modified to change the die gap from 0.8 mm to 2.5 mm to eliminate the high flow restriction and the melt fracture problem. Figure 4.23 shows the original die (D = 250 mm and δ = 0.8 mm) before the mandrel modification. Figure 4.24 shows the same die after the mandrel modification (D = 250 mm and δ = 2.5 mm). It should be noted that the screw was designed to deliver the maximum specific mass output for the modified spiral mandrel die. The final results, in this case, were a 50% increase in output, elimination of the melt fracture problem, and good film quality.

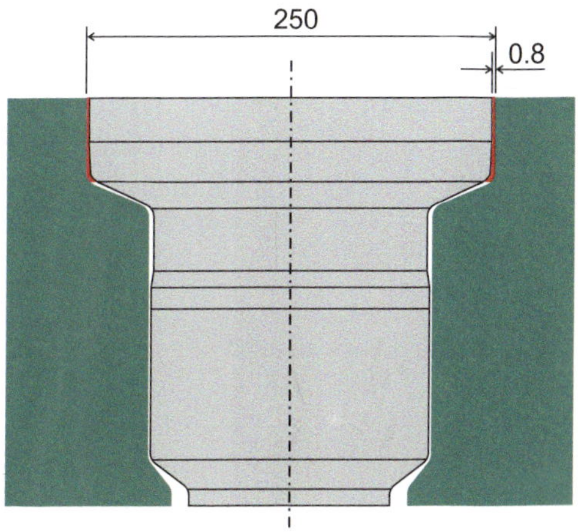

Figure 4.23 Original spiral mandrel die

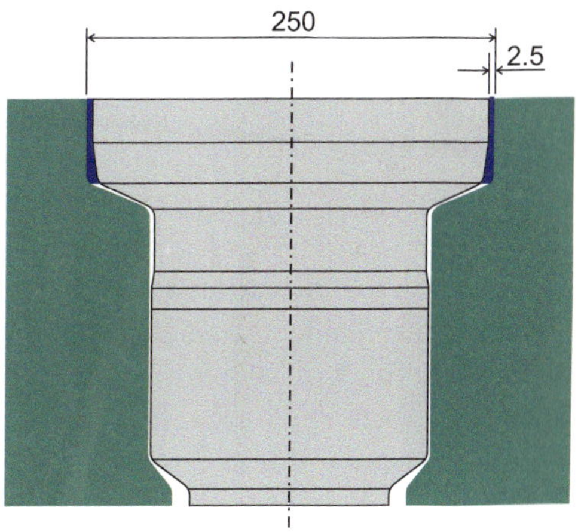

Figure 4.24 Modified spiral mandrel die

■ 4.12 Masterbatch Selection

4.12.1 Description of the Problem

The problem was very strong color variations in a final extruded product (an extrusion coated product) due to the presence of agglomerates and striations. The color concentrate was prepared in-house. Two formulations were established for the color concentrate (CC), and the request was to select the best formulation for the current extrusion coating process.

The raw materials for the formulations are described in Table 4.16, and the established formulations are shown in Table 4.17. Formulation 1 provided a masterbatch with a high melt index material (MFI = 8) in a rheologically similar substrate (MFI = 10). Formulation 2 provided masterbatch with a very low melt index material (MFI = 1) in a substrate with very high flow ability (MFI = 54).

Table 4.16 Raw Material Specifications

	Material 1	Material 2	Material 3	Material 4
Polymer type	EAA Copolymer	LDPE + 50% TiO_2	LDPE	HDPE + 50% TiO_2
MFI g/10 min 190 °C/2.16 kg	10	8	54	1

Table 4.17 Formulation 1 and 2

Formulation 1	Formulation 2
Material 1: 85%	Material 3: 85%
Material 2: 15%	Material 4: 15%

4.12.2 Analysis of the Problem

Each formulation was blended in a torque rheometer for 10 minutes. Samples of the polymer melt were taken after 1 minute and after 10 minutes of mixing, and then pressed into thin plaques. The pressed film samples were analyzed for color variation using an optical microscope and transmitted light. The torque rheometer was a Haake Rheocord 90-200, equipped with a Rheomix 600 mixer, and the optical microscope used was a Leica Laborlux 12 Pol S, equipped with polishing and microtome units.

Figure 4.25 presents the torque rheometer curves for both formulations during the 10 minutes of mixing. The viscosity of Formulation 1 is higher than the viscosity of Formulation 2. Figure 4.26 shows micrographs of both formulations after 1 minute and after 10 minutes of mixing. The upper two pictures show Formulation 1 and the lower two pictures show Formulation 2.

The micrographs suggest that the ingredients of Formulation 2 (lower two micrographs) are incompatible and do not produce a homogeneous mix. After both 1 minute and 10 minutes of mixing, the quality of the mixing is not good at all, and the agglomerates and striations are quite noticeable. On the other hand, Formulation 1 (upper two micrographs) is compatible and produces a homogeneous mix. The quality of the mixing after 10 minutes is quite good and there are no agglomerates or striations. Good dispersion and good distribution can be observed.

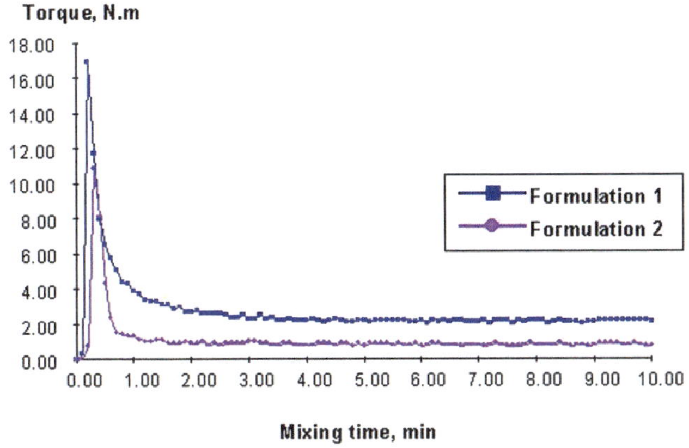

Figure 4.25 Torque rheometer curves for both formulations

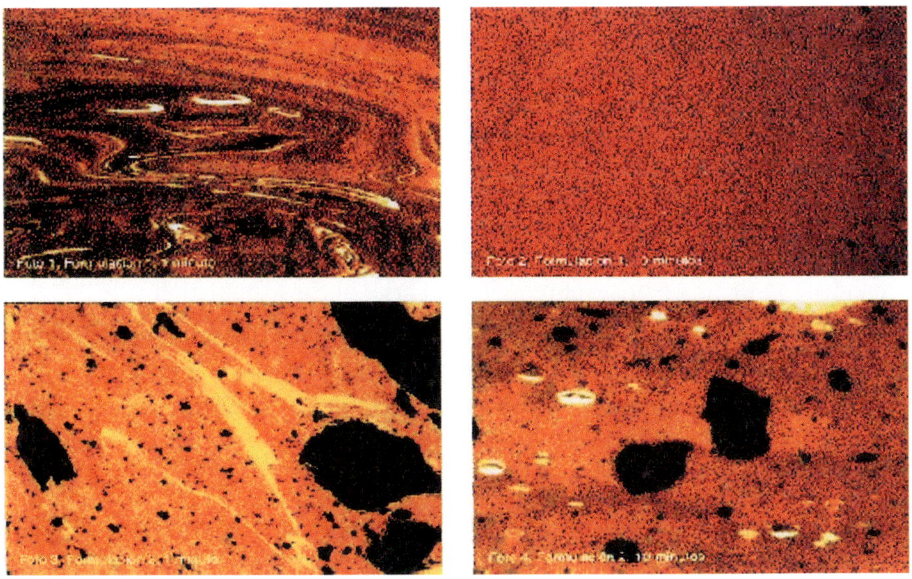

Figure 4.26 Micrographs of samples (crosscuts) from both formulations. Left: Formula 1 after 1 minute (top) and Formula 2 after 1 minute (bottom). Right: Formula 1 after 10 minutes (top) and Formula 2 after 10 minutes (bottom)

4.12.3 Solution

In view of the results, Formulation 1 was selected for the extrusion coating process. This formulation was rheologically compatible because the viscosities of both substrate and concentrate were quite similar. Formulation 1 was run in the extrusion line with excellent mixing results. Agglomerates and striations were not noticeable in the final product.

It is important to point out that the extruder screw was appropriate for mixing. The screw had two mixers of $2D$ each.

As a rule of thumb, based on Grace theory [161], in order to achieve good dispersion, the viscosity ratio between concentrate and substrate should not exceed 3.8 at the specific operating conditions of temperature and shear rate. For in-house manufacturing of a masterbatch that will produce excellent mixing results, it is recommended that a slightly less viscous concentrate on a slightly more viscous substrate or matrix be selected.

4.13 Pipe Extrusion Problem

4.13.1 Description of the Problem

The problem concerned plastic pipes with very poor impact and sustained pressure test results. The pipes were made from a rigid PVC compound with some amount of recycled PVC material. The pipes were pigmented to a light yellow color. It was determined in the manufacturing facility that the problem was strongly dependent on the percentage of recycled material in the formulation.

The pipe extrusion line included a single-screw extruder (D = 60 mm, L/D = 20), calibrator, cooling bath, haul-off, and cutting unit. The screw geometry was a conventional three-zone screw with square pitch and without mixing units. This extruder setup is typical for processing of hard PVC because it accommodates the shear sensitivity of the compound.

It was observed that there were no problems when large amounts of recycled material were used in the formulation. However, with small amounts of recycled material the extrusion process became very problematic and calibration of the pipes was very difficult. The recycled material was made from a PVC compound for a blister film application.

4.13.2 Analysis of the Problem

A pipe sample with the sustained pressure failure is shown in Figure 4.27 (outer surface) and Figure 4.28 (inner surface). The outer diameter of the pipe was 82.7 mm and its wall thickness was 3.4 mm. The pipe failure has a typical fishbone form reported in several studies in the literature.

Figure 4.27 Sustained pressure failure in a pipe (outer surface)

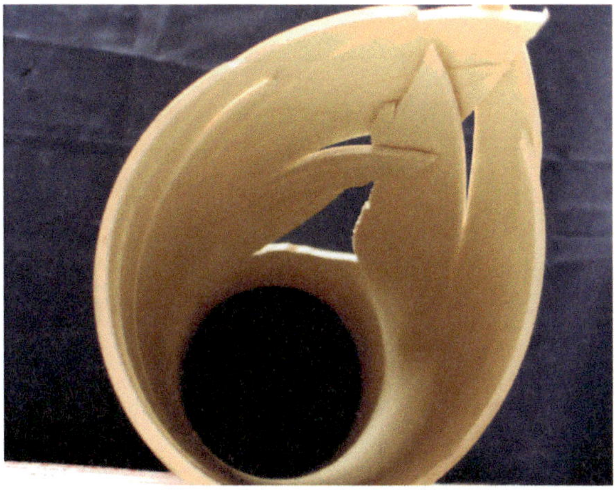

Figure 4.28 Sustained pressure failure in a pipe (inner surface)

4.13.3 Solution

In order to find a strategy for the solution to the problem, the following topics were considered:

- PVC formulation
- PVC recycled material
- Blend of pure PVC compound and recycled material
- Operating conditions in extrusion
- Extruder configuration

The first step in the troubleshooting process was to check the influence of the proportion of recycled material used on properties of the final product. The formulations in Table 4.18 were proposed and run in the extrusion plant. The laboratory results for the extruded pipes produced under identical extrusion conditions for all the formulations are shown in Table 4.19.

Table 4.18 Proposed Formulations

Formulations of pure PVC and recycled material	Percentage of pure PVC compound (%)	Percentage of recycled material (%)
No. 1	100	0
No. 2	75	25
No. 3	50	50
No. 4	25	75
No. 5	0	100

Table 4.19 Laboratory Results of Extruded Pipes

Formulations of pure PVC and recycled material	Impact test (Min. 35 J/m)	Sustained pressure test (Min. 2.76 MPa)
No. 1	Passed	Passed
No. 2	Passed	Passed
No. 3	Passed	Failed
No. 4	Failed	Failed
No. 5	Failed	Failed

From the test results it was clear that to produce pipes satisfying the quality standards for both impact and sustained pressure, the formulation should not include more than 25% recycled material. This finding was accepted without any problems.

Formulation No. 2 could still be processed in the 60-mm extruder, but the machine was running at the torque limit. This result was in agreement with the initial observation (see Section 4.13.1). Since the recycled material was less viscous than the pure PVC compound, processing was easier when a high percentage of recycled material was used.

The troubleshooter proposed modifying the extruder drive system to make it possible to process pure PVC compounds. This would ensure that the pipe extrusion line would be able to process Formulations No. 1 and No. 2 while meeting good quality standards. This proposal was accepted, and the drive system was modified by increasing the capacity of motor installed.

4.14 Gel Formation in a Coextruded Film

4.14.1 Description of the Problem

A five-layer coextruded blown film exhibited gel formation. The film was of natural color with a layer structure of nylon–tie layer–EVOH–tie layer–LLDPE. The total film thickness was 100 microns intended for packaging application for cold cured meat. The defects looked like small globules or so-called fish eyes. The objective in this case was to determine the polymer material and the layer causing the gel problem.

The EVOH was the most temperature sensitive polymer in the mentioned multi-layer structure. The operating conditions of this coextrusion blown film line looked inappropriate for EVOH processing. High melt temperature could be observed.

It is known from literature [163] that the formation of hydrogen bonds between the hydroxyl groups in vinyl alcohol parts of EVOH copolymers with high vinyl alcohol fraction and the formation of hydrophobic interaction between the methylene groups in ethylene parts of EVOH copolymers with high ethylene fraction contributes to gel formation during processing.

EVOH decomposition may occur, causing gels or voids in the extrudate at temperatures above 240 °C and/or extended residence time [164].

4.14.2 Analysis of the Problem

A film sample was microtomed from a part containing gels and the cross section was examined in detail by optical microscopy (see Figure 4.29). Colorimetry was used to differentiate and reveal the different layers. Iodine in low concentration was used to color the nylon red and the EVOH brown. Figure 4.29 shows the micrographs at 20 × (left) and 40 × (right) magnification.

The micrographs indicated that the gel problem was located in the EVOH barrier layer (brown tinted). The recommended extrusion temperature for EVOH ranges between 220 °C and 230 °C, because its melting point is approx. 180 °C. The actual temperature extrusion temperature was approx. 250 °C, measured with an infrared thermometer.

Figure 4.29 Micrograph of coextruded 5-layer film, 20 × (left) and 40 × (right)

4.14.3 Solution

The solution was visualized based on previous information, raw material data sheets, and the fishbone diagram shown in Figure 4.17. Each of the causes was checked, focusing on the polymer material. Both the barrel temperatures of the extruder delivering the EVOH material and the die temperatures of the coextrusion line were lowered to achieve a melt temperature at the die exit below 230 °C. Once the melt temperature was reduced, the gel problem of the barrier layer disappeared. In this case, the colorimetric method was crucial to determine the location of the gel problem and to decide how to correct it.

4.15 Agglomerates and Grammage Variation in a PP Sheet

4.15.1 Description of the Problem

The industrial case concerned the double wall sheet extrusion of a PP compound comprised of PP homopolymer, PP copolymer, and a masterbatch containing calcium carbonate ($CaCO_3$) and color pigments. The problem was related to the following issues:

- the original screw was a conventional 3-zone screw (Ø 90 mm, L/D = 30), delivering a melt with physical and thermal inhomogeneities,
- the output and quality of the product were deficient due to mixing problems and strong sheet thickness and grammage variations, and

- the PP material required a preceding compounding and pelletizing operation caused by the lack of mixing capacity of the conventional 3-zone screw.

During the manufacturing process for continuously extruded products, such as sheet extrusion, thickness and grammage uniformity are key quality factors and they are directly linked to extruder flow oscillations and/or flow disturbances at the die [165].

It is known from literature that morphology analysis at high shear stress showed that large $CaCO_3$ agglomerates are randomly dispersed and are not oriented in flow direction, suggesting poor mixing [166]. Therefore, dispersive or elongational mixing is required for successful extrusion of this type of PP compound.

4.15.2 Analysis of the Problem

This industrial case involved two complicated problems: A grammage variation that seemed to be related to an output problem and the presence of agglomerates in a sheet or a dispersive mixing problem.

The solution was visualized based on the fishbone diagram in Figure 4.30, checking each one of the possible causes. Solids conveying problems were checked and quickly discarded. The original 3-zone screw (D = 90 mm, L/D = 30) did not have mixers to accomplish dispersion of agglomerates. Wear of screw flights was noticeable.

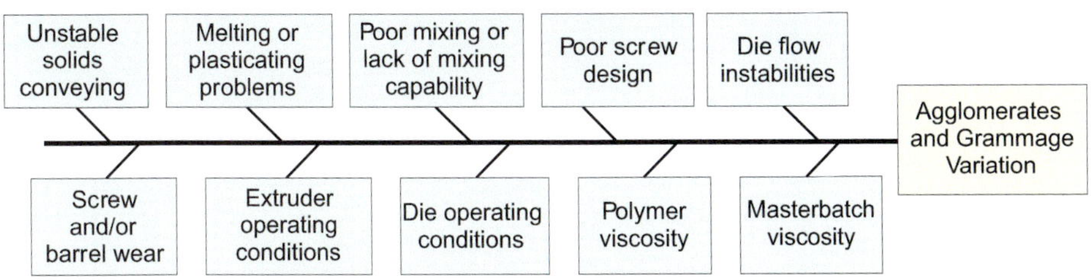

Figure 4.30 Fishbone diagram for agglomerates and grammage variation

A capillary rheometer was used to obtain complete viscosity curves at three temperatures. Measurements for polypropylene were carried out at 200, 220, and 240 °C. The viscosity curves could be obtained using circular capillaries, while a rectangular one was used to obtain data at low shear rates. Figure 4.31 presents the corrected viscosity curves at three temperatures for the PP compound.

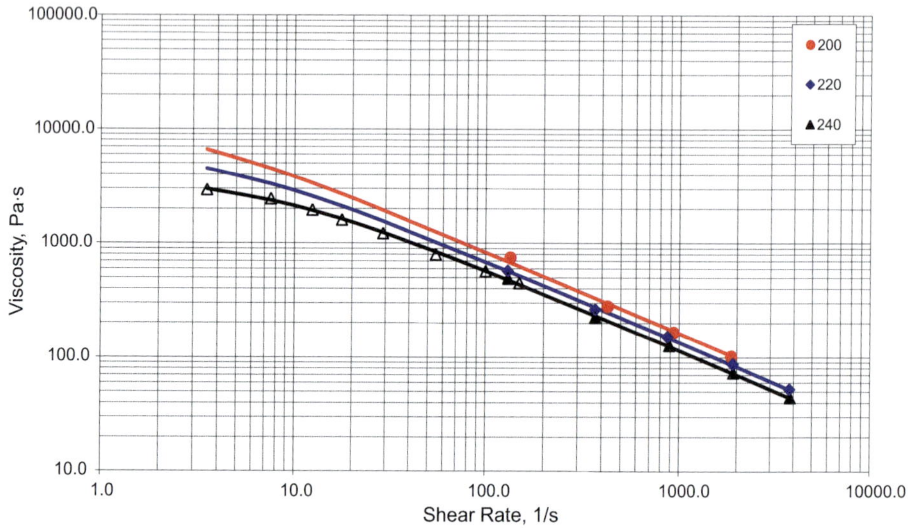

Figure 4.31 Viscosity curves of PP compound

Figure 4.32 presents the calculated melting profile (SBP, solid bed width profile after Tadmor and Pearson models) of the original screw at the original operating conditions: 100 kg/h output at 60 rpm and 185 bar backpressure. The barrel temperature profile was ascending from 200 °C in the feeding zone up to 220 °C in the metering zone. Figure 4.33 shows the geometry of the original 3-zone screw with square pitch and a compression ratio 2.2 : 1, which is low for conventional extrusion of PP.

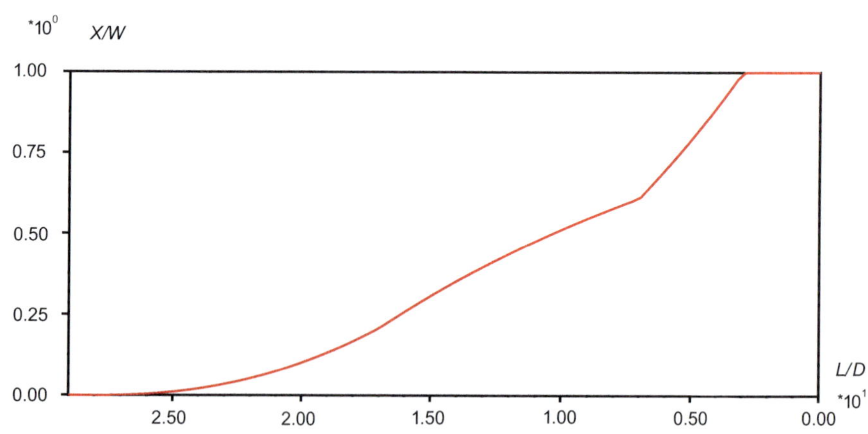

Figure 4.32 Calculated melting profile of the original 90-mm screw

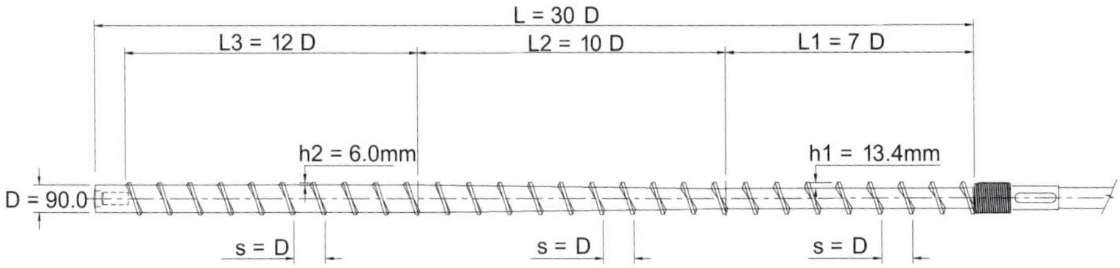

Figure 4.33 Original 90-mm screw

It was possible to confirm a delayed melting, leading to homogeneity problems observed in this extrusion line. The calculated melt temperature was higher than 220 °C, in agreement with the extrusion line observations.

In double wall sheet extrusion, higher melt temperatures are undesirable because of subsequent difficulties in product shaping and cooling in order to obtain the exact product dimensions and tolerances, in particular thicknesses and grammage.

4.15.3 Solution

The solution was visualized based on previous information, computer simulations, extrusion line trials, and the fishbone diagram shown in Figure 4.30. Each of the causes was checked, focusing on screw design.

The new single-screw to be designed needed to address melting and mixing to avoid the above described problems. Therefore, a barrier screw was designed (D = 90 mm, L/D = 30), including a 12D barrier section and a 2D elongational mixer. Figure 4.34 shows the new optimized design delivering, both in simulations and final trials, a higher output of 125 kg/h at 60 rpm (operating condition) and lower melt temperature (200 °C) and backpressure (162 bar). The barrel temperature profile was also ascending from 200 °C in the feeding zone up to 220 °C in the metering zone. Figure 4.35 presents a picture of the intensive elongational mixer included in this design. The melt homogeneity of the polymer compound was quantified by means of morphology analysis (optical microscopy), confirming that $CaCO_3$ agglomerates were no longer present and that good thermal and physical homogeneity was achieved.

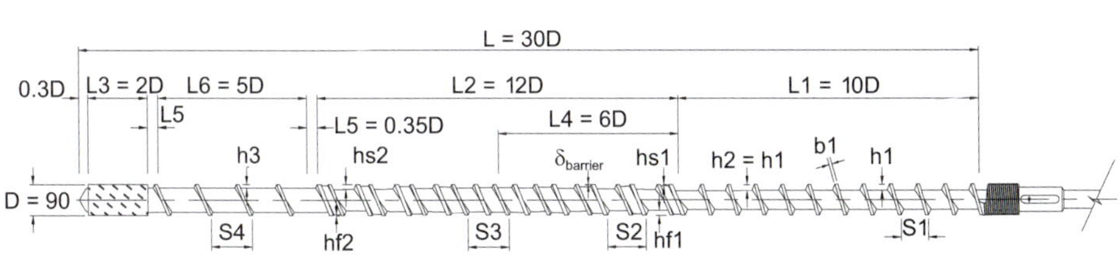

Figure 4.34 90-mm barrier-mixing screw

Figure 4.35 Intensive elongational mixer

With the new barrier-mixing screw, the agglomerates and grammage variation problems disappeared. A good single-screw simulation tool, together with a measured and corrected viscosity curve, was key in solving these extrusion problems.

■ 4.16 Insufficient Melting and Mixing in a Plasticating Unit

4.16.1 Description of the Problem

This manufacturing case involved the plasticating unit of a 150-ton injection molding machine. Injection of a high melt flow index PP (MFI: 45 and 60 g/10 min) with a masterbatch containing thermally sensitive colors into the mold was deficient.

The problem included the following issues:

- the selected machine was not able to process low viscosity PP to produce thin-wall containers in various colors,
- the original screw design of the plasticating unit (D = 50.8 mm, L/D = 17.5) was short and showed insufficient melting for low viscosity PP, and
- output, reproducibility, and quality of the product were deficient, presenting homogeneity problems with critical colors, particularly striation and presence of agglomerates.

In order to reduce the flow resistance during injection molding, a decrease in polymer viscosity (higher melt index) is required. But it is known from extrusion processing that low viscosity materials generate less viscous dissipation and low viscous dissipation reduces the amount of heat generated affecting polymer melting [167].

The plasticating unit in injection molding is a single-screw extruder and its screw design is crucial to guarantee polymer melting, particularly for high melt index polymeric materials. Barrier screws are becoming more common in plasticating units because they provide better melt homogeneity and mixing [168].

4.16.2 Analysis of the Problem

In this particular case, two processes had to be considered: melting and melt conveying taking place in the plasticating unit and during injection molding. The 50.8-mm screw of the plasticating unit was checked in operation during the production of thin-wall containers.

The solution was visualized based on the fishbone diagram in Figure 4.36, checking each one of the possible causes. The original 3-zone screw (D = 50.8 mm, L/D = 17.5) was short and did not have mixers to achieve mixing of thermally sensitive colors.

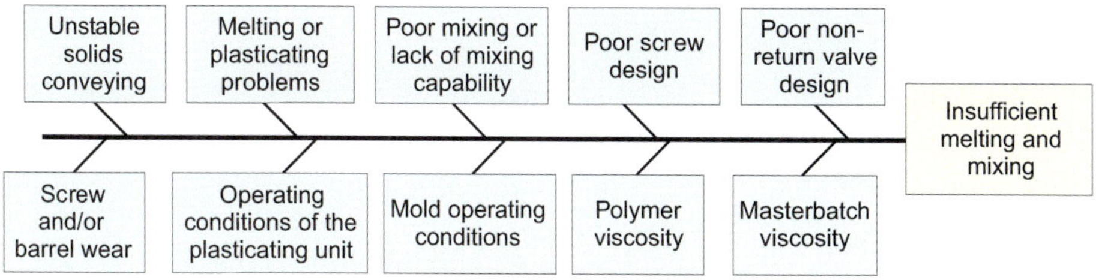

Figure 4.36 Fishbone diagram for insufficient melting and mixing

A capillary rheometer was used to obtain complete viscosity curves at three temperatures. Measurements for polypropylene were carried out at 200, 220, and 240 °C. The viscosity curves could be obtained using circular capillaries. Figure 4.37 presents the corrected viscosity curves at three temperatures for PP (MFI, melt flow index: 60 g/10 min at 2.16 kg and 230 °C).

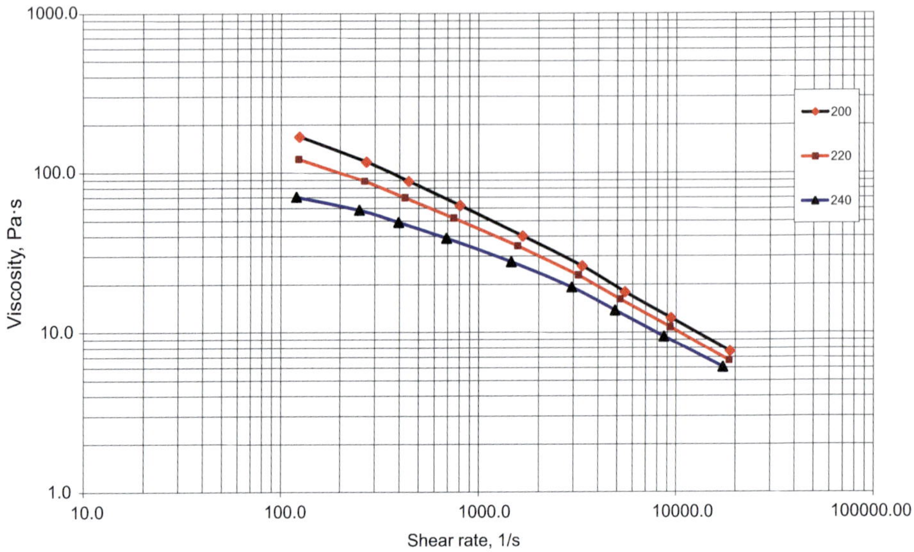

Figure 4.37 Viscosity curves of PP MFI 60 g/10 min

The calculated melt temperature profile of the original 3-zone screw at a plasticating rate of 29 g/s and a screw rotation speed of 170 rpm delivered a very high viscous dissipation, 50 °C above the set temperature. Figure 4.38 shows the geometry of the original 3-zone screw with square pitch and a compression ratio of 2.3:1, which is low for PP processing.

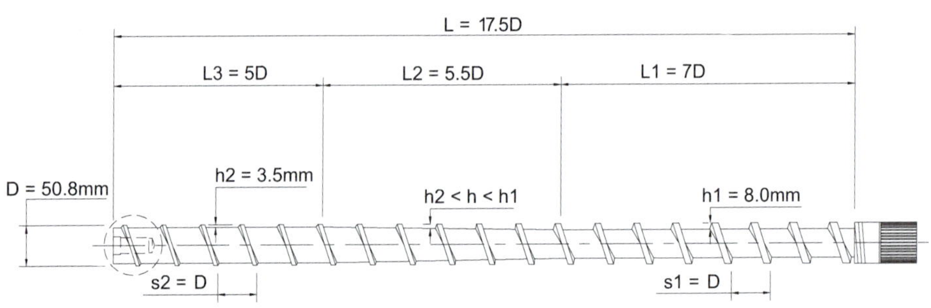

Figure 4.38 Original 50.8-mm screw

The problem was observed and confirmed during the operation of this injection molding machine. On the other hand, significant agglomerates and inhomogeneities were detected in the thin-wall plastic parts.

Thin-wall injection molding is a preferred option for reducing part weight and size. This process allows the rapid manufacture of less expensive, more compact parts, because of fast cooling [169]. However, thermal and physical homogeneity of the polymer melt is required for injection into the mold cavity.

4.16.3 Solution

The solution was visualized based on previous information, computer simulations, injection molding trials, and the fishbone diagram shown in Figure 4.36. Each one of the causes was checked. Particular attention was paid to the screw design of the plasticating unit.

The newly designed screw needed to address melting capacity, melting length, and mixing to avoid the original problems. Therefore, the screw length needed to be increased by $2D$ (D = 50.8 mm, L/D = 19.5), including a $10D$ barrier section and a $2D$ distributive mixer. Figure 4.39 shows the new optimized design. It delivered an identical plasticating rate of 29 g/s and a screw rotation speed of 170 rpm, both in simulations and in final injection molding trials, together with insignificant viscous dissipation. This last issue was important to preserve the critical thermally sensitive colors. The melt homogeneity of the polymer material was quantified by means of morphology analysis (optical microscopy), confirming that agglomerates were no longer present and that good thermal and physical homogeneity was achieved.

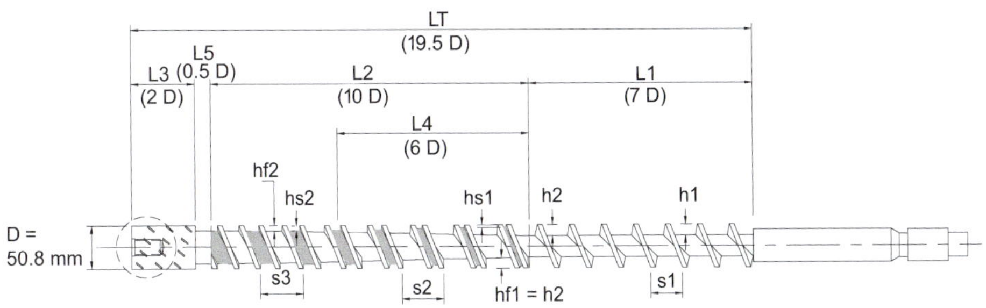

Figure 4.39 50.8-mm barrier screw for plasticating unit

Figure 4.40 shows a Saxton type distributive mixer with an improved design of a ring non-return valve to guarantee part reproducibility (weight vs. number of parts over time) and no leakages during high speed injection molding of high melt index polymer materials.

With the new barrier-mixing screw and a new non-return valve, the insufficient melting and melt homogeneity problems disappeared. A good simulation tool together with measured and corrected viscosity curves and in-situ trials were crucial in solving this problem.

Figure 4.40 Distributive mixer and non-return valve

4.17 High Melt Temperature and Insufficient Output in Coextrusion

4.17.1 Description of the Problem

This industrial case involved the 3-layer film coextrusion of a PE blend comprised of LDPE, LLDPE, and m-PE. The problem was related to the extruders of external layers A and C. Both extruders had the same screw geometry but delivered PE blends in different compositions. The problem presented the following issues:

- the original 88.9-mm screw showed excessive wear,
- the output of 92.9 kg/h at 69 rpm was insufficient for the intended multi-layer structure,
- surging could be observed, and
- high melt temperature was measured at LLDPE and/or m-PE contents higher than 70%, negatively affecting blend homogeneity and the mechanical film properties over time.

This multilayer structure was used for thin and transparent film packaging and agricultural applications. An increase of at least 25% in extruder output was required to manufacture the desired 3-layer film structure. With this new output, it was necessary to avoid high melt temperature to guarantee multilayer film specification and good quality.

It is known from literature that inadequate shear stress and too high melt temperature can also cause mixing problems by lowering the melt viscosity and generating inadequate shear stress. High temperatures can also cause thermal degradation of the polymer, which can lead to yellowing of transparent films [170].

Surging in plasticating extrusion can be avoided with barrier screws or screw cooling [171]. The main reason for surging is solids breakup. Barrier screws solve this problem by separating the solids from the melt pool with an additional screw flight.

4.17.2 Analysis of the Problem

This film extrusion case presents two difficult problems: high melt temperature in extruders A and C of the coextrusion line and insufficient output in both extruders to produce the intended multi-layer structure. Using higher screw rotation speeds in order to increase output leads to increased melt temperatures so that the process went out of control in the subsequent operations such as film cooling and calibration.

The solution was visualized based on the fishbone diagram in Figure 4.41, analyzing each one of the probable causes. Setup and operating conditions were checked. The original screw (D = 88.9 mm, L/D = 25.5) had an intensive dispersive mixer (3*D* Maddock Mixer) close to the screw tip, contributing to the increase in melt temperature. Wear of the screw flights was evident.

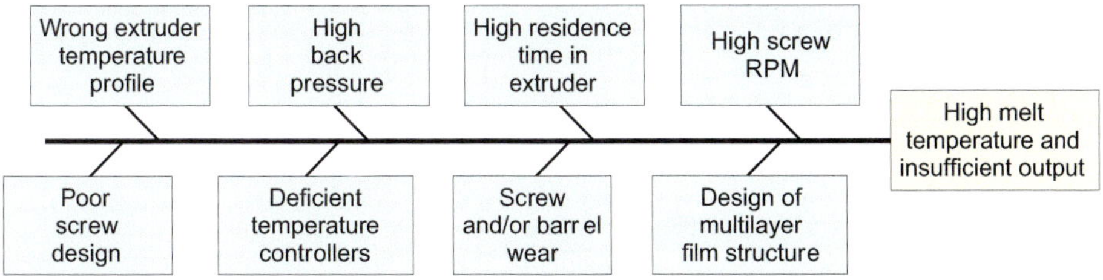

Figure 4.41 Fishbone diagram for high melt temperature and insufficient output

Figure 4.42 presents the calculated melting profile (SBP after Tadmor model) of the original screw at typical operating conditions (92.9 kg/h output at 69 rpm and 129 bar backpressure). Figure 4.43 shows the geometry of the original 3-zone screw with square pitch and a compression ratio of 2.6:1, which is low for conventional PE extrusion.

Delayed melting was confirmed, which led to blend homogeneity problems observed in the coextrusion line. The calculated melt temperature was higher than 230 °C in conformity with the trial observations. This melt temperature was too high for PE film extrusion, because LLDPE has a melting point at approx.125 °C.

In blown film coextrusion, high melt temperatures are unwanted because of subsequent difficulties in bubble cooling and calibration in order to obtain the required film thickness and tolerance.

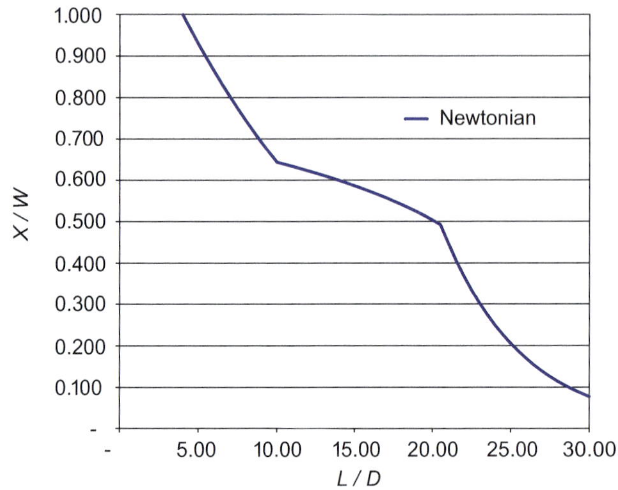

Figure 4.42 Calculated melting profile of the original 88.9-mm screw

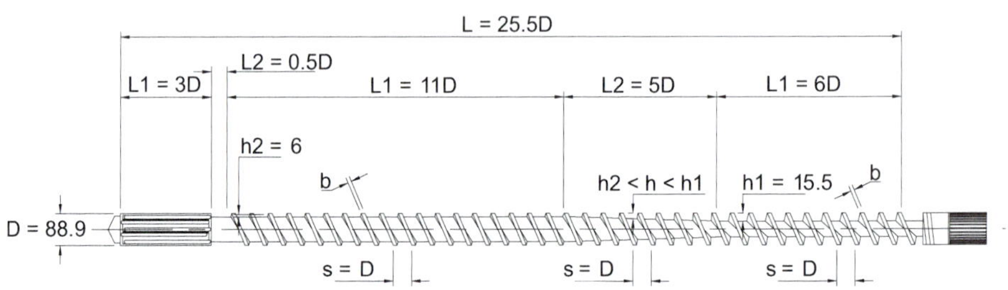

Figure 4.43 Geometry of original 88.9-mm screw

4.17.3 Solution

The solution was visualized based on previous information, computer simulations, coextrusion line trials, and the fishbone diagram shown in Figure 4.41. Each of the causes was considered, focusing on a new screw design for the extruders of layers A and C.

The newly designed single-screw needed to address higher specific output, defined as mass flow rate per unit screw rpm, and mixing to avoid the above described problems. Therefore, a barrier screw was designed (D = 88.9 mm, L/D = 25.5) including a 13D barrier section and a 2D Saxton type distributive mixer. Figure 4.44

shows the new optimized design delivering, both in simulations and final trials, a higher output of 120 kg/h at the required screw rpm as well as lower melt temperatures. Figure 4.45 shows a detail of the barrier section and Figure 4.46 shows the Saxton type distributive mixer included in this design.

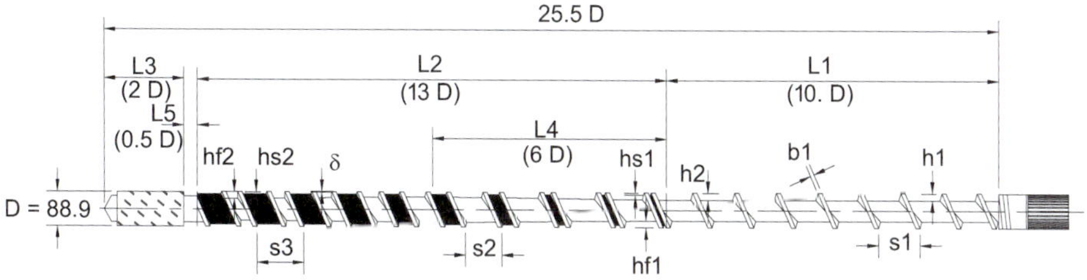

Figure 4.44 New 88.9-mm barrier screw

Figure 4.45 Detail of barrier section

Figure 4.46 Saxton type mixer

The melt homogeneity of the polymer blend was verified indirectly by means of mechanical and physical tests of the thin films, such as tensile test, tear test, and drop impact test, among others. Figure 4.47 shows tensile test results for the multi-layer film carried out in a universal testing machine following the ASTM D882 (Standard Test Method for Tensile Properties of Thin Plastic Sheeting). The left picture shows how the grips hold the ends of the test specimen for procedure and the right picture shows the film elongation.

Figure 4.47 Tensile test of 3-layer film

The variation of film properties over time disappeared and the required multi-layer structure could be manufactured without high melt temperature and overheating problems.

Figure 4.48 shows a micrograph of the obtained 3-layer film taken at 500× magnification under illumination with transmitted and polarized light, using an optical microscope. The film sample was embedded in epoxy resin for subsequent polishing.

Figure 4.48 Micrograph of 3-layer film, 500×

4.18 Deficient Solids Conveying and Dispersion

4.18.1 Description of the Problem

This case involved the mono-layer blown film extrusion of a PE blend comprised of LDPE and LLDPE (C4 or C8). The extruder operated with a shallow grooved barrel surface in the feed section. The extrusion line manufactured films from PE blends in different compositions and colors. The problem presented the following characteristics:

- the original 55-mm screw showed wear,
- the output of maximum 55 kg/h was unsatisfactory for the intended production program,
- a solids conveying problem was present, and
- the presence of agglomerates and other mixing problems were noticeable for several color masterbatches.

These films were used for thin and colorful T-shirt style bags in wide variations. Subsequent film conversion was required.

An output increase of at least 30% was necessary for the planned production program. It was necessary to avoid solids conveying problems and deficient dispersion of agglomerates to guarantee product specification and good color quality of the film for the new output.

Grooves in the internal barrel surface of the feed section increase the effective external friction and can improve solids conveying considerably. This is due to the fact that in the inner area portion of the grooves the internal coefficient of friction is effective and always higher than the external coefficient of friction [171].

The solids conveying mechanism in a grooved feed section is the so-called “nut-on-screw” conveyance, where the solids move in the groove direction (in this case in axial direction), like a nut on a rotating screw. This mechanism is of great importance, because the extruder output is at a maximum and completely independent of backpressure over a wide range [171].

4.18.2 Analysis of the Problem

This PE film extrusion case presented two difficulties: a solids conveying problem related to the operation of the grooved feed section and deficient dispersion of agglomerates introduced by color masterbatches.

The solution was visualized based on the fishbone diagram in Figure 4.49, examining each one of the tentative causes. Setup and operating conditions were checked. The original screw with $D = 55$ mm, $L/D = 30$ was a 3-zone screw with inefficient mixers for dispersion of agglomerates. The screw design in the feed section (channel depth and pitch) was not correct to operate with the shallow grooved feed section. Wear of the screw flights was evident as well.

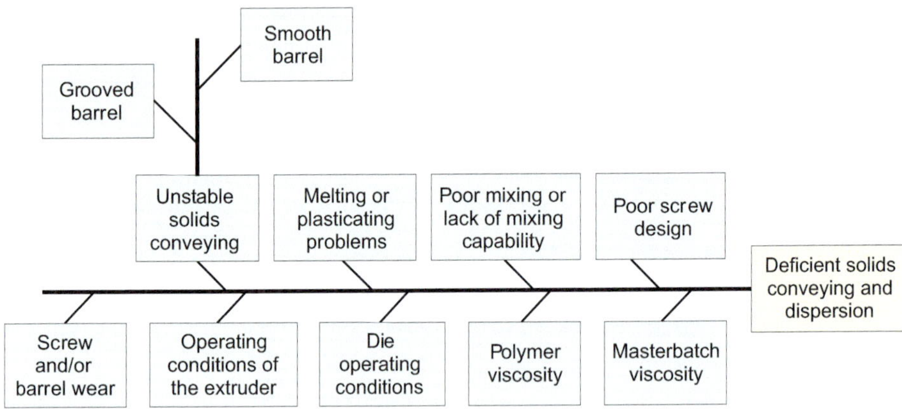

Figure 4.49 Fishbone diagram for deficient solids conveying and dispersion

Figure 4.50 presents a micrograph of the polymer melt analyzed for color variation. Optical microscopy with transmitted light was used at a magnification of 100 ×. The micrograph shows the presence of striations and agglomerates, a strong indication of mixing problems.

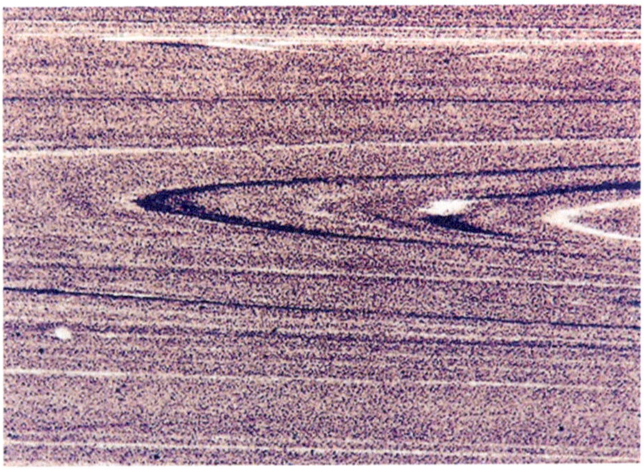

Figure 4.50 Micrograph of PE film

4.18.3 Solution

The solution was visualized based on previous information, computer simulations, extrusion line trials, and the fishbone diagram shown in Figure 4.49. Each of the causes was checked, focusing on a screw design suitable for the use of a grooved feed section and providing efficient mixing.

The newly designed single-screw needed to address 30% more output and sufficient mixing capability to avoid the above described problems. Therefore, a variable pitch screw was designed (D = 55 mm, L/D = 30), including a $4D$ dispersive mixer of Maillefer type. Figure 4.51 shows the new optimized design delivering, both in simulations and final trials, a higher output of 80 kg/h at 120 rpm and 200 bar backpressure. Melt temperature was approx. 200 °C as expected and appropriate for PE blends. Figure 4.52 presents the calculated melting profile (SBP after Tadmor model) of the new screw at the typical operating conditions: 80 kg/h output at 120 rpm. The remaining melting takes place in the Maillefer mixing unit. Figure 4.53 shows the dispersive mixing section of Maillefer type included in this design.

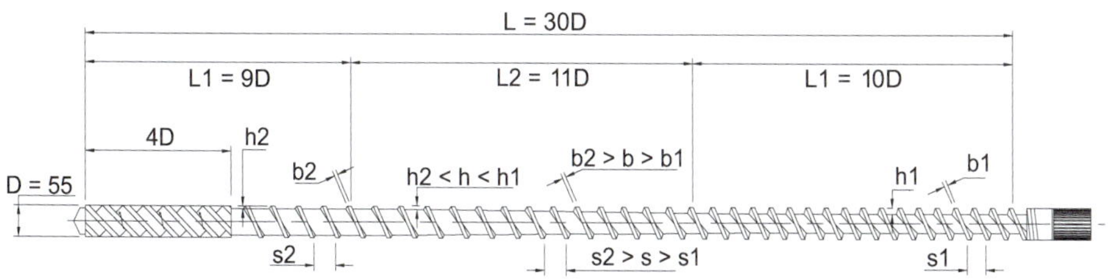

Figure 4.51 New 55-mm variable pitch screw

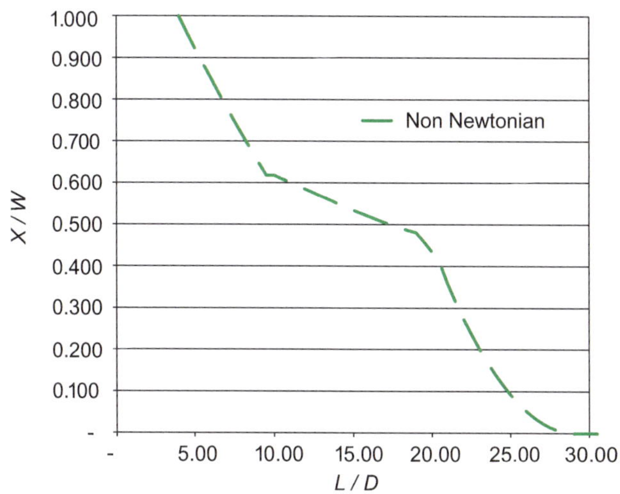

Figure 4.52 Calculated melting profile of new 55-mm screw

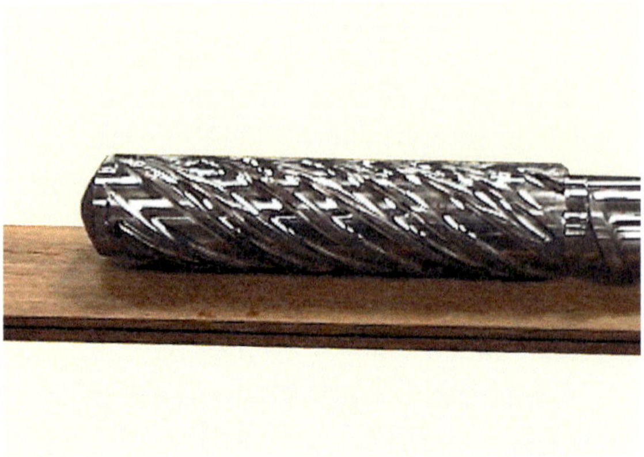

Figure 4.53
Maillefer type dispersive mixer

The melt homogeneity of the polymer blend and the film color were verified by means of optical microscopy and color spectrometer. Agglomerates and striations disappeared and the expected output was obtained for the required production program.

4.19 Instability of Formation at the Die

4.19.1 Description of the Problem

This problem concerned the single-screw extrusion of a PP blend comprised of recycled post-consumer PP and virgin PP material for industrial strapping and banding. The extruder operated with a smooth barrel surface in the feed section. The extrusion line manufactured heavy duty strapping tapes from PP blends in different compositions and colors. The recycled post-consumer PP content in the blend varied between approximately 40 and 60%, depending on product requirements.

The problem presented the following characteristics:

- the extrusion process was unstable,
- instability of formation at the die was noticeable,
- presence of inhomogeneities, e.g., agglomerates visible for several blend compositions, especially at higher amounts of recycled PP, and
- variations in mechanical resistance of the tapes.

These tapes were produced in 0.5 mm and 0.8 mm thickness and 13 mm width. A simultaneous extrusion of two tapes was required (e.g., two-cavity die). A subsequent drawing process was necessary to achieve the desired mechanical strength and stiffness of the tapes.

A minimum extruder output of 60 kg/h was considered. A melt temperature below 220 °C was a condition for a successful subsequent drawing process (e.g., drawing ratio 10:1).

The performance of an extrusion line is strongly dependent on the interaction between extruder and extrusion die, in particular when using a conventional extruder with a smooth barrel. With regard to the design of an extrusion die, following key criteria should be taken into account [172]:

- The pressure drop is an important parameter, because it is related directly to output and melt temperature difference.
- Ensuring that the melt emerges from the die at the same average rate all across the outlet cross section.
- The surface of the extrudate and/or the interfaces of the melt layers should remain smooth at operating conditions (e.g., within the processing window). Flow anomalies and/or stagnations should be avoided along the die length.

4.19.2 Analysis of the Problem

This tape extrusion case presented an interaction of extruder and extrusion die, where the operation of the main components, screw and die, suffered from a deficient design. In addition, there was a source of visible inhomogeneities of various types caused by recycled post-consumer PP.

The solution was visualized based on the problem tree showed in Figure 4.54, considering each one of the potential causes. A problem tree is a problem analysis tool that illustrates the cause and effect relationship of problem(s) using a hierarchical tree diagram.

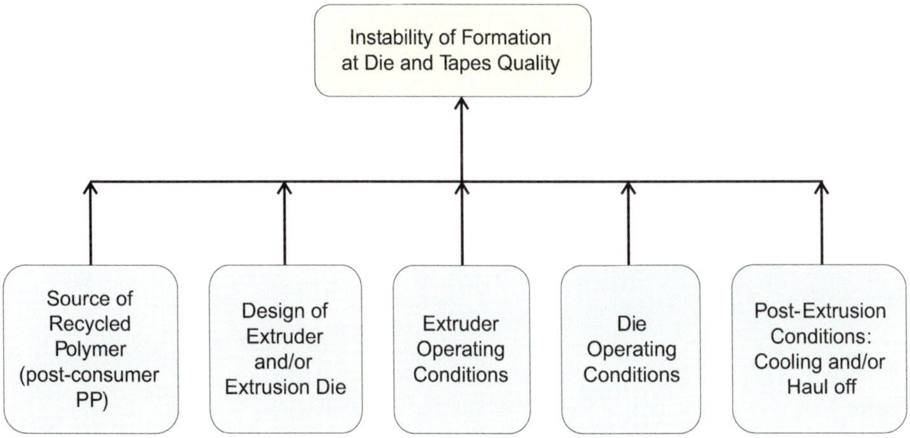

Figure 4.54 Problem tree for instability of formation at die

Setup and operating conditions were checked; it was discovered that the melt temperatures were too high (above 240 °C), affecting extrudate cooling and other post-extrusion processes. The original screw (D = 63.5 mm, L/D = 30.9) was a conventional screw with square pitch and a compression ratio of 2.8:1. This screw had an inefficient mixer for dispersion of agglomerates located at the end of the compression zone. Figure 4.55 shows the geometry of the original screw.

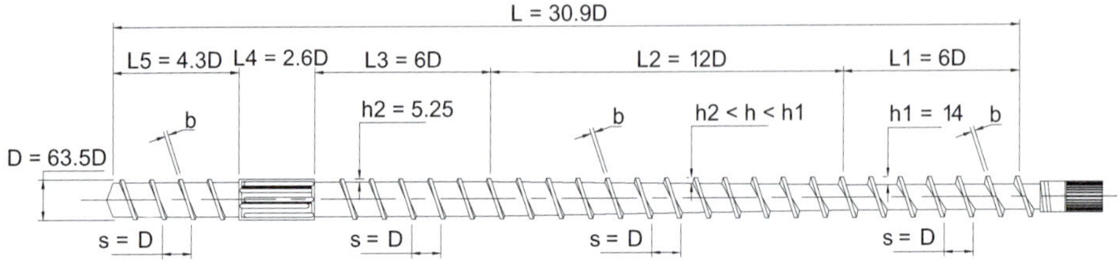

Figure 4.55 Original 63.5-mm screw

A capillary rheometer was used to obtain complete viscosity curves at three temperatures. Measurements for a polymer blend comprising recycled post-consumer PP were carried out at 200, 220, and 240 °C. Figure 4.56 presents the corrected viscosity curves at three temperatures for a specific PP blend composition. These viscosity curves were strongly affected by the content of recycled PP.

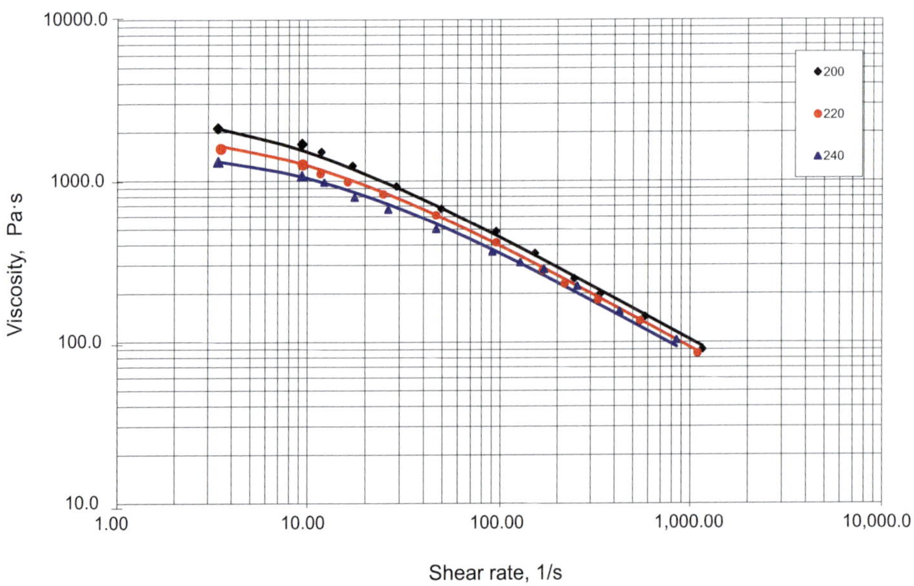

Figure 4.56 Viscosity curves of blend with recycled PP content

4.19.3 Solution

The solution was visualized based on previous information, computer simulations, extrusion line trials, IR thermography, and the problem tree shown in Figure 4.54. Each of the causes was analyzed, focusing on a new screw design suitable for a new two-cavity die.

The newly designed single-screw needed to achieve 60 kg/h output and sufficient mixing to avoid the presence of inhomogeneities or agglomerates caused by recycled PP. Therefore, a variable pitch screw was designed (D = 63.5 mm, L/D = 30.9), including a 4D dispersive mixer of Maillefer type and a 2D distributive mixer of Saxton type. Figure 4.57 shows the new optimized design delivering, both in simulations and final trials, a higher output of 60 kg/h at 112 rpm. A uniform rate of molten material could be conveyed to the extrusion die.

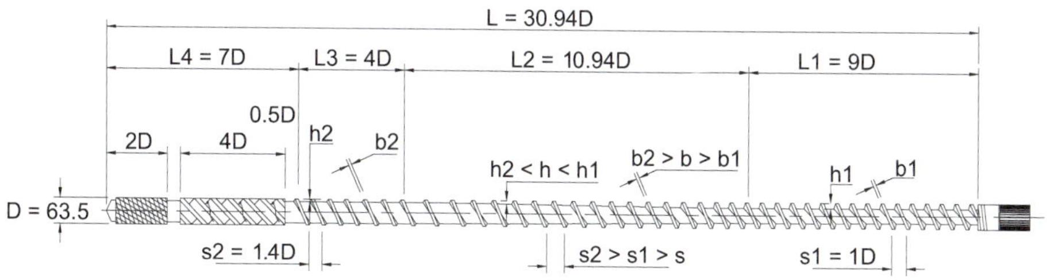

Figure 4.57 New 63.5-mm variable pitch screw

The new two-cavity die was designed to achieve the same 60 kg/h output with a 2.3 mm die gap for manufacturing both tape sizes (0.5 mm and 0.8 mm thickness and 13 mm width). Figure 4.58 shows the calculated pressure drop along the die length. A pressure drop of approx. 65 bar was predicted and obtained in final trials with small deviations. A low melt temperature increase (less than 5 °C) was observed because of low viscous dissipation.

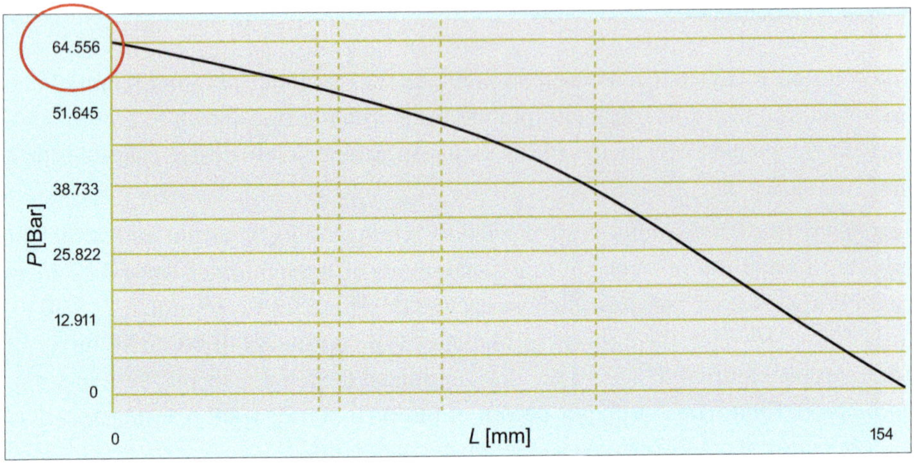

Figure 4.58 Calculated pressure drop along the die length

Figure 4.59 provides important information with regard to flow front velocity at the die exit. It shows that the melt front velocity was the same for the two die cavities. The simultaneous and uniform tape extrusion was a requirement in this case. The final trials showed a successful uniform extrusion of tapes.

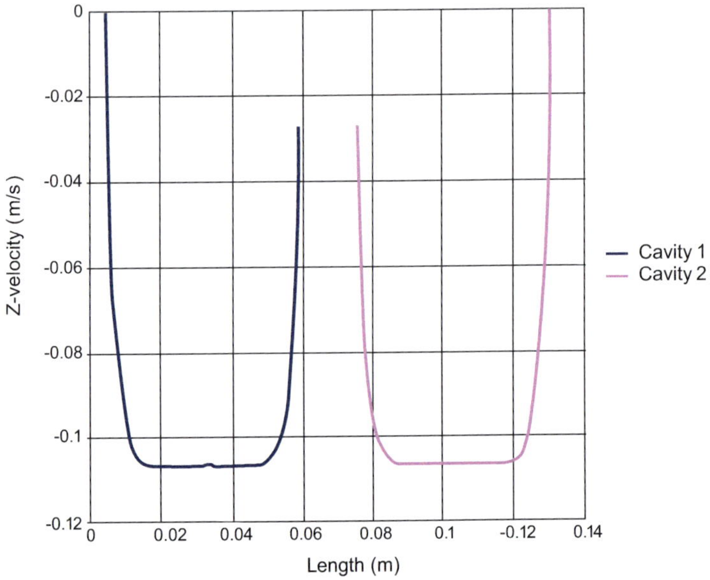

Figure 4.59 Flow front velocity for two-cavity die

Figure 4.60 presents the calculated shear stress at the wall along the die length. The obtained values did not generate flow anomalies or extrudate distortions, which was confirmed in final trials.

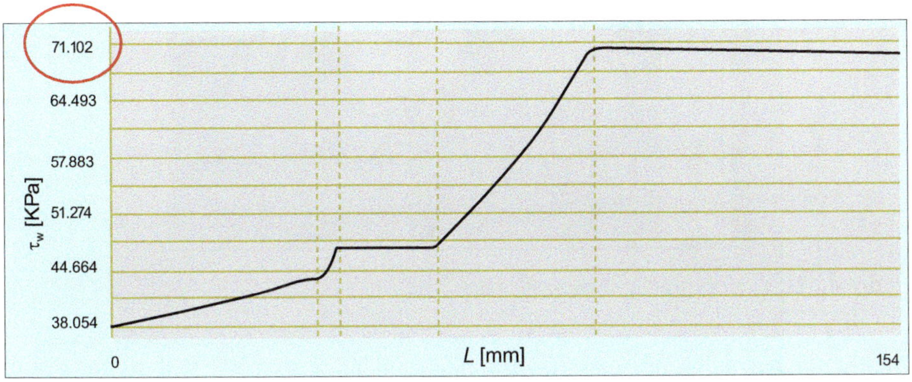

Figure 4.60 Calculated shear stress along the die length

Figure 4.61 shows the calculated shear rate at the wall along the die length. The obtained values of shear rate were within the expected range and they did not contribute to any flow anomalies or extrudate distortions, which was confirmed in final trials.

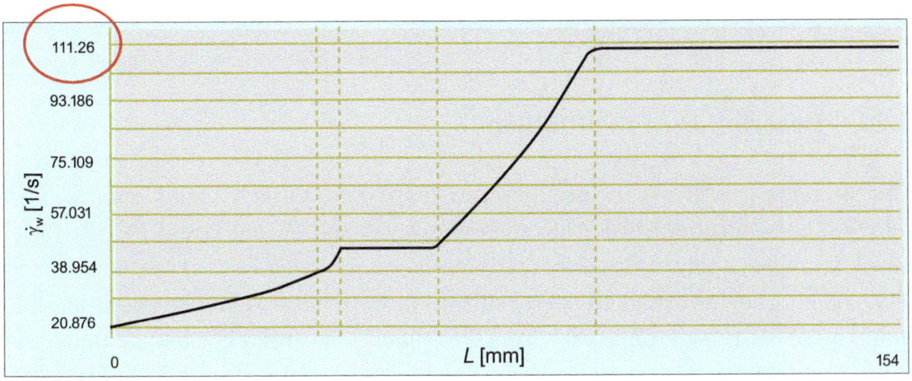

Figure 4.61 Calculated shear rate along the die length

Melt temperature was approx. 223 °C, optimized by IR thermography measurements setting the right temperature profiles in extruder and extrusion die. Figure 4.62 presents the IR thermography of the new die in operation at the typical extrusion line conditions.

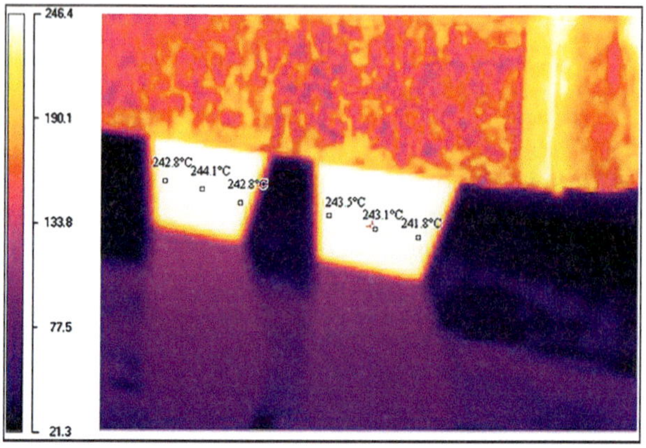

Figure 4.62 IR thermography of two-cavity die

Finally, the extrusion process with the new components (screw and die) was able to achieve stable conditions with a good extrudate formation at the die. The presence of agglomerates disappeared due to the Maillefer dispersive unit in the screw and melt filtration. The strapping tapes reached the required values of mechanical resistance. Changes in the drawing process were not necessary.

4.20 Intermittent Pumping in a Vented Extruder

4.20.1 Description of the Problem

This problem concerned the single-screw extrusion of a PET blend comprised of recycled post-consumer PET from soda and water bottles and virgin PET material for plastic sheet applications. The extruder operated with a 75-mm vented screw or a two-stage screw and smooth barrel surface in the feed section. The extrusion line manufactured transparent sheet from PET blends in different compositions and translucent colors. The recycled post-consumer PET content in the blend varied, depending on product requirements.

The problem presented the following characteristics:

- the extrusion process was unstable showing intermittent pumping,
- material flow out of the vent port was observed at several operating conditions,
- presence of inhomogeneities, e.g., gels and dark specks, were visible for several blend compositions, in particular at higher amounts of recycled PET, and
- the sheet quality did not meet requirements.

Subsequent processes (e.g., thermoforming and conversion) were necessary to achieve the different product specifications.

A minimum extruder output of 100 kg/h in stable extrusion conditions was considered. The maximum screw rotation speed of the extruder was 100 rpm. Pumping was noticeable in the range between 70 and 100 rpm. A melt temperature of approx. 285 °C was observed.

The performance of a vented screw or a two-stage screw is strongly dependent on the right screw balance between the two stages. Special attention should be paid to the screw conveying capacity in the second stage, which must be higher than the one in the first stage [173].

Other issues are important in a vented screw extruder [173]:

- The polymer has to be completely molten in the first stage before reaching the vent port, where degasification or extraction of volatiles takes place.
- The vent port geometry plays an important role. The vent opening should be neither too small nor too big.
- The operating conditions in both screw stages should be set looking for reducing conveying capacity in the first stage and increasing conveying capacity in the second stage.

Gear pumps are useful when using vented extruders to avoid polymer flow out of the vent. They can be used for most common thermoplastics, such as PET, nylon, polyolefins, polystyrene, and others [174].

4.20.2 Analysis of the Problem

The issue at hand was a vented extruder processing recycled PET material in non-uniform flakes delivered by a shredding process. The two-stage screw had a deficient design, because it was originally designed for virgin PET granules. In addition, there was a source of visible inhomogeneities such as gels and dark specks caused by recycled post-consumer PET.

The solution was visualized based on the problem tree showed in Figure 4.63, examining each one of the probable causes and emphasizing on a screw design suitable for post-consumer PET resin.

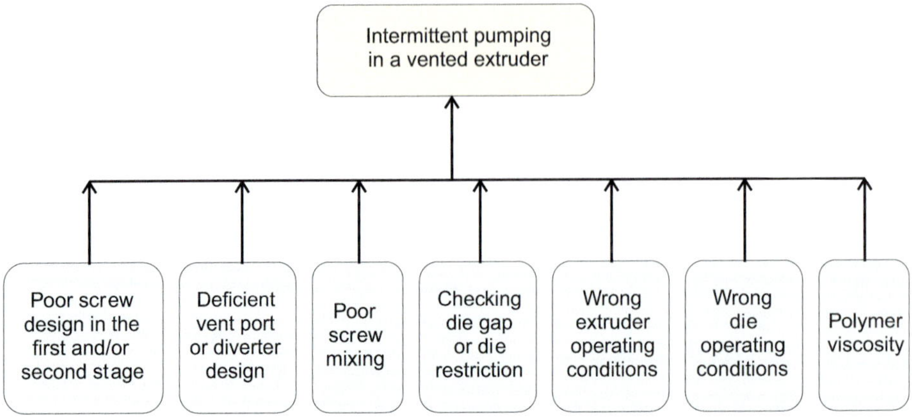

Figure 4.63 Problem tree for intermittent pumping

Setup and operating conditions were checked. The original screw ($D = 75$ mm, $L/D = 35$) was a vented screw with square pitch and a compression ratio of 4:1 in the first stage. The second stage did not have enough channel depth for higher melt conveying than the first stage and the screw pitch was inadequate, $S/D < 1D$. This caused material to flow out of the vent port at various operating conditions, because conveying capacity in the second stage was lower than conveying capacity in the first stage. This screw had an inefficient mixer for the dispersion of gels, located closer to the screw tip. Figure 4.64 shows the geometry of the original two-stage screw.

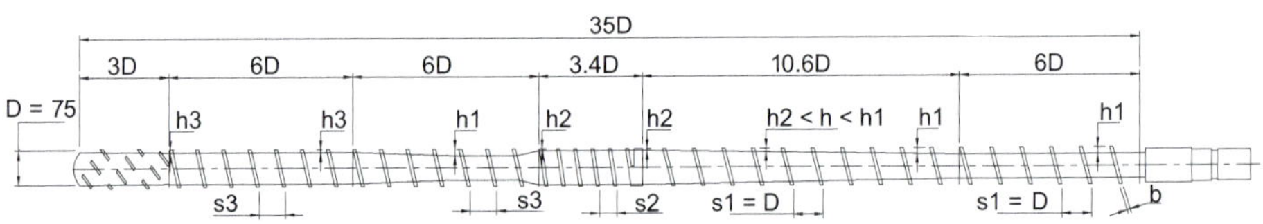

Figure 4.64 Geometry of original 75-mm two-stage screw

The backpressure of this extrusion line was less than 50 bar at 100 kg/h output and 70 screw rpm. A specific output of 1.4 (kg·min/h·rev) was obtained (mass flow rate per unit screw rpm). Intermittent pumping was observable. Sheet thickness was not within the set quality standards.

The intrinsic viscosity (IV) of the recycled PET material was measured before and after processing, obtaining an initial IV value of 0.8 dl/g and a final IV value of 0.7 dl/g after extrusion. This value was still suitable for plastic sheet applications.

The melting point of the recycled PET material was measured using DSC equipment. Figure 4.65 shows the obtained DSC curve, presenting a melting point of 246 °C. Rheological and thermal characterization of the post-consumer PET showed suitability for sheet extrusion.

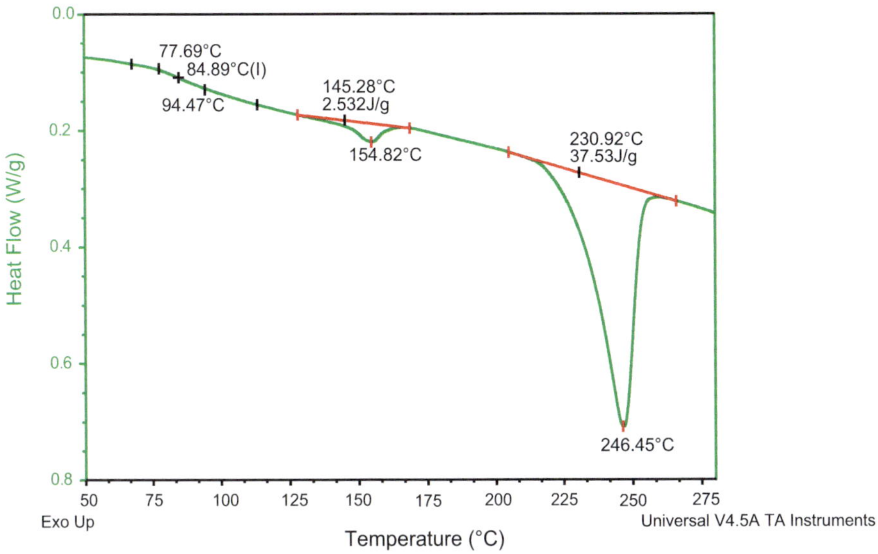

Figure 4.65 DSC of recycled PET

4.20.3 Solution

The solution was visualized based on previous information, computer simulations, extrusion line trials, IR thermography, and the problem tree shown in Figure 4.63. Each of the causes was checked, focusing on a new screw design adequate for post-consumer PET flakes to be operated with a gear pump to guarantee highest sheet quality, in particular thickness tolerances. The gear pump was acquired from a well-known international manufacturer.

The newly designed single-screw had to provide at least 100 kg/h output and sufficient mixing to avoid the presence of inhomogeneities (gels and dark specks) caused by recycled PET. Therefore, a barrier screw was designed (D = 75 mm, L/D = 35) including a 14D barrier zone and 4D distributive mixer of Saxton type. The new optimized design delivered, both in simulations and final trials, a higher output of 100 kg/h at 50 rpm in operation with a gear pump at 70 rpm. A higher specific output of 2 (kg·min/h·rev) was achieved. Higher outputs were achieved up to 145 kg/h at 70 screw rpm.

A uniform rate of molten material could be conveyed to the screen changer and to the extrusion die. Obtained sheet thickness variation was less than 3%.

Figure 4.66 schematically illustrates the operation of an extruder with a gear pump [174]. In this case, the melt pump increased sheet line efficiency by assuming the pressure buildup function and relieving the extruder of this task.

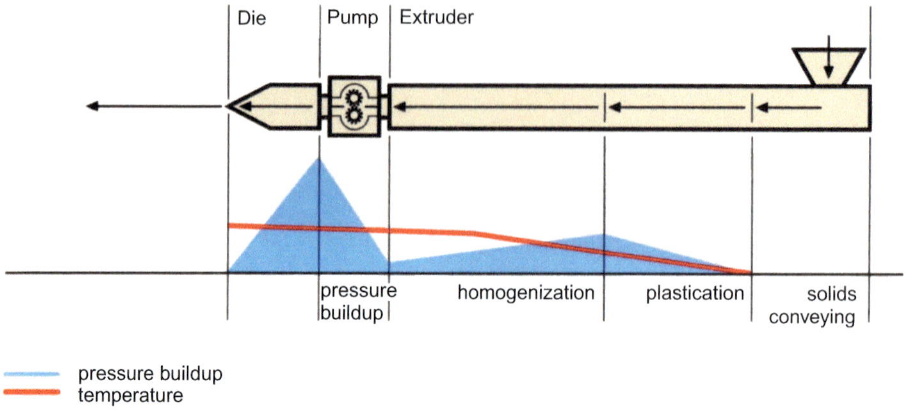

Figure 4.66 Operation of extruder and gear pump

Melt temperature was at approx. 292 °C, optimized by IR thermography measurements setting the right temperature profiles in the sheet extrusion line. Figure 4.67 presents the IR thermography of the extrudate at the die exit. Melt temperature was measured along the extrudate width to assure sheet thickness.

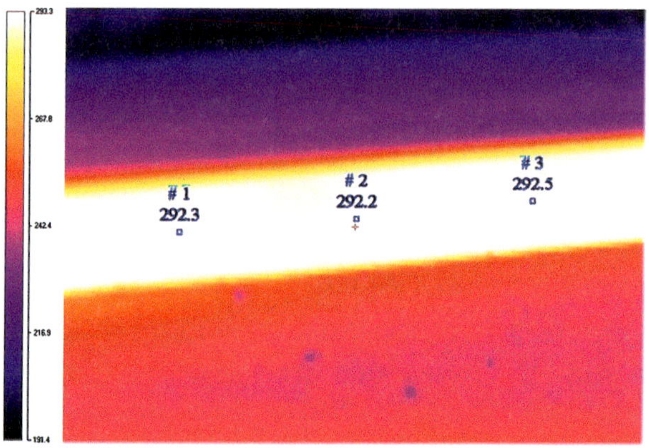

Figure 4.67 IR thermography of extrudate

Finally, the sheet extrusion process with the new components (barrier screw and gear pump) achieved stable conditions with good extrudate and product quality. The presence of gels disappeared due to the barrier-mixing screw and screen changer. The transparent and translucent sheets met the quality requirements. Changes in the subsequent process (e.g., thermoforming and conversion) were not necessary.

4.21 Melt Fracture or Sharkskin in m-PE

4.21.1 Description of the Problem

This case involved extrusion of an m-PE blend in a 45-mm single-screw extruder for strand pelletizing. The strands were white-colored with a 15% masterbatch (MB) containing 50% titanium dioxide pigment (TiO_2). The defects on the strands resembled extrudate surface roughness.

The extrusion of m-PE involves higher output, higher back pressure and torque than extrusion of branched LDPE. High melt temperature could be observed during extrusion trials.

It is known from literature that melt fracture is a flow instability related to the wall adhesion of the polymer melt. If the wall shear stress is larger than a critical shear stress, an unstable region appears in the vicinity of the die orifice, in which the flowing melt changes from adhering to slipping [172]. In the unstable region, both adhesion and slipping can occur.

This phenomenon is the so-called stick-slip effect, which leads to periodically rough extrudate surfaces and can be predicted by rheological measurements using a capillary rheometer.

4.21.2 Analysis of the Problem

The solution was visualized based on the fishbone diagram shown in Figure 4.68, examining each of the probable causes and focusing on polymer material and operating conditions of the strand extrusion line.

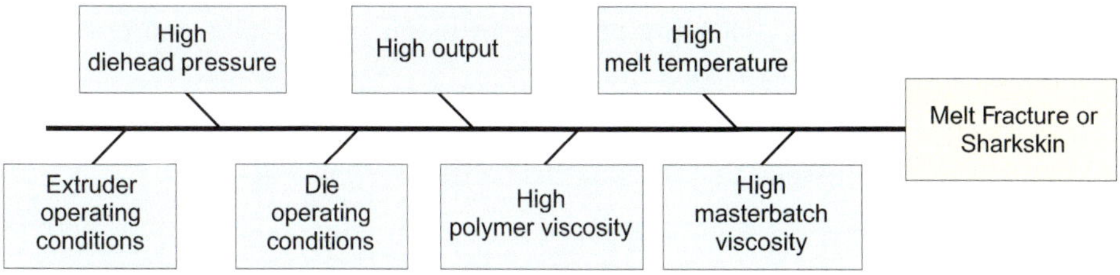

Figure 4.68 Fishbone diagram for melt fracture in m-PE

The original screw (D = 45 mm, L/D = 25) was a barrier screw with an 11D barrier section but without mixers. Setup and operating conditions were checked and determined to be not optimal because of high melt temperature (up to 268 °C) and high die-head pressure (up to 197 bar). The extruder temperature profile was ascending from 230 to 270 °C.

Strand samples were obtained at different extruder outputs or different screw rotation speeds (rpm) and at different masterbatch concentrations. Trials were carried out at 30, 60, and 90 screw rpm and at 5, 10, and 15% masterbatch content.

Strands were examined in detail by optical microscopy (Figure 4.69). The moderate magnification micrographs are shown for the blend of m-PE with 15% masterbatch and at three screw rpm: 30, 60, and 90 rpm. At higher screw rpm (e.g., shear rate), the extrudate surface roughness or strand surface was greater.

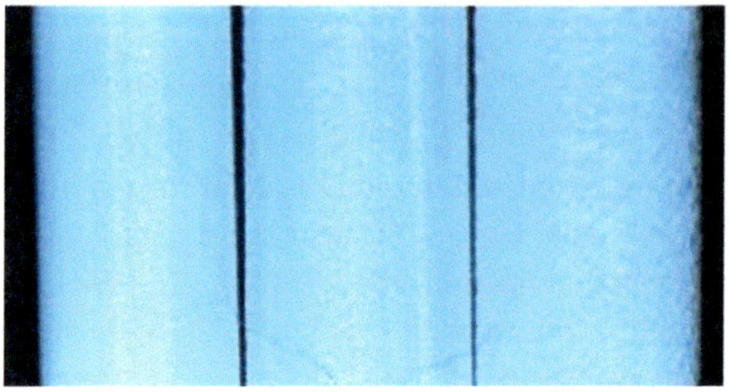

Figure 4.69 Micrographs of strands at different screw rpm

Other micrographs showed that at lower masterbatch content the extrudate surface roughness or strand surface was less noticeable.

4.21.3 Solution

The solution was visualized based on previous information, raw material data sheets, extrusion trials, and the fishbone diagram shown in Figure 4.68. Each of the causes was analyzed, focusing on polymer blend composition and operating conditions of the strand extrusion line.

It was possible to lower the die-head pressure to 109 bar by reducing the die restriction, which increased the output of this pelletizing line.

The melt temperature was slightly reduced with a new temperature profile, both in the extruder and in the die. The blend composition was changed, reducing masterbatch concentration without losing the desired white color. This was relevant because the polymer blend viscosity could be reduced.

In this case, the problem could be corrected because of a significant reduction in melt viscosity and die-head pressure. The micrographs were useful to state the progress of the problem at the different operating conditions and blend compositions.

4.22 Scale-up of LLDPE Single-Screw Extruder

4.22.1 Description of the Problem

This industrial case involved an extruder of 55 mm with a length of 25.5 *L/D* delivering LLDPE for an external layer in a 3-layer film coextrusion for thin and transparent film packaging and agricultural use. An important increased output was required to produce the desired 3-layer film structure. With the actual 55-mm extruder with barrier screw it was not possible to reach the higher mass throughput without high melt temperatures affecting negatively the film quality. The extruder was operating with a conventional feeding zone.

On the other hand, the plant operators were familiar with the operating windows (i.e., different conditions) of the 55-mm extruder.

4.22.2 Analysis of the Problem

This industrial case presents two difficulties for the existing extruder: higher output requirements, more than double, and avoiding high melt temperatures that affect film specifications and quality.

In film coextrusion, high melt temperatures are undesirable because of subsequent problems in bubble cooling and calibration to achieve the right film thickness and tolerance.

The solution was visualized in an acquisition of a new coextrusion line comprising an extruder, translating the good performance of the 55-mm extruder to a new larger extruder maintaining constant *L/D*, adjusting solids and melt conveying, and checking for appropriate melting of LLDPE. This was a typical case for an extrusion scale-up.

The Fisher-Potente scaling factors for single-screw extruders will be used for the solution of this industrial case [193, 194]. This technique is based on the following assumptions:

1. The melt temperature of the extruded material for all the extruder sizes should remain constant
2. The average thermal and flow properties of the material for a given extruder size are assumed constant, i.e., density, heat capacity, viscosity, among others
3. The ratio between pressure flow and drag flow is assumed to be constant
4. The extrusion process should run quasi-adiabatically; this means that the input heat in the feeding zone should be extracted with the output

5. The principle of similarity underlies the scale-up analysis of extruders, for example, constant length *L/D* and constant screw pitch *S/D*

The value of ψ (scale-up exponent) was obtained experimentally from measurements of production extruders in different sizes, conventional and grooved-feed extruders. The measurements were done for three different polymers: HDPE, HIPS (high impact polystyrene), and HMPE (high molecular weight PE). The obtained values of ψ averaged 0.7.

4.22.3 Solution

The solution was based on a scaling-up of a 55-mm single barrier screw (*D*1) to a new 90-mm barrier screw (*D*2) maintaining a constant length of 25.5 *L/D*.

Table 4.20 shows the estimated screw dimensions and operating parameters.

Table 4.20 Dimensions of the Original and Scaled-Up Screws

		D1	*D2*	Equation
Parameters Dimensions	Units	55 mm	90 mm	
N	rpm	100.0	74.4	$N2 = N1(D2/D1)^{2\psi-2}$
Mp	kg/h	80	225.0	$Mp2 = Mp1(D2/D1)^{3\psi}$
h1	mm	11	15.5	$h12 = h11(D2/D1)^{\psi}$
h2	mm	11	15.5	$h22 = h21(D2/D1)^{\psi}$
hf1	mm	11	15.5	$hf12 = hf11(D2/D1)^{\psi}$
hf2	mm	3	4.2	$hf22 = hf21(D2/D1)^{\psi}$
hs1	mm	3	4.2	$hs12 = hs11(D2/D1)^{\psi}$
hs2	mm	10	14.1	$hs22 = hs21(D2/D1)^{\psi}$
δ	mm	0.7	1.0	$\delta 2 = \delta 1(D2/D1)^{\psi}$
L/D	-	25	25	-

Average ψ = 0.7

Figure 4.70 shows the type of single barrier screw, where dimensions are visualized.

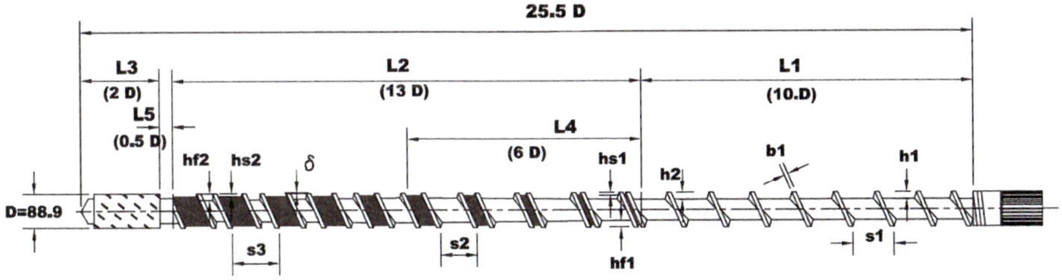

Figure 4.70 Barrier screw type

The nomenclature of Table 4.20 is the following:

N: screw rpm

Mp: mass throughput

h1, h2: channel depths in the feeding zone

hf1, hf2: channel depths of the solids channel in the barrier section

hs1, hs2: channel depths of the melt channel in the barrier section

δ: barrier gap

ψ: scale-up index

The different screw lengths were kept constant as indicated in Table 4.21.

Table 4.21 Screw Lengths of the Original and Scaled-Up Screws

Extruder zone	*L/D*
Feeding zone	10
Barrier section	13
Mixing section	2.5
Total screw length	25.5

The new scaled-up 90-mm barrier screw showed the good performance of the 55-mm extruder delivering the desired film properties over time. The 3-layer film structure could be produced with adequate melt temperatures avoiding overheating problems. The new 90-mm extruder with 25.5 *L/D* for a conventional feeding zone was checked for solids and melt conveying and for appropriate melting of LLDPE at approximately 200 kg/h mass throughput.

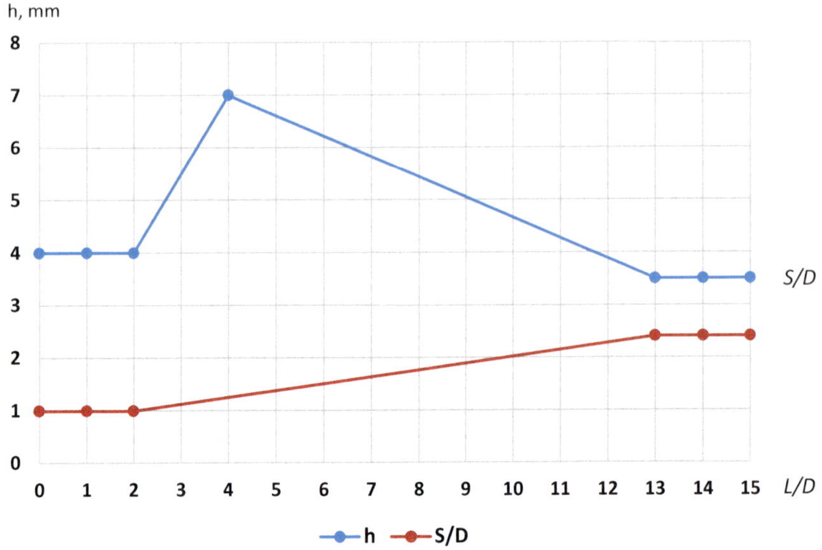

Figure 4.72 Channel depth and screw pitch profiles of new 45-mm screw in the feeding and melting zones

The mixing zone of the new 45-mm extruder with 24 *L*/*D* was comprised of a distributive pre-mixer of 2*D*, a Maddock dispersive mixer of 4*D*, and a distributive mixer of 2*D* to guarantee the melt and temperature homogeneity.

The important improvement in melt and temperature homogeneity required for the manufacturing of HDPE bottles was obtained; furthermore, stable parison formation, calibration, and extruder output were achieved.

4.24 High Melt Temperature in Sheet Extrusion

The case study involves a sheet extrusion company that contacted the author to conduct in-house training for operators and process engineers. The company extrudes PP and HDPE sheets. In the extrusion class one of the experienced operators brought up a concern about die temperature zones overriding. Figure 4.73 shows a screen from the data acquisition system showing the die temperatures, pressures, melt temperature, and screw speed.

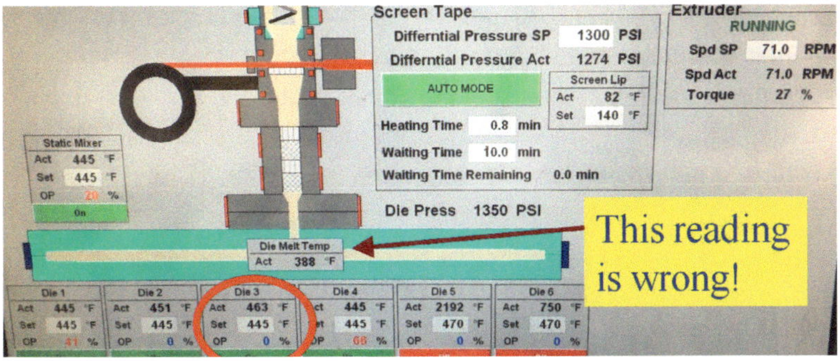

Figure 4.73 Screen showing temperatures, pressures, and screw speed

The extruder is equipped with a continuous belt screen changer, a static mixer, and a sheet die with six temperature zones. The DAS screen shows that die zones 2 and 3 are overriding, in other words, the actual temperature is higher than the setpoint temperature. The setpoint for die zone 3 is 445 °F (229.4 °C) with the actual temperature 463 °F (239.4 °C). No heating occurs in die zones 2 and 3. Die zones 5 and 6 are not used. The melt temperature in the die is measured at 388 °F (197.8 °C).

The actual die zone temperature is above the setpoint with zero heating of the die heater. This indicates that the die is heated by the polymer melt to a temperature above the setpoint. That must mean that the actual melt temperature in the die is well above 463 °F (239.4 °C). Figure 4.74 shows the actual readouts of die zones 2, 3, and 4. Die zone 4 is at setpoint and the die zone 4 heater is operating at 41% power. This indicates that there are enough heat losses in die zone 4 so that some heating is required to maintain setpoint. Die zone 4 is at the edge of the die where more heat losses occur than in the middle of the die; the same is true for die zone 1.

Figure 4.74 Readouts of die zones 2, 3, and 4

Clearly, the die melt temperature reading in Figure 4.73 cannot be correct. We know that the melt temperature in the die has to be higher than 463 °F (239.4 °C);

therefore, the 388 °F (197.8 °C) "melt" temperature reading cannot be correct. It turns out that the die melt temperature was measured with a combination pressure/temperature probe, see Figure 4.75. These P/T probes are commonly used in the extrusion industry. Unfortunately, these probes are notoriously unreliable when it comes to measuring melt temperature.

Figure 4.75 Combination pressure/temperature transducer

The temperature sensor in a P/T transducer is located inside the transducer; it is not in direct contact with the polymer melt. As a result, the temperature measured is the temperature of the transducer and not the temperature of the polymer melt. The melt temperature can be measured with an immersion temperature sensor or with an infrared thermometer. In this case the temperature of the melt emerging from the sheet die was measured with an infrared camera; the photo is shown in Figure 4.76.

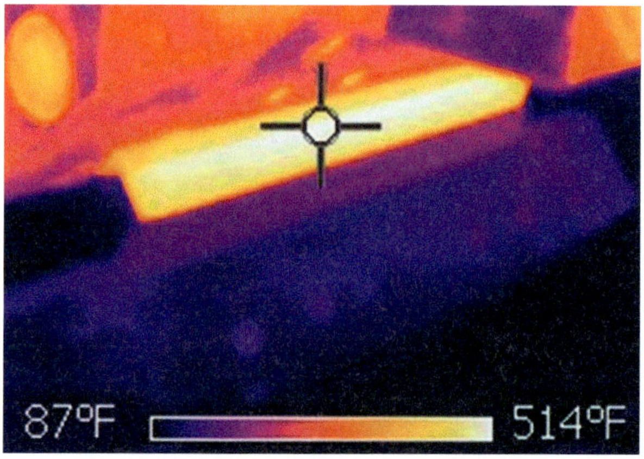

Figure 4.76 Infrared photograph of melt emerging from sheet die

The temperature measured at the spot size was 513 °F (267.2 °C). The "melt" temperature shown on the DAS screw, see Figure 4.73, was 388 °F (197.8 °C). This shows that the temperature measured with the P/T transducer was 125 °F (69.4 °C) lower than the melt temperature measured with the infrared camera. This example demonstrates clearly that the temperature reading from a P/T transducer can provide completely incorrect information regarding melt temperature.

This company had been running several extrusion lines with melt temperatures significantly higher than the indicated "melt" temperature for a long period of time without knowing it. The high melt temperatures caused degradation and resulted in high scrap rates over a long period of time, costing the company a substantial amount of money. Fortunately, this problem could be corrected rather easily and inexpensively by measuring the melt temperature with an immersion probe. Figure 4.77 shows an immersion melt temperature probe with adjustable probe depth.

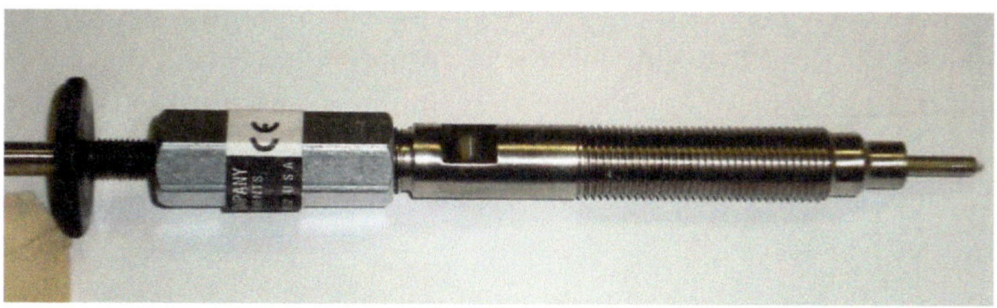

Figure 4.77 Immersion melt temperature probe with adjustable depth

The melt temperature probe shown in Figure 4.77 has a sensing stem with adjustable depth of immersion. One benefit of such a probe is that the melt temperature can be measured at different locations. Another benefit is that the stem can be completely retracted when the extruder starts up or before shutdown so that the chance of damage of the probe is minimized.

Conclusions

The three most critical measurements in the extrusion process are melt pressure, melt temperature, and motor load. These can be considered the vital signs of the extrusion process. If one of these vital signs is missing or not measured correctly, it severely handicaps proper control of the extrusion process. P/T transducers are used in many extrusion operations. Unfortunately, these P/T transducers do not provide a good measurement of melt temperature.

In this case study the indicated "melt" temperature from the P/T transducer was about 70 °C lower than the actual melt temperature – a huge difference! As a result, significant polymer degradation occurred in the extrusion process, causing high scrap rates. This case study amply demonstrates that incorrect melt temperature measurement can be very costly to an extrusion company, especially when the problem occurs over a long period of time. The use of P/T transducers to measure "melt" temperature is one of the most common mistakes made in the extrusion industry.

4.25 Gear Pump Speed Variation in Sheet Extrusion

This case study involved a sheet extrusion operation with a 125-mm single-screw extruder connected to a gear pump. The inlet pressure of the gear pump was controlled by the screw speed using pressure feedback control. It was found that the gear pump speed was varying but the plant personnel were not aware that this was an abnormal situation – the technical expertise at the plant was not at a high level.

In gear pump assisted extrusion, the gear pump speed is generally set at a constant value and the screw speed is adjusted to maintain constant inlet pressure at the gear pump. This adjustment is done automatically when pressure feedback control is used. When the gear pump speed varies, pressure feedback control cannot work properly.

This extrusion line was not equipped with a data acquisition system. Therefore, the dynamic behavior of the extruder could not be easily analyzed. It is quite common that older extrusion lines do not have data acquisition capability. Unfortunately, this makes process analysis and troubleshooting much more difficult. In order to determine the short-term process variation, a one-minute video was taken of the control panel so that the following process parameters could be monitored: drive current, screw speed, gear pump speed, extruder discharge pressure, and gear pump inlet pressure. The video was taken under normal production conditions.

The information from the video was used to allow determination of short-term process variation by taking data every video frame. Each video frame represents about 34 milliseconds. Therefore, a one minute video contains about 1800 frames. Clearly this is a very tedious and time-consuming method of analyzing the process. However, in the absence of a good data acquisition system (DAS) this is one way to assess short-term process variation. It took more than five hours to capture all the data and enter the data into an Excel file. If a DAS had been available, the process data over an extended period of time would have been available within seconds.

Figure 4.78 shows the gear pump speed versus time. There is a significant variation in gear pump speed, about 2 rpm variation on an average speed of about 39.4 rpm. This corresponds to about 5% speed variation. Since the extruder output is determined by the gear pump, a speed variation of 5% will cause an output variation of 5%. This will result in a 5% dimensional variation of the extruded sheet. Such variation is generally unacceptable.

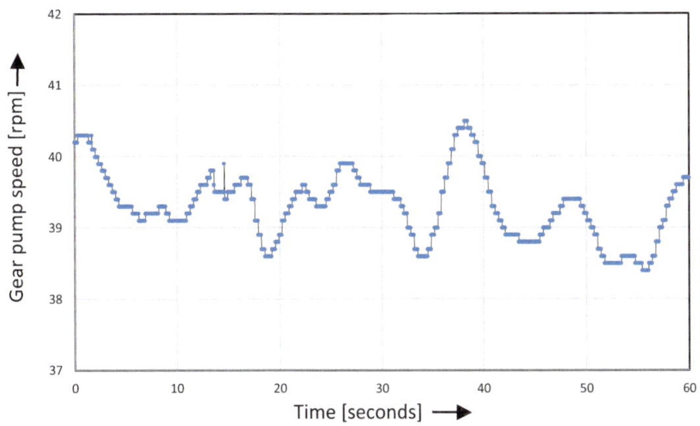

Figure 4.78 Gear pump speed versus time

When an engineer from a different plant assisted in analyzing the extrusion line it was found that the actual gear pump speed was different from the speed indicated on the control panel readout. The readout showed 48 rpm while the actual speed was 20 rpm – a very large discrepancy! It was also found that the actual gear pump speed was constant. Therefore, the large speed variation noted from the readout was actually nothing but measurement error.

It was remarkable that this extrusion line had been running with a faulty gear pump speed readout for a long time. This happened because nobody at the plant realized that this level of gear pump speed variation is abnormal and unacceptable.

Figure 4.79 shows the screw speed versus time. The apparent screw speed variation is about 4 rpm on an average screw speed of 83 rpm. This corresponds to about 5% screw speed variation. The screw speed graph shows a severe step pattern. Such a pattern is typical of poor resolution. In this case, the screw speed can only be displayed as whole numbers. That means that for a screw speed of 50 rpm, the actual screw speed can be anywhere between 50.5 and 49.5 rpm. As a result, at 50 rpm there can be a 2% change in screw speed without any change in the value shown on the readout.

The lack of resolution results in an unnatural variation in the screw speed graph. The screw speed should be shown to at least one decimal point. Unfortunately, it is quite common that screw speed readouts only show whole numbers. This is a mistake made by many extruder manufacturers. For proper process analysis, the screw speed needs to be known to at least one decimal point.

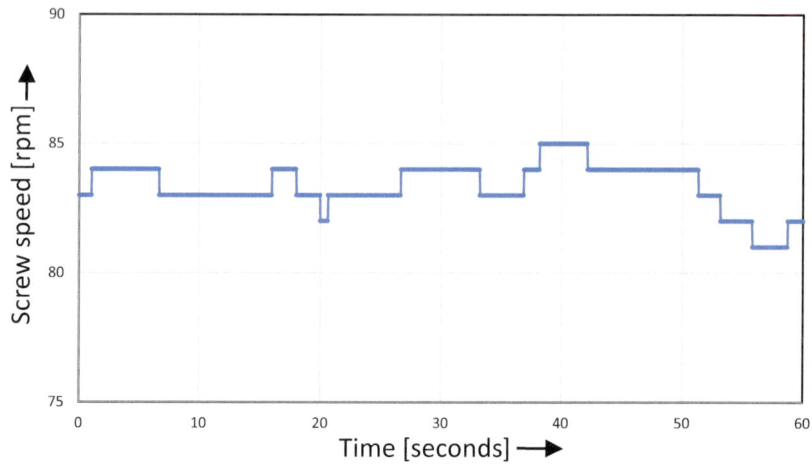

Figure 4.79 Screw speed versus time

To illustrate how lack of resolution can distort the process patterns Figure 4.80 shows the gear pump speed in two ways. The first data, in blue, is based on actual readings to one decimal point – this is the same as the graph shown in Figure 4.78. The second data, in red, is based on readings that are rounded to zero decimal points. The figure clearly shows that the rounding of the readings changes the actual variation significantly; it creates a step pattern and the actual variation is distorted.

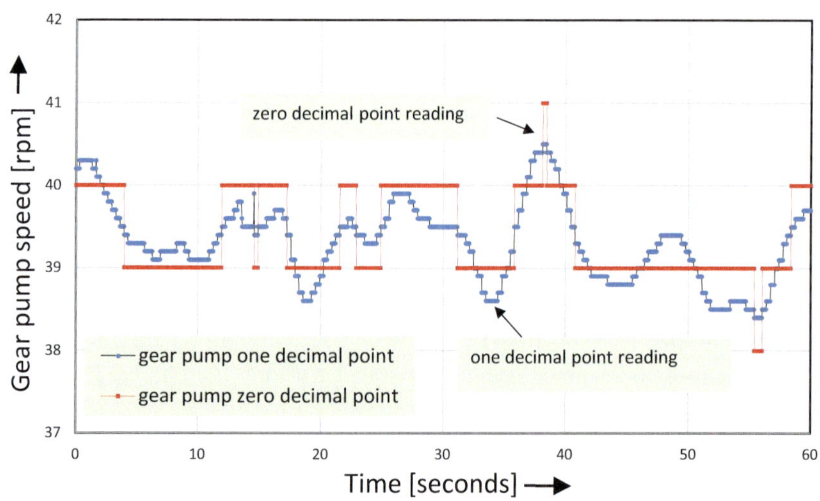

Figure 4.80 Gear pump speed versus time actual and rounded numbers

Figure 4.81 shows the extruder discharge pressure and the gear pump inlet pressure. Generally, the gear pump inlet (suction) pressure should not vary more than ± 50 psi (0.34 MPa). The gear pump discharge pressure should not vary more than ± 10 psi (0.07 MPa).

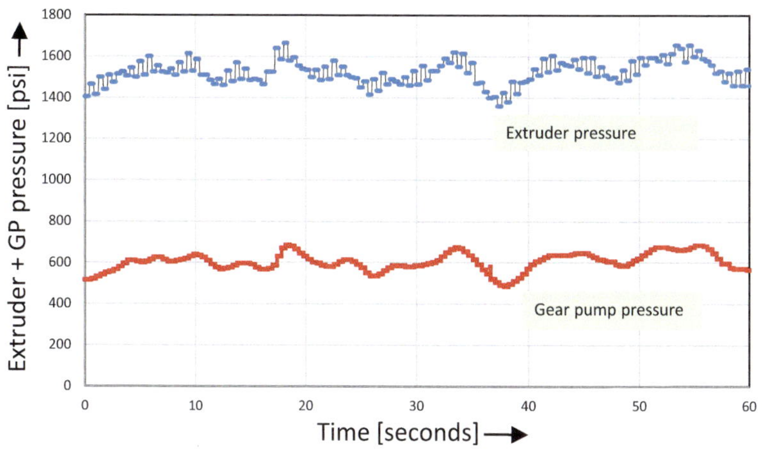

Figure 4.81 Extruder discharge pressure and the gear pump inlet pressure

The extruder pressure varies from about 1350 to 1700 psi; the pressure variation is about 23%. The gear pump pressure varies from about 500 to 700 psi – this variation is more than twice as high as normal. The gear pump pressure varies about 33%. The extruder pressure shows a stairs pattern similar to the screw speed graph in Figure 4.81. This indicates insufficient resolution in the extruder pressure data. One way to get a better idea of the actual variation is to use a moving average trend line; this is shown in Figure 4.82.

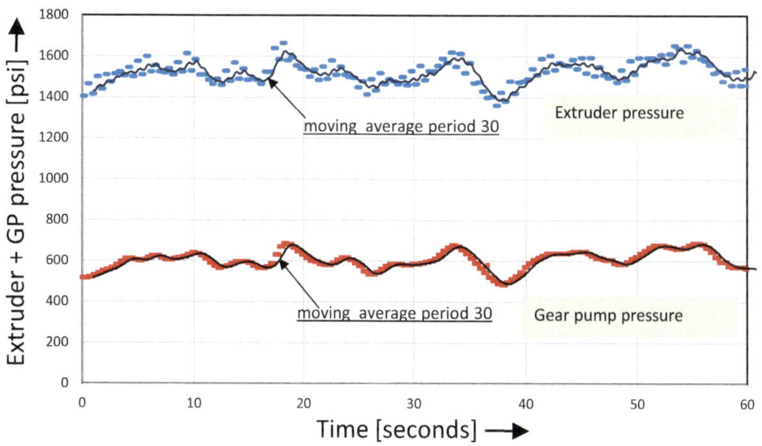

Figure 4.82 Extruder and gear pump pressure versus time with moving average trend lines

When the data is plotted in Excel it is very easy to create a moving average trend line. The period of the moving average trend line can be easily adjusted. It is one of many useful tools that Excel offers.

Figure 4.83 shows the drive current versus time.

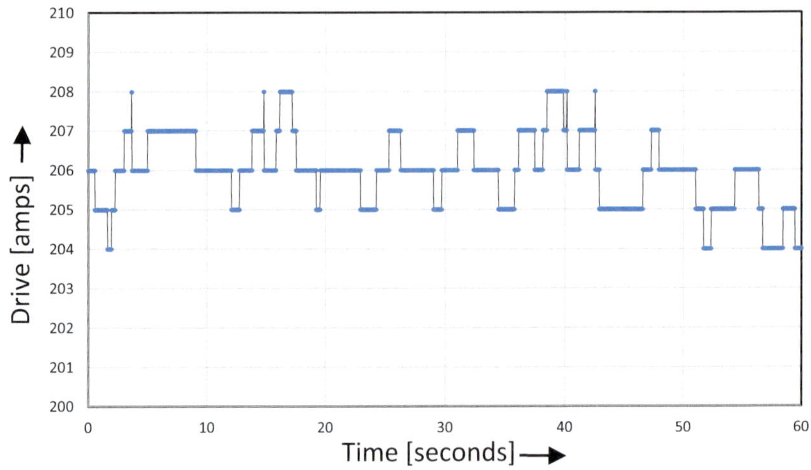

Figure 4.83 Drive current versus time

The drive current graph shows a strong stairs pattern. This results from the fact that the current in amps can only be read as whole numbers. This is similar to the graph of screw speed and extruder pressure. The apparent current variation is 4 amps on an average of about 206 amps; this corresponds to a current variation of about 2%.

The instrument panel did not have a readout of the power consumption of the drive. In this case, the power consumption could not be determined accurately because the drive voltage was not measured. Without a reading of motor power it is not possible to determine the specific energy consumption or SEC; this is one of the most important process parameters. Absence of data on specific energy consumption limits the ability to analyze the process.

Conclusions

In this particular extrusion plant there appears to be a lack of technical expertise. The process engineer responsible for this extrusion line had been alerted on multiple occasions that the gear pump speed variation was abnormal and needed to be investigated. However, no action was taken over a period of several months. This situation may be the result of lack of training or simply a lack of competency. A competent extruder operator or process engineer would have noticed the difference between the actual gear pump speed (20 rpm) and the indicated speed (48 rpm). This is a very large difference that could have been noticed by simple visual observation.

In a well-run extrusion operation all critical process measurements should be checked and calibrated on a regular basis. This should be done at least on an annual basis. Maintenance records should be available so that one can verify that the instrumentation was checked and calibrated. In this plant the extruder instrumentation was not tested on a regular basis and no records of testing and calibration were available. When data shown on the extruder instrument panel is incorrect, it is impossible to properly run the extruder, to analyze the process, and to troubleshoot the process. When one process measurement is completely incorrect, it puts all the process measurements into question.

In this extrusion line several process parameters (screw speed, drive current, extruder discharge pressure) were not measured with sufficient resolution. Insufficient resolution results in distorted apparent dynamic behavior of the extruder. As a result, the true dynamic behavior of the extruder cannot be properly analyzed. Furthermore, without information on motor power consumption the specific energy consumption cannot be determined. This limits the ability to analyze the process.

The variation in gear pump inlet pressure in this case study is larger than normal. It is possible that a retuning of the parameters of the pressure feedback control would reduce the gear pump inlet pressure variation.

The lack of a data acquisition system causes serious problems. The dynamic behavior of the extruder can be analyzed by taking a video of the instrument panel and the readings can be analyzed on a frame by frame basis. However, this is extremely tedious and time-consuming. In this case, it took more than five hours to analyze a one-minute video. This is simply not practical if it needs to be done repeatedly. Poor instrumentation and lack of DAS makes it hard to do the following:

- Control the extrusion process
- Know what is happening inside the extruder
- Optimize the extrusion process
- Detect a problem in a timely fashion
- Troubleshoot the extrusion process
- Have a profitable extrusion operation

■ 4.26 Melt Temperature Variation in Tubing Extrusion

This case study involved a 75-mm 28*D* grooved feed, single-screw extruder used to produce nylon 11 tubing [196]. The tubing has an OD of 14 mm and a wall thickness of 1.25 mm. The nylon is dried for 8 hours at 70 °C and vacuum-conveyed to the extruder feed hopper. The temperature of the tubing emerging from the die is measured with an infrared thermometer. Figure 4.84 shows the infrared temperature measurement of tubing.

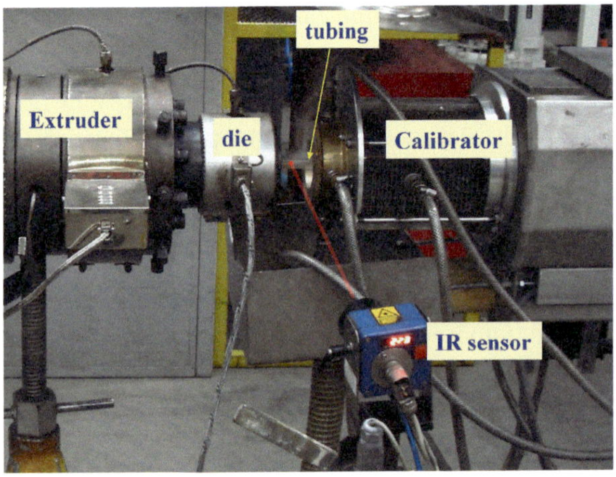

Figure 4.84 Infrared temperature measurement of tubing

The extrusion line has a data acquisition system with trending capability. It was found that several process temperatures were varying in a regular sinusoidal pattern with a cycle time of about 6 minutes. Figure 4.85 shows 11 process temperatures and the torque. The following temperatures were measured and displayed:

1. Melt temperature measured with immersion probe
2. Melt temperature measured with infrared thermometer
3. Extruder zone 1, actual temperature
4. Extruder zone 2, actual temperature
5. Extruder zone 3, actual temperature
6. Extruder zone 4, actual temperature
7. Die head zone 1, actual temperature
8. Die head zone 2, actual temperature
9. Die head zone 3, actual temperature

10. Die head zone 4, actual temperature
11. Die head zone 5, actual temperature

All temperatures are measured in degrees centigrade.

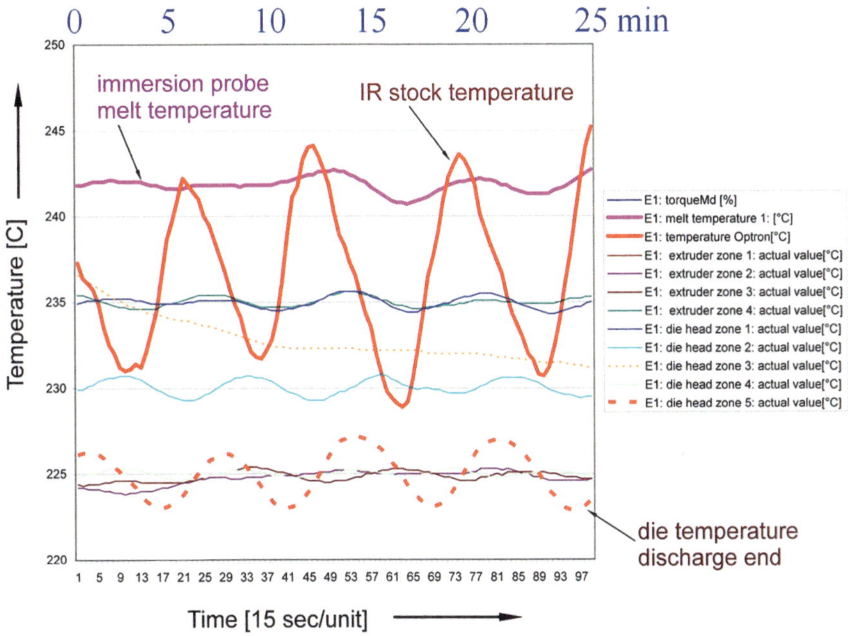

Figure 4.85 Process temperatures versus time

Several barrel and die zone temperatures are oscillating; also the melt temperature measured with an immersion probe. The temperature variation of die head zone 5 is about 4 °C. The melt temperature variation with an immersion probe is about 2 °C. The greatest variation occurs in the melt temperature measured with the infrared thermometer. The IR melt temperature variation is about 15 °C – this is a very large variation!

The following questions can be asked:

1. Is the IR melt temperature variation real?
2. If it is real, do we need to worry about it?
3. What can be the cause of the IR melt temperature variation?
4. How can we reduce the IR melt temperature variation?

The following process variables are relatively constant: motor load, barrel temperatures, screw speed, and take-up speed. There is also no significant variation in the dimensions of the extruded tubing.

The IR melt temperature variation is most likely real because several other process temperatures are varying in a similar pattern. A melt temperature variation of 15 °C is certainly something to worry about. Even if the product can be made to the required dimensions, the variation in melt temperature will affects the stresses in the extruded product. With non-uniform stresses in the product, the relaxation of these stresses may produce changes in the product over time.

The die temperature at the discharge end varies about 4 °C and it would be tempting to conclude that the die temperature variation causes the tubing temperature variation. However, this is not possible because the tubing temperature variation is about 3–4 times greater than the die temperature variation. A die temperature variation of 4 °C cannot cause a tubing temperature variation of 15 °C. Clearly, the die temperature variation is caused by the melt temperature variation and not the other way around. A similar argument can be used for the variation in barrel temperatures.

There is little variation in motor load and no variation in screw speed. As a result, the tubing temperature variation cannot be caused by changes in viscous dissipation. This leaves one other possible cause of the problem and that is variation in the temperature of the feed material entering the extruder. Unfortunately, this extruder was not equipped with a pellet temperature sensor in the feed port. However, there was strong indirect evidence of pellet temperature variation. The polymer was dried before it was conveyed to the extruder. It turned out that the feed hopper was filled with dry and hot material approximately every six minutes.

The feed hopper was not heated. Therefore, the hot pellets from the drier would cool down in the feed hopper as they made their way to the feed opening of the extruder. The hopper was filled at six minutes intervals. This created a six minute cycle in the temperature of the pellets entering the extruder. Once the cause of the problem is known, the solutions are obvious. One possible solution is to install a hopper drier on the extruder so that the pellets stay dry and at constant temperature in the hopper.

It should be noted that the problem would have been noticed immediately if the extruder had been equipped with a pellet temperature sensor at the feed opening of the extruder. Such a temperature sensor provides important information. Unfortunately, extruder manufacturers generally do not provide a temperature sensor in this part of the machine.

4.27 Black Specks in Tubing Extrusion

The next case study deals with black specks in extruded tubing made out of nylon 11. The extruder used in this case study is the same extruder described in Section 4.26. The tubing has an OD of 14 mm and a wall thickness of 1.25 mm. The customer specified that the tubing could have no more than one black speck in 10 meters of tubing. This requirement created a problem because it was found that some virgin pellets contained black specks. The resin supplier tested their pellets and found that in a 2 kg sample there were 124 pellets with black specks. Figure 4.86 shows examples of discolored specks in pellets.

Figure 4.86 Discolored specks in pellets

Discolored specks are a common problem in extrusion. They are similar to gels, another common defect. Like gels, discolored specks are formed not only during processing but also in resin polymerization. So in order to get a handle on the problem, processors first need to figure out how many specks are in the incoming raw material, or the so-called p-specks. Then they need to solve specks caused during the extrusion process, or what we will refer to here as "e-specks".

It is not easy to test for black specks in virgin resin pellets because the pellets are quite small. A typical plastic pellet may have a diameter and length of about 3 mm, with a volume of about 20 mm^3. Assuming a density of 1 g/cm^3, the mass of a typical pellet will be about 20 mg. This means 2 kg of resin (about 4.4 lb) will consist of about 100,000 pellets. If a processor runs its extruder at 300 kg/h (about 660 lb/h), that means approximately 1.5 million pellets will pass through the extruder every hour.

Commercial instruments are available today that can analyze millions of pellets. Optical Control Systems (OCS) GmbH in Witten, Germany, is one company that produces such systems. OCS pellet defect detection systems are used by many resin producers. Therefore, most resin companies know how many defects occur in the pellets they produce. Figure 4.87 shows a pellet analyzer (PS-25C) from Optical Control Systems [198].

Figure 4.87 Pellet analyzer from Optical Control Systems [198]

If we know that we have one pellet with a discolored speck for every N pellets, it is possible to determine the average incidence of specks in the extruded product. If the cross-sectional area of the extruded product is A_e and the volume of the pellet or powder particle is V_p, the average length over which a p-speck will occur in the extruded product is:

$$L_{ps} = NV_p/A_e$$

More specifically, consider the extrusion of tubing with an outside diameter of 12 mm and a wall thickness of 1.25 mm. The cross-sectional area of this product is:

$$A_e = 0.25\pi(D_o^2 - D_i^2) = 0.25 \cdot 3.14 \cdot (144 - 90.25) = 42.22\ \text{mm}^2$$

The volume of the pellet is 20 mm^3. If we consider a plastic raw material that contains 124 pellets with a discolored p-speck for every 100,000 pellets (2 kg), that corresponds to 1 black speck pellet for every 806 pellets.

$$L_{ps} = 806 \cdot 20/42.22 = 382\ \text{mm}$$

With this input data, we can determine that the average length over which a p-speck will occur is 382 mm or 0.382 meters. If additional specks are generated in the extrusion process, the average length over which a speck occurs in the extruded product will be even less than 0.382 meters. Clearly, in this case it would not be possible to produce tubing with an average length between black specks of more than 10 meters.

In order to produce tubing with a spacing of black specks of more than 10 meters the number of p-specks in the raw material should be less than 1 black speck pellets in 4000 pellets. Since resin producers can scan millions of pellets for specks, they can remove pellets with specks from the pellet stream during production. In other words, the processor can specify the ppm level of pellets with specks that can be tolerated by the process, and the resin supplier can set up its parameters accordingly.

Clearly, sorting pellets this way will cost the resin producer money, which will almost certainly be passed on to you. Alternatively, processors can do this sorting in-house by buying a pellet scanning and sorting system on their own. Evaluate the overall cost; if in-house scanning and sorting is less expensive than the cost of buying pre-sorted pellets, it makes sense to go that route.

Pellets can be scanned and sorted not only for discolored specks but also for irregularities such as tails, see Figure 4.88. Such pellets can cause problems in extrusion and should be eliminated from the feed stream. De-dusting equipment can be used to eliminate tails, angel hair, and dust.

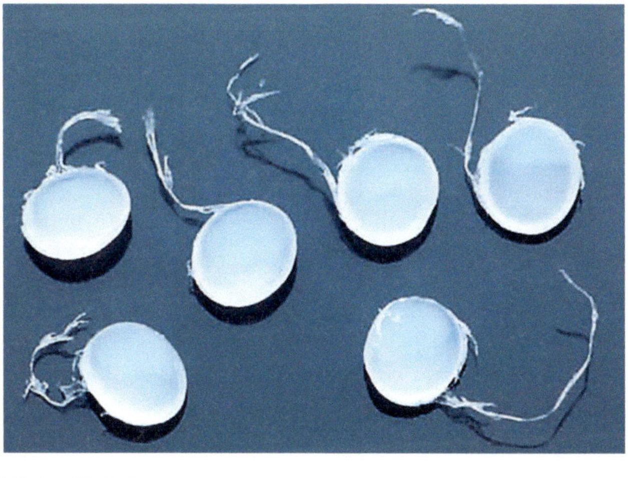

Figure 4.88 Pellets with tails

How to Avoid Specks Created in the Extrusion Process

Discoloration of the plastic inside the extruder can be caused by degradation, contamination, and several other things. Figure 4.89 shows a list of possible causes of specks generated in the extrusion process. By no means is this a complete listing. We will discuss one significant cause in more detail – screw wear.

Screw wear is a fact of life. The question is not if screw wear will occur, but when. The key question is at what point the outside diameter (OD) of the screw is being reduced by wear. More importantly, we need to know when wear has progressed to the point where it starts causing unacceptable problems. At this point the worn screw needs to be replaced with a new or refurbished one.

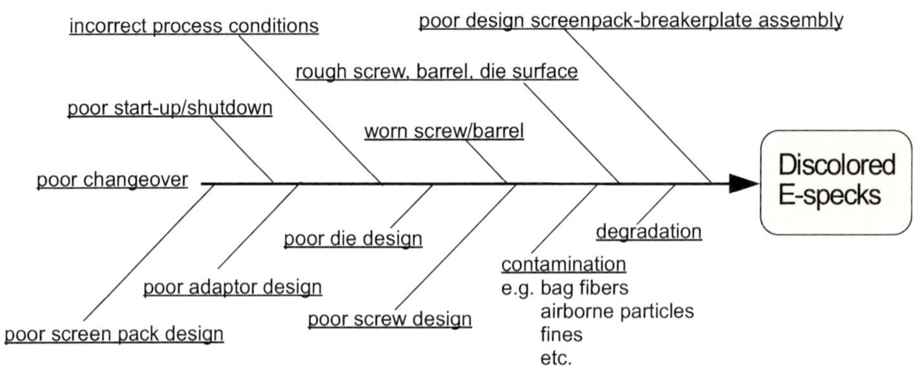

Figure 4.89 Fishbone chart for discolored specks formed in extrusion

In order to determine at what point the screw needs to be replaced, you will need to regularly measure the OD of the screw over its entire length. Special tools are available to do this; Figure 4.90 shows a screw flight micrometer. In a typical extrusion operation the screw and barrel should be measured at least once a year. Some processors do not monitor screw and barrel wear at all. These companies are putting themselves at risk, as they will have no idea how badly wear is affecting their process and product. When product quality starts deteriorating because of wear, they will not realize the appropriate corrective action.

This is being penny wise and pound foolish, because if you put yourself in this situation things will quickly get costly. A problem that could have been solved by installing a $15,000 spare screw may wind up costing you 10 to 100 times more if the solution to the problem is not clear.

Figure 4.90 Screw flight micrometer

Careful process monitoring will often allow processors to detect the effects of wear. The process parameters that are affected by wear are melt temperature, output, and motor load. As wear progresses, the quality of the extruded product tends to deteriorate. This may manifest itself as discoloration, streaks, gels, discolored specks, holes, etc. It is very important to monitor the specific energy consumption (SEC) and specific extruder throughput (SET) because changes in these parameters often correlate with wear.

SEC is the ratio of motor power divided by the throughput. It is the mechanical power consumed per unit mass of plastic extruded. The SEC tends to correlate with the melt temperature. SEC is normally expressed in kWh/kg. A typical value of the SEC is 0.25 (kWh/kg) for extrusion of polyolefins. SET is the output divided by screw speed; it is often expressed in kg/h/rpm. For instance, for a 75-mm extruder the SET may be 2 kg/h/rpm; this corresponds to 0.033 kg/rev. The SET is the amount of resin extruded per screw revolution.

Both SEC and SET are normalized performance parameters that allow comparison of data achieved at different process conditions, such as screw speed. As wear progresses the SEC values tend to increase and the SET values tend to decrease.

Besides losing energy efficiency, why should processors be concerned about screw and barrel wear? As wear progresses, the gap between the screw and barrel increases. This will reduce the pumping capability of the screw. However, the more critical issue may be the fact that the thickness of the stagnant melt layer on the barrel increases with wear. This will increase degradation inside the extruder and reduce extrudate quality.

Screw wear will create a thicker insulating melt layer at the barrel surface. This will inhibit heat transfer between the barrel and the melt in the screw channel. As a result, the control of melt temperature will be diminished and excessively high melt temperatures are more likely. This further increases the chance of degradation.

There are additional adverse effects of screw and barrel wear. Increases in clearance will reduce the ability of the screw flights to wipe the barrel surface. This will increase the average residence time and broaden the residence time distribution. As a result, changeover from one resin to another will take longer. Also, purging will take longer, as will startup and shutdown. This means that more scrap will be produced, downtime will increase, and there will be a corresponding reduction in uptime. This can have a significant negative effect on production cost.

Discolored specks are ubiquitous in extrusion. Controlling discolored specks starts with quantifying p-specks in the incoming raw material. Once the p-specks are controlled to an acceptable level, the e-specks can be addressed. Important factors that affect e-specks have been identified. One factor, screw wear, has been discussed in detail here.

■ 4.28 Mechanical Degradation in TPE Extrusion

Mechanical degradation is the reduction in molecular weight resulting from stresses imposed on the plastic. An interesting study of degradation in an extruder was published by Paakinaho et al [195]. Figure 4.91 shows the change in molecular weight (MW) along the length of a 19-mm laboratory single-screw extruder.

In this study three polylactic acid (PLA) resins were extruded on a small lab extruder. Polylactic acid or polylactide (PLA) is a biodegradable and bioactive thermoplastic aliphatic polyester derived from renewable resources. Samples were taken from the screw along the length of the extruder.

The figure shows that the low molecular weight PLA22 experiences only a small loss in MW. The medium MW PLA48 experiences a greater MW loss. The high MW PLA63 experiences the greatest loss in MW. In fact, the final MW of PLA63 is lower than the final MW of PLA48.

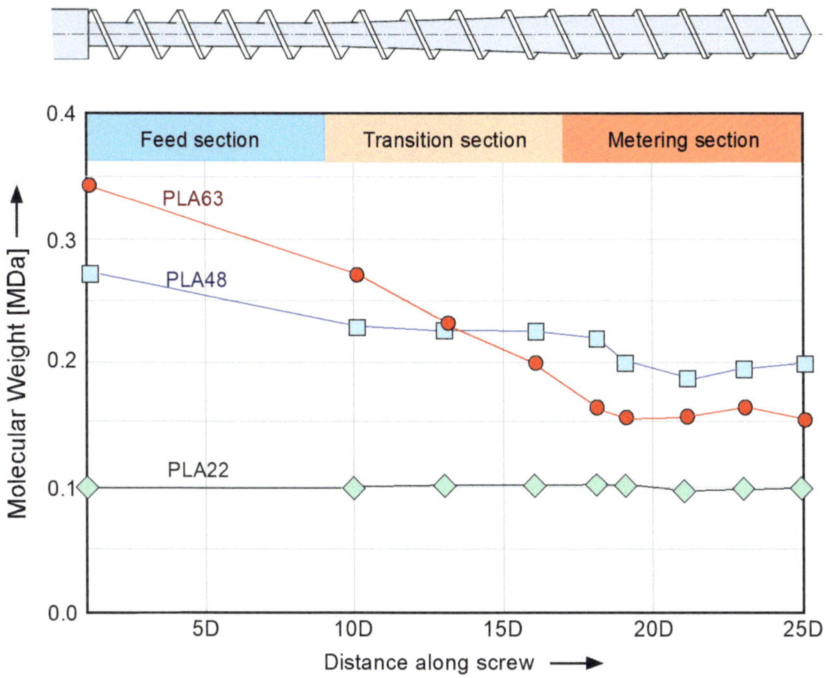

Figure 4.91 Molecular weight versus length of laboratory single-screw extruder [195]

Mechanical degradation can occur in single-screw extruders when the polymer molecules are exposed to high stresses [200, 201]. High stresses can occur in barrier screws and in fluted mixing sections. These screws have barrier flights with a small undercut. These screws are designed such that all the polymer melt has to flow over the barrier flight. Figure 4.92 shows a cross section of a barrier screw showing the solids channel, the melt channel, the main flight, the barrier flight, and the barrier undercut.

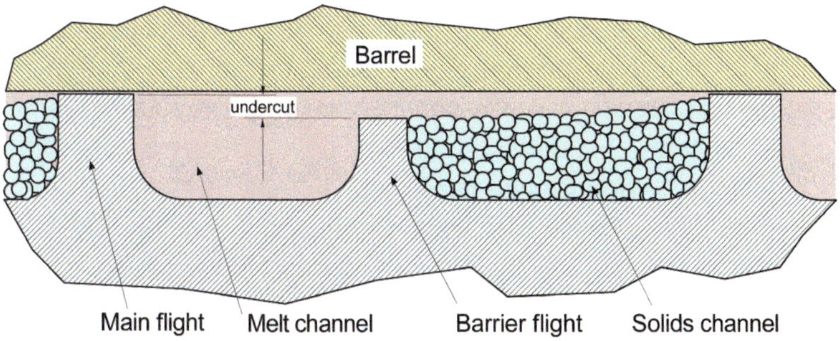

Figure 4.92 Cross section of barrier screw

The barrier undercut typically ranges from 0.5 mm to 2.0 mm depending on the size of the extruder. In the barrier gap the polymer melt is exposed to high shear stresses. In a 75-mm extruder with a channel depth of 5 mm and a barrier undercut of 1 mm the shear rate in the barrier gap will be five times greater than the shear rate in the channel. If the screw speed is 100 rpm, the barrier gap shear rate is 392.7 s^{-1} and in the channel 78.5 s^{-1}.

The shear rate in the flight clearance will be much higher. If the 75-mm extruder has a radial flight clearance of 0.10 mm, the shear rate in the flight clearance will be 3927 s^{-1}. This means that significant mechanical degradation is likely to occur in the flight clearance. However, the effect of this degradation will be small since only a small fraction of the polymer will be affected by this degradation. The reason for this is that only a small fraction of the polymer melt will actually pass through the flight clearance. On the other hand, all of the polymer melt will pass over the barrier flight.

A medical company was developing a new product made out of a thermoplastic elastomer. This company found that in their extrusion process they experienced significant degradation with the existing extruder screws to the point that the required molecular weight in the product could not be achieved. One existing screw was a barrier type screw; the other was a screw with a LeRoy/Maddock fluted mixing section.

The company contacted Rauwendaal Extrusion Engineering (REE) with a request to provide a screw that would minimize degradation. REE supplied a screw with a CRD422 mixing section; this is a multi-flighted slotted mixer with tapered slots. Figure 3.9 in Section 3.2.4 shows a picture of the CRD422 mixer. The taper of the slots create elongational flow as the polymer melt flows through the slots. It was found that only the CRD mixing screw was able to produce product with a molecular weight higher than the minimum requirement. Figure 4.93 shows the molecular weight versus screw design for three extruder screws.

One process variable that can provide information on molecular weight reduction in extrusion is the head pressure. The head pressure is dependent on the melt viscosity. The melt viscosity is strongly dependent on the polymer molecular weight. If the head pressure reduces at constant melt temperature, output, die temperatures, and die configuration, the most likely cause is reduced viscosity. If the melt viscosity reduces at constant melt temperature, output, die temperatures, and die configuration, the likely cause is reduction in molecular weight. Therefore, head pressure can be an indication of the polymer melt molecular weight.

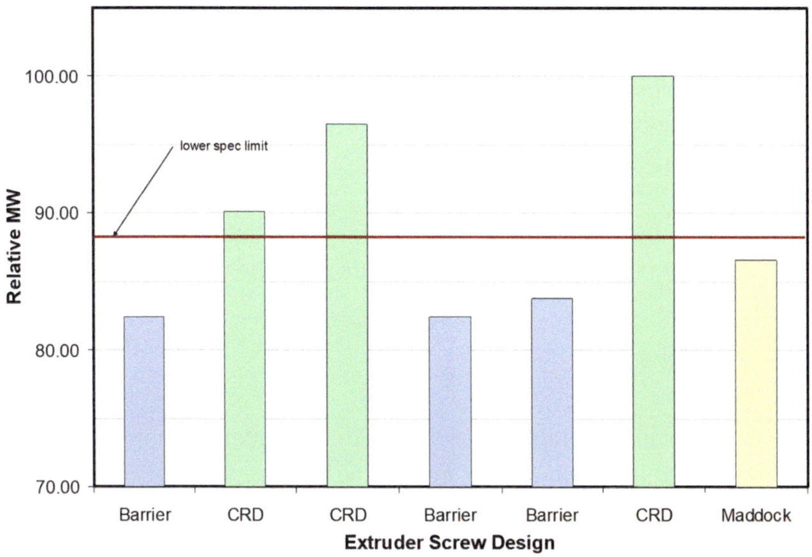

Figure 4.93 Molecular weight versus screw design

4.29 Degradation in a Long Adapter

This case study concerns a sheet coextrusion operation with a long adapter between the extruder and the sheet die. The output of the extruder is 26 kg/h, the internal diameter of the adapter is 40 mm, and the length of the adapter 2500 mm. The company had experienced a high level of black specks in the extruded sheet. The question that was asked was the following: Is it possible that long residence times in the adapter can cause black specks?

We can take a first look at the residence times by determining the mean residence time; this is the ratio of the volume occupied by the polymer melt divided by the volumetric flow rate. The volume can be calculated from $\pi R^2 L$, where R is internal radius (20 mm) and L the length of the adapter (2500 mm). The volume of the adapter is 3140 cm^3. The volumetric flow rate is the mass flow rate divided by the melt density. The mass flow rate is 26 kg/h or 7.22 g/s. If we take the melt density as 1 g/cm^3, the volumetric flow rate is 7.22 cm^3/s.

The mean residence time is volume/flow rate = 3140/7.22 = 434.9 s = 7.25 min. This is a long mean residence time, especially when we consider that a substantial fraction of the volume will have a longer residence time. If the induction time for thermal degradation is known, it is possible to predict the amount of polymer that will be degraded. We will use the variable t_{ind} for the induction time.

The velocities in the adapter can be determined from the flow rate and the viscosity of the polymer melt. We will assume that the melt viscosity can be described by the power law expression. This expression relates the melt viscosity (η) to the shear rate ($\dot{\gamma}$), the consistency index (m), and the power law index (n). The power law can be written as [176]:

$$\eta = m\dot{\gamma}^{n-1}$$

The velocities in a circular channel with radius R can be written as [176]:

$$v(r) = \frac{3n+1}{n+1}\bar{v}\left[1-\left(\frac{r}{R}\right)^{1+1/n}\right]$$

In this expression r is the radial coordinate, R is the internal radius, $\bar{v}$ is the average velocity, and n is the power law index. The average velocity is the volumetric flow rate ($\dot{V}$) divided by the cross-sectional area of the channel (A). For a circular channel the cross-sectional area is: $A = \pi R^2$. In this example the volumetric flow rate is 7.22 cm^3/s and the cross-sectional area is 12.566 cm^2. The average velocity becomes 7.22/12.566 = 0.575 cm/s.

The critical velocity is the velocity that produces a residence time equal to the induction time. The critical velocity can be written as:

$$v_{crit} = L / t_{ind}$$

In this example the critical velocity is 2500/1800 = 1.39 mm/s, if we assume that the induction time is 30 minutes (1800 seconds). We now want to determine the critical radius (r_{crit}); this is the radial position where the velocity equals the critical velocity. The critical radius can be expressed as follows:

$$r_{crit} = R\left[1-\frac{n+1}{3n+1}\frac{v_{crit}}{\bar{v}}\right]^{\frac{n}{n+1}}$$

At this point we can calculate the critical radius. If the power law index n = 0.3, the critical radius becomes r_{crit} = 1.9185 cm. For radial positions $r > r_{crit}$ the velocities will be lower than the critical velocity and the residence times will be longer than the induction time. That means that from r = 1.9185 to 2.0 cm the polymer will be degraded. Thus, a layer with thickness 0.0815 cm (0.032 inch) will degrade. This is a rather thick layer and cause for concern. We can determine how the degraded layer thickness changes with the radius of the adapter; this is shown in Figure 4.94.

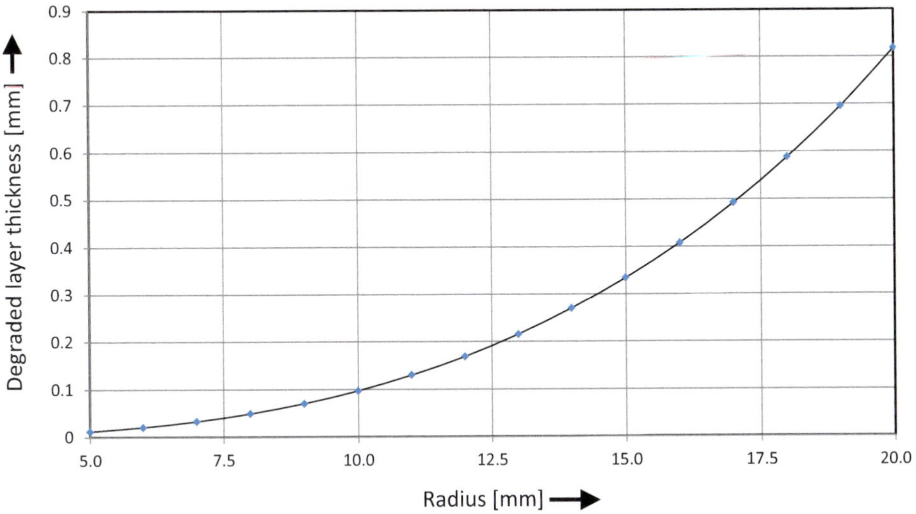

Figure 4.94 Degraded layer thickness versus adapter radius

The degraded layer thickness drops rapidly when the adapter radius is reduced. When the radius is reduced from 20 mm to 10 mm the degraded layer thickness reduces from 0.81 mm to just under 0.10 mm – a very significant reduction! These results suggest that a 10 mm adapter radius would be a much better choice than the 20 mm radius.

There is another concern that needs to be addressed. When the adapter radius is reduced, the pressure drop will increase. When we reduce the adapter radius we need to make sure that the pressure drop does not become excessively high. The pressure drop can be determined from the following expression [176]:

$$\Delta P = \frac{2mL}{R}\left[\frac{\left(3+\frac{1}{n}\right)\dot{V}}{\pi R^3}\right]^n$$

Figure 4.95 shows the pressure drop versus adapter radius. We assume that the consistency index m = 1000 Pa·s^n.

We see that the pressure drop at 20 mm radius is 0.3 MPa. When the adapter radius is reduced to 10 mm, the pressure drop increases to 1.1 MPa. This is still a low pressure drop and this would indicate that the adapter radius of 10 mm is a good choice. The pressure drop is directly proportional to the consistency index. Therefore, for different values of the consistency index the pressure drop can be found easily by multiplying with the ratio of consistency indices.

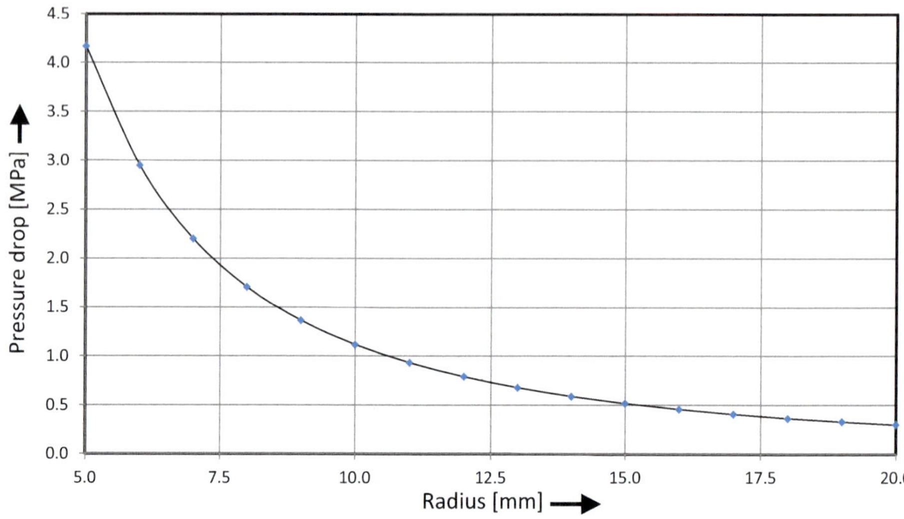

Figure 4.95 Pressure drop versus adapter radius

Conclusions

Longer adapters have to be sized carefully to avoid excessive degradation. The internal diameter is the most critical design parameter. In this section we have developed a method to predict the thickness of the degraded polymer layer. In this example, the thickness of the degraded polymer layer was 0.815 mm with an induction time of 30 minutes. A degraded layer thickness of about one millimeter is too large and will result in defects in the extruded product. In a properly designed adapter the degraded polymer layer thickness should be less than 0.1 mm. The design must also take into account the pressure drop in the adapter.

4.30 Shrink Voids in Rod Extrusion

This case study involved problems with shrink voids in solid rod extrusion. In this case a solid rod was extruded out of several polymers such as polyethylene, polypropylene, and polyether ether ketone. The rod diameter was 6 mm. It was found that shrink voids limited the line speed that could be achieved in the extrusion operation. Shrink voids are caused by the cooling process. For this reason, the cooling process was analyzed in detail to determine how to optimize cooling with respect to shrink voids.

In extrusion of thick-walled products, such as rods and solid profiles, shrinks voids occur readily because of the non-uniform temperatures and shrinkage that occur in the cooling process. The outer wall of the profile cools rapidly while the inside is still hot and molten. This creates a solid rigid outer wall with the inside still hot, see Figure 4.96. As the inside of the profile cools and increases in density the stiff outer wall will pull the material from the inside to the outer wall. This will create one or multiple shrink voids in the center of the profile.

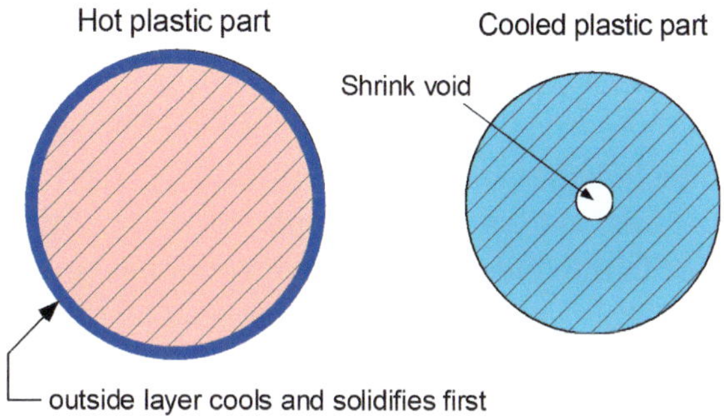

Figure 4.96 Cooling and formation of shrink void

Shrink voids can be avoided by reducing the line speed. However, reducing the line speed reduces the production rate and, therefore, this is an unattractive solution of the shrink void problem. A more effective way of solving the problem is to modify the cooling process such that shrink voids are avoided. One way of inhibiting shrink void formation is by using staged cooling rather than using a long, continuous cooling bath. In staged cooling multiple cooling baths are used, allowing adjustment of the distance between the individual cooling units. It is also possible to maintain the various cooling baths at different temperatures. This allows greater control of the cooling process and provides significant flexibility.

When a short first cooling tank is used followed by a second cooling bath, the outer wall of the extrudate will be cool when it exits the first cooling tank. However, the outer wall will heat up very quickly as heat from the inside starts heating the outside. This creates a temperature inversion that reduces the temperature differences within the extrudate – it is like an annealing step. As the outer wall heats up it will soften and reduce the chance of shrink void formation.

In this section we will perform an analysis of the cooling process in rod extrusion with the following objectives:

1. To study the cooling process using one long cooling bath
2. To define cooling conditions that can lead to shrink voids
3. To study alternate layouts of the cooling line to prevent shrink voids

The goal is to achieve optimal design of the cooling line so that shrink voids can be avoided at normal line speeds – in other words, without having to sacrifice production rates.

Shrink Void Formation in Rod Extrusion

Shrink voids are formed when the extrudate cools after exiting the extrusion die [204, 205]. As soon as the strand enters the water bath the strand surface temperature drops quickly. Within a few centimeters the surface temperature will drop below the melting point of the polymer and a solid skin will form. The thickness of the solid skin grows as cooling continues along the length of the water bath. As the solid skin grows thicker it becomes rigid and resists further reduction in diameter as cooling continues.

The inner region of the extrudate will still be molten and significant volume reduction will occur as the inner region cools down. The outer solid skin will cause the polymer in the inner region to shrink away from the center. This can lead to a single large shrink void in the center or multiple shrink voids near the center. Multiple shrink voids are likely to form when the melt temperature distribution in the center region is not uniform.

The onset of shrink void formation will occur when the solid skin becomes thick and rigid enough to resist deformation due to shrink stresses acting on the skin. It may be possible to define a critical solid skin thickness for preventing voids formation. The critical skin thickness will likely be of the order of 1 mm. If we assume a critical solid skin thickness of 1 mm, the time it takes to reach this thickness will be about 1 second given a typical value of the thermal diffusivity of 10^{-7} m^2/s. At a line speed of 30 m/min (0.5 m/s) this critical skin thickness is reached about 1 meter from the entrance of the cooling bath.

The goal in cooling is to avoid reaching the critical solid skin thickness early in the cooling process. One way to do this is by using air cooling. This method is used sometimes in extrusion of high temperature polymers such as PFA, PPS, PES, PEEK, and others. Another way is to use staged water cooling. This involves using multiple cooling baths rather than one long one. The cooling baths are separated from each other by non-cooled sections. It is also possible to use hot water instead of water at room temperature or chilled water.

In staged cooling of the non-cooled sections, the cold skin will heat up rapidly because the hot inner region will transfer heat to the outer region. At the same time, the solid skin will reduce in thickness. There is a temperature reversal taking place in the outer region in the non-cooled section. It is like a short-term annealing process where the heat from the inner region is used to anneal the outer region. At the end of the non-cooled section the temperature distribution in the rod will be more uniform than at the start. The result is more uniform cooling and reduced thermal stresses in the product.

The cooling process can be studied using software developed for this purpose by SHS Plus in Oberhausen, Germany; this software is called chillWARE® [189–191]. This simulation software allows detailed analysis of the progress of cooling along the length of the cooling line. The temperature distribution at each cross section can be calculated knowing the cooling conditions and the thermal properties of the polymer.

It would make sense to simulate the current cooling process. This will tell us the growth of the solid skin along the cooling bath. With this information we can analyze a staged cooling process designed to keep the solid skin thickness below the critical value in the first cooling bath. The first non-cooled section (annealing section) should be long enough to reduce the solid skin thickness sufficiently and increase the surface temperature to a high enough level to soften the polymer. Another variable that can be controlled is the water temperature of each cooling bath. It may be beneficial to maintain the first cooling bath a high temperature (50–80 °C) to slow down the growth of the solid skin.

The first cooling bath and annealing zone are the most critical part of the cooling process when it comes to the formation of shrink voids. We can simulate a number of different configurations of the staged cooling process. When the most promising configuration has been determined by simulation, it makes sense to do a number of actual trials on the extrusion line.

Simulation with chillWARE®

The cooling simulation software chillWARE® was used to simulate the cooling process [189–191, 203–205]. This software can simulate cooling in various extrusion operations, such as pipe, sheet, profile, and rod extrusion. Such a simulation program can be helpful in understanding, troubleshooting, and optimizing the cooling process. Figure 4.97 shows the rod temperatures along the length of the cooling line.

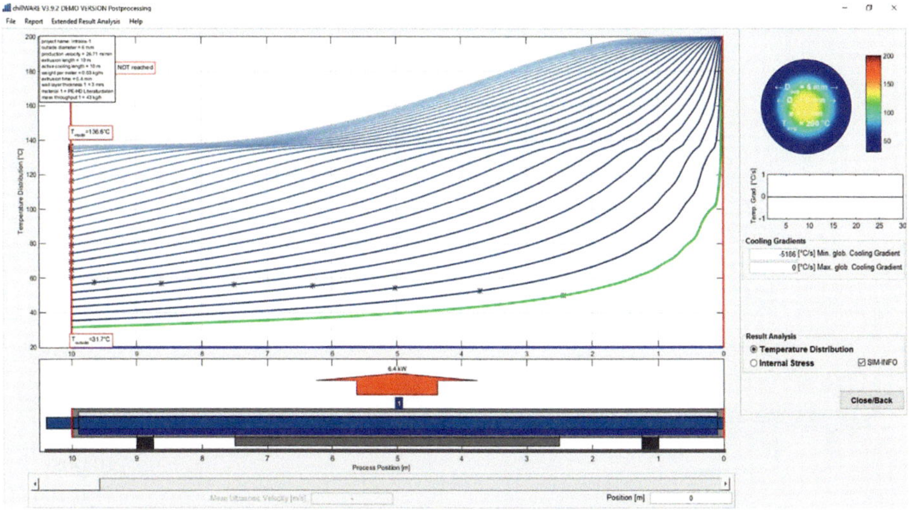

Figure 4.97 Rod temperatures along the length of the cooling bath for HDPE

The results are for a 6-mm HDPE rod extruded at a rate of 43 kg/h using a 10 meter long cooling bath. The output rate of 43 kg/h corresponds to a line speed of 26.7 m/min. The initial rod temperature is 200 °C; it is assumed that the initial rod temperatures are uniform. In a real extrusion operation the initial temperatures may not be uniform because of the low thermal conductivity of molten polymers.

The graph shows the temperature of different layers of the rod along the length of the cooling bath. The green curve represents the temperatures at the outside wall; the top curve represents the temperatures in the center of the rod. Clearly, the rod temperatures are not uniform at the end of the cooling line.

The inlet temperatures are shown on the right and the exit temperatures on the left. The color contour plot, top right, shows the temperatures in the rod cross section at the end of the cooling bath. The wall temperature (green line) drops quickly in the first 100 mm from 200 °C to 100 °C. The final wall temperature is about 31.7 °C. The temperature in the center does not change until after about 1 meter. Beyond this point the temperature drops slowly to reach a final temperature of 136.6 °C.

If we take the melting point of HDPE at 125 °C, it is clear that at the end of the cooling line the inner region is still molten. Figure 4.98 shows the rod temperatures along the length of the cooling line for polypropylene (PP). The initial rod temperature is 220 °C. The final rod temperature is 25.8 °C at the wall and 120.1 °C in the center. Clearly, the PP rod cools more quickly than HDPE. Because PP is extruded at higher temperatures than HDPE, there will be greater temperature differences in the cooling of PP, causing faster cooling. Also, HDPE has a greater enthalpy rise from 20 °C to 200 °C than PP.

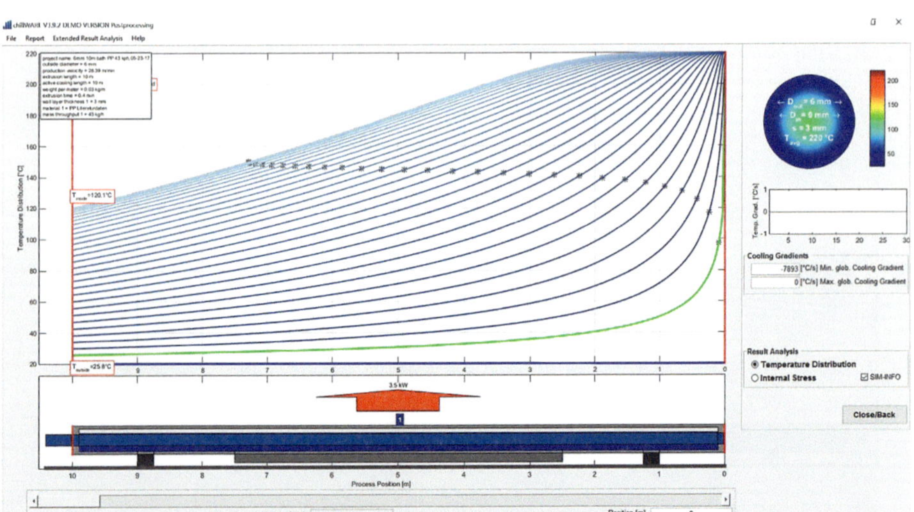

Figure 4.98 PP rod temperatures along a single long cooling bath

If we take the melting point of PP at 160 °C, it is clear that the PP rod is completely solidified at the end of the cooling line. In fact, the rod has solidified completely after about 6 meters in the cooling tank. The results from chillWARE® can be exported to Excel so that the data can be graphed in many different ways, see Figure 4.99.

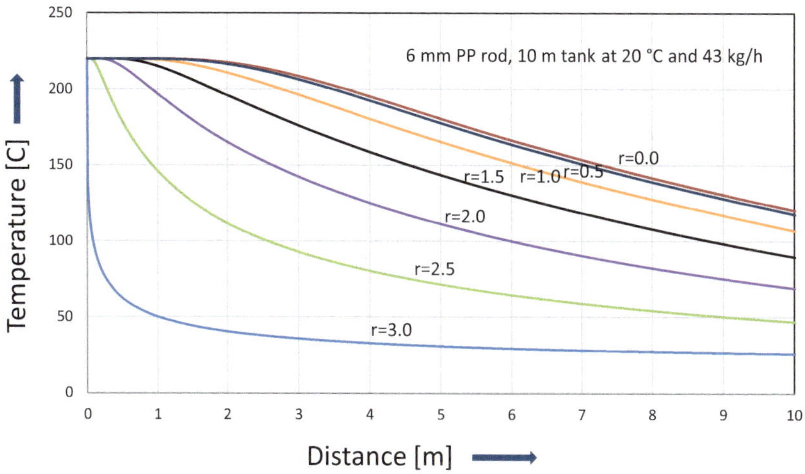

Figure 4.99 PP rod temperatures along a single 10 m long cooling bath

Figure 4.99 shows the outer (r = 3 mm) temperatures in blue, other radial positions are shown in different colors. The temperature differences diminish toward the center of the rod.

Figure 4.100 shows the time to solidification for the different layers; the solidification temperature is taken as 160 °C.

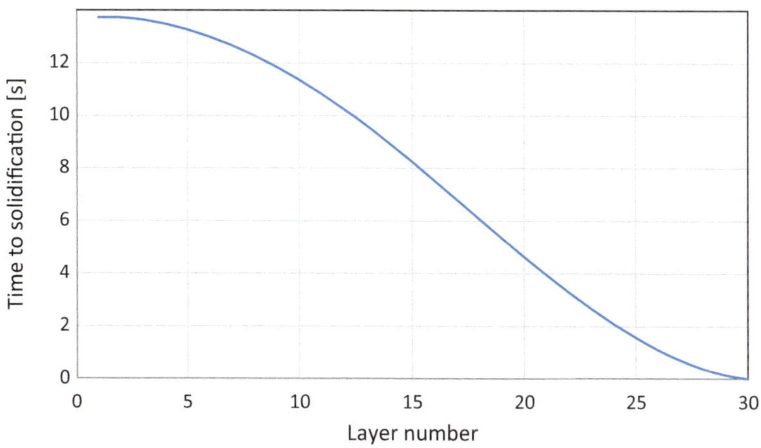

Figure 4.100 Time to solidification for the different layers

The outer layer (number 30) cools down to 160 °C almost immediately, in milliseconds. The inner layers cool down more slowly. The layers close to the center take about 14 seconds to cool down to 160 °C. At 30 m/min line speed a cooling time of 14 seconds corresponds to a distance of 7 meters.

Figure 4.101 shows the radial temperature profiles for 10 axial positions starting at 1 meter from the start of cooling and ending at 10 meters at the end of the cooling bath. The layers close to the center, number 1 through 10, cool down very slowly. The first ten layers (1 mm) do not change within the first 1 meter of the cooling bath. The layers close to the outside diameter cool down rapidly. Almost the entire temperature drop occurs within the first 1 meter of the cooling bath and there is little change in temperature over the next 9 meters.

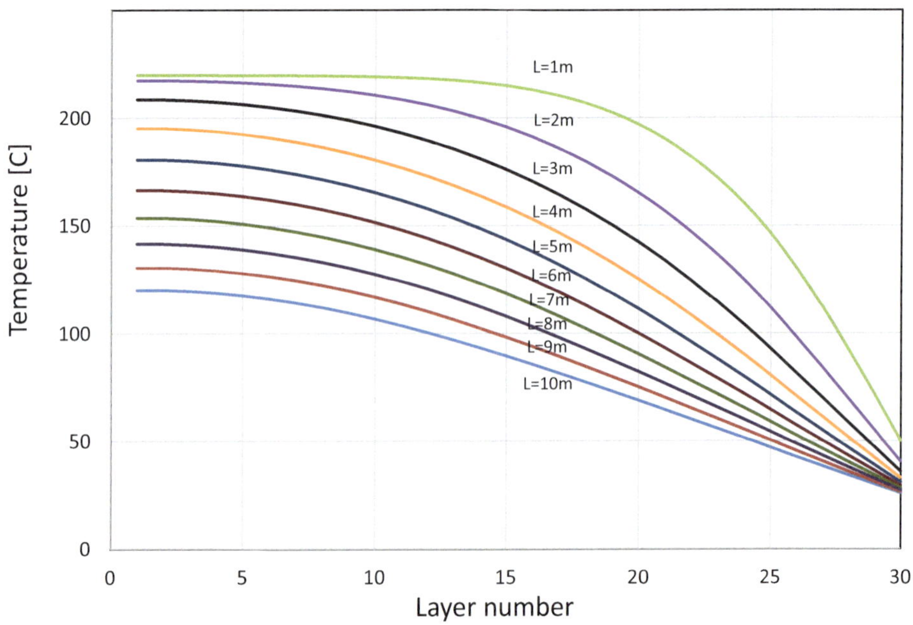

Figure 4.101 Radial temperature profiles for 10 axial positions

Figure 4.102 shows a cooling arrangement with two cooling baths. The first cooling bath is 2.9 m long and the second cooling bath is 7.65 m long. The air space between the first and second bath is 0.8 m. With two cooling baths separated by an air space the rod temperatures are significantly different from those in a single long cooling bath. The final rod temperature is 25.3 °C at the wall and 109.6 °C in the center. When the rod exits the first cooling tank, the wall temperature quickly increases from 36 °C to almost 90 °C. The hot inner layers heat up the outer layers by conduction. The air space between the first and second cooling bath acts like an annealing section.

It should be noted that the increase in wall temperature in the air space is not high enough to reach the melting point. This is significant with regard to shrink void formation. As the outer layers solidify they form a rigid skin. As the inner layers cool down they will shrink toward the outer rigid skin, thus creating conditions for shrink voids to form. In order to avoid shrink voids the initial cooling process has to be retarded. It should also be noted that the predicted wall temperatures in the air space between the water baths can be verified experimentally by using infrared temperature measurement.

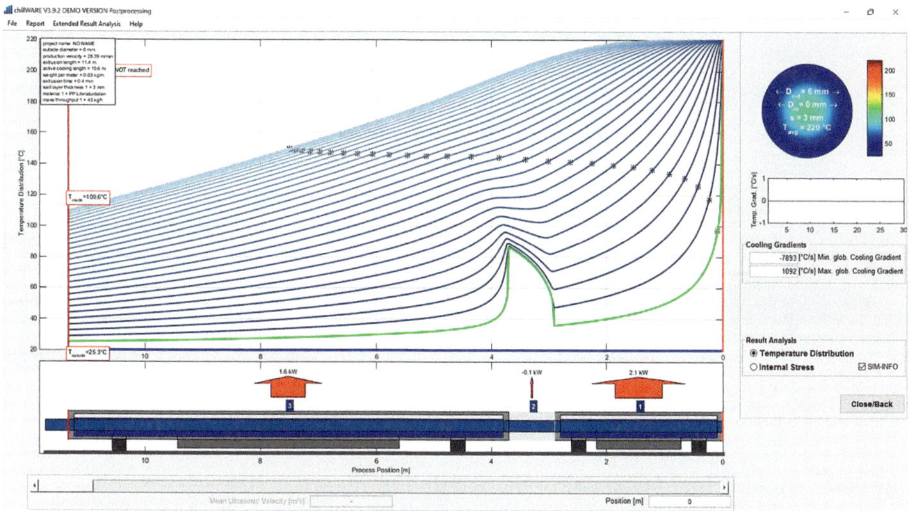

Figure 4.102 PP rod temperatures along two cooling baths (2.9–0.8–7.65)

After 1 meter the solid layer is 0.6 mm thick. Actual extrusion experiments have shown that the shrink voids form around 500 to 700 mm from the entrance of the first cooling bath. After 0.5 meters in the first cooling bath the solid layer is 0.4 mm thick, after 1 meter the solid layer is 0.6 mm thick. At the end of the first cooling bath the solid layer is 1.2 mm thick. These results suggest that the critical solid layer thickness is around 0.4 mm.

Slower initial cooling can be achieved by using a shorter first cooling bath and also by maintaining the water at a temperature above room temperature. Figure 4.103 shows the rod temperatures using a 1 meter first cooling bath followed by a 1 meter air space and 9 meter second cooling bath. Both water baths are maintained at 20 °C.

The solid layer is 0.6 mm at the end of the first cooling bath. If we assume the critical solid layer thickness to be 0.4 mm, the first cooling bath is too long to keep shrink voids from forming. Figure 4.103 shows that the maximum wall temperature in the air space is now much higher – about 130 °C. A higher wall temperature will result in a softer outer layer of the rod reducing the chance for voids to form. The modulus of polypropylene can reduce as much as an order of magnitude when the temperature increases from 20 to 100 °C.

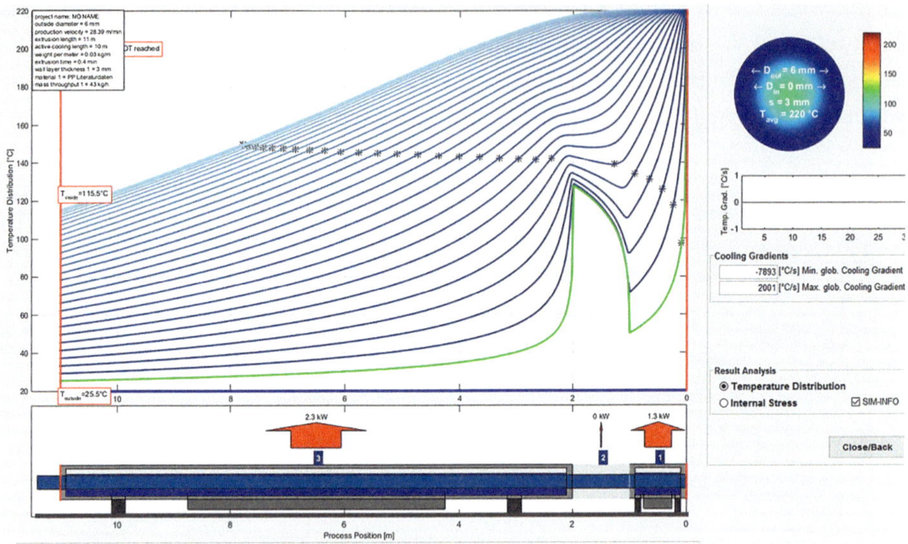

Figure 4.103 PP rod temperatures along two cooling baths (1.0–1.0–9.0)

The color contour plot is shown in more detail in Figure 4.104.

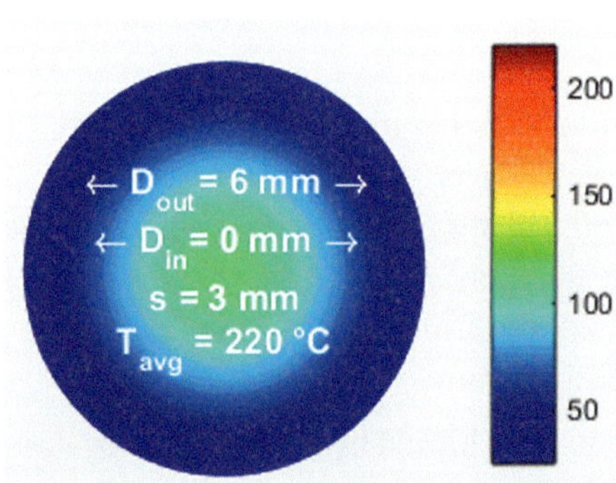

Figure 4.104 Color contour plot from Figure 4.103

Figure 4.105 shows the rod temperatures when the first cooling bath is maintained at 60 °C and the second at 20 °C. In this case the solid wall temperature at the end of the first cooling bath is 0.45 mm – slightly thicker than the assumed critical solid wall thickness. For that reason the length of the first bath should be reduced.

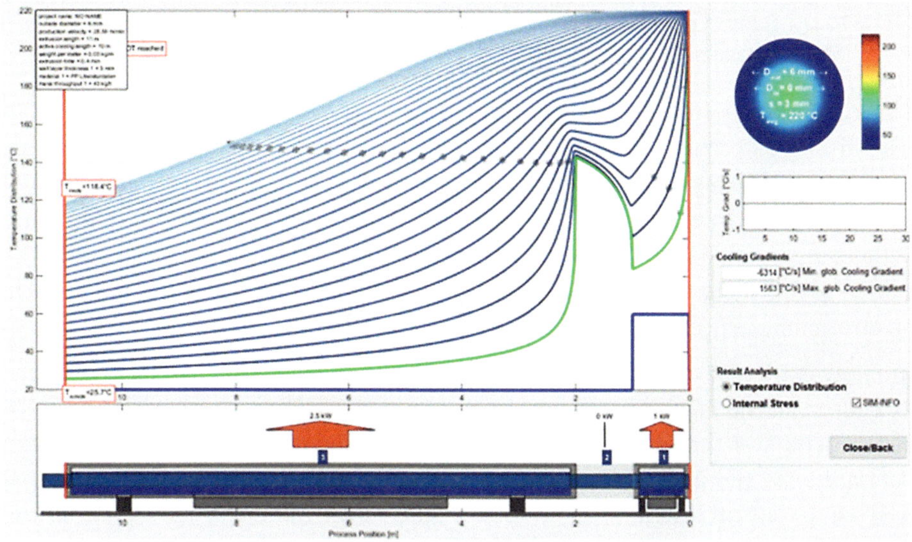

Figure 4.105 PP rod temperatures along two cooling baths (1.0–1.0–9.0)

As expected, with the higher water temperature (60 °C) the wall temperature at the exit of the first water bath is now higher. The wall temperature is now about 85 °C compared to about 50 °C when the water is maintained at 20 °C. The maximum wall temperature in the air space is now about 145 °C – this approaches the melting point. At these temperatures it is likely that the outer skin is soft enough to keep shrink voids from forming.

It is important to make sure that in the air space between the water baths the wall temperature does not increase too close to the melting point of the polymer. This would result in stretching of the rod and causing a change in rod diameter. It is also possible that the rod would break.

Conclusions

These results indicate that a short first water bath maintained at high water temperatures can create cooling conditions that reduce the chance of shrink void formation. In fact, with these conditions no more shrink voids could be observed. In the actual process several cooling parameters can be made adjustable. The distance between the water baths can be easily adjusted if the baths are located on tracks and equipped with wheels. The cooling water temperature can also be adjusted. The cooling bath can be made with removable partitions so that the length of the bath can be adjusted. It is even possible to make one water bath into two, or more, shorter water baths separated by an air space. A cooling line with a flexible configuration is useful to perform cooling experiments so that the cooling process can be optimized with respect to avoidance of shrink voids.

4.31 Improper Preheating of Extruders

This case study pertains to preheating of extruders [203]. Improper preheating can lead to serious problems, such as breaking the extruder screw upon startup. This study was undertaken as a result of work with a company that had encountered multiple screw breaks during startup. This company was running polyether ether ketone, a very high temperature polymer that is processed close to 400 °C. In the course of the study, it was found that the actual preheating time was too short to ensure complete preheating of the extruder.

In many extrusion operations the extruder is shut down with polymer still in the extruder and the machine cools down to room temperature. Often, the slide valve at the bottom of the hopper is closed and the screw turns slowly until no more polymer exits the discharge end of the extruder. This procedure may empty the feed section of the extruder; however, in most cases the metering section of the extruder remains partially full with molten polymer.

Improper preheating of an extruder after a shutdown can cause a number of problems. Insufficient preheating can cause damage to the screw or the die can blow off the extruder. It is one of many reasons why it is critical for operators to watch the extruder very carefully during startup. A few seconds of inattention can cause serious damage to the machine and, even worse, serious accidents.

During cooldown the polymer can pull away from the barrel. When this happens, the required preheating times can increase several times the normal preheat times. As a result, problems with improper preheating can occur quite easily. Excessive torsional load on the screw can damage the coating and hard-facing; especially when the coating or hard-facing is harder than the base material of the screw – this is often the case.

When the polymer is in contact with a layer of high temperature air for an extended period of time, this will lead to significant oxidative degradation. In fact, this situation presents ideal conditions for oxidative degradation. As a result, when the extruder is restarted, degraded polymer will be extruded for a considerable period of time – this can continue over a period of several hours!

This problem can be mitigated by not shutting down power to the extruder completely and leaving the barrel and die temperatures at a temperature between room temperature and process temperatures. This procedure will reduce the chance of air gaps forming and degradation of the polymer. In addition, the preheat times will reduce significantly if the barrel and die temperatures are kept below process temperatures at a level where no significant degradation occurs.

This section will analyze the heating process of extruders from room temperature so that the required heating time can be calculated. Measures that can be taken to minimize the chance of problems on startup are listed. Also, results of computer simulations of the extruder heating process will be presented.

What are Appropriate Extruder Shutdown Procedures?

Extruders generally can be stopped for a short period of time without negative consequences upon restarting. A short time period is defined here as five to ten minutes. Longer time periods are more likely to be detrimental. One important issue is that stopping the extruder will change the thermal conditions in the machine. When screw rotation stops, there are no more solid polymer pellets entering the extruder – ending the cooling effect of the cold pellets. As a result, the screw and pellet temperatures in the feed section will increase because heat from the discharge end of the extruder will travel to the feed end.

When the feed section of the screw heats up, a melt film can form where previously there was no melt film. When the extruder restarts and cold pellets enter again into the screw channel, the feed section of the screw will cool down and the melt film that was formed on the screw during the stoppage will solidify. This will create high friction conditions on the screw surface and this will reduce the solids conveying capability.

High friction at the screw surface can create conveying problems. In some cases, the problem will correct itself after a short time as regular thermal conditions are reestablished in the extruder. In other cases, the problem persists and it may be necessary to pull the screw to remove the solidified layer on the screw in the feed section. Obviously, this can cause long downtimes and significant scrap.

Another issue with a stoppage is long exposure of the polymer to high temperatures. For thermally stable polymers the extruder can be stopped for more than 30 minutes without serious consequences. An example of such a polymer is low density polyethylene (LDPE). However, for a less thermally stable polymer a 30 minute stoppage will create sufficient thermal degradation so that problems occur upon restarting the extruder. Even with LLDPE, a close family member of LDPE, a 30 minute stoppage will create problems with degradation.

In some cases, rather than stopping screw rotation completely, it may be better to set the screw speed at a very low value so that the polymer continues to move. For this situation it can be helpful to have a diverter valve between the die and extruder. Such a valve would allow work on the die to be done while the screw is still turning. With a diverter valve the flow through the die can be stopped completely.

Preheating the Barrel and Screw

The barrel heating time depends primarily on the heating capacity of the barrel heaters, the mass of the barrel, and the presence of insulation. The time to heat the barrel can be 15 to 30 minutes on smaller extruders (less than 50 mm) and over one hour for larger extruders (larger than 100 mm).

If the total barrel heater capacity is Q and the mass of the barrel M_b, then the barrel heating time will be:

$$t_b = \frac{C_p M_b \Delta T}{Q} \tag{4.1}$$

where C_p is the specific heat of the barrel material and ΔT the temperature rise.

As an example we will use a 90-mm extruder with a barrel 2700 mm long and a thickness of 50 mm. In this case the volume of the barrel will be 0.0594 m^3. If the density of the steel is 8000 kg/m^3, the mass of the barrel will be 475.2 kg. The specific heat of steel is about 0.5 kW·s/kg°C. If the temperature rise ΔT = 200°C and the total heating capacity is 30 kW, the barrel heating time will be t_b = 1584 s (26.4 minutes). This barrel heating time assumes no heat losses. In reality, of course, there will be heat losses and the actual barrel heating time will be about 40–45 minutes.

The time to heat the screw can be determined from the same expression except that the mass of the barrel has to be replaced by the mass of the screw (M_s). If we take the screw length as 2700 mm and the root diameter 70 mm, the volume of the screw will be about 0.01 m^3. With a density of 8000 kg/m^3 the mass of the screw is 83 kg. With a temperature rise of 200 °C and a heat capacity of 30 kW, the screw heating time will be t_s = 276.7 s (4.6 minutes).

Without thermal losses the total time to heat the barrel and the screw will be about 30 minutes. With losses the heating time will be about 45–60 minutes.

Preheating the Polymer without Air Gap

During normal preheating there is contact between the barrel and the polymer inside the barrel. The time required to bring the polymer up to the barrel temperature can be calculated from the Fourier number. When the Fourier number is one, the polymer will have reached relatively uniform temperatures. The Fourier number can be expressed as:

$$N_{Fo} = \frac{\alpha t}{H^2} \tag{4.2}$$

where α is the thermal diffusivity, t the heating or cooling time, and H the thickness of the part being heated or cooled.

In single-sided heating, the time needed to achieve more or less uniform temperatures becomes:

$$t_{eq} = H^2 / \alpha \tag{4.3}$$

A typical value of the thermal diffusivity is 10^{-7} m^2/s. Figure 4.106 shows how the time to thermal equilibrium depends on the thickness. Clearly, there is a power law relationship between time and thickness – this is expressed by Equation 4.3.

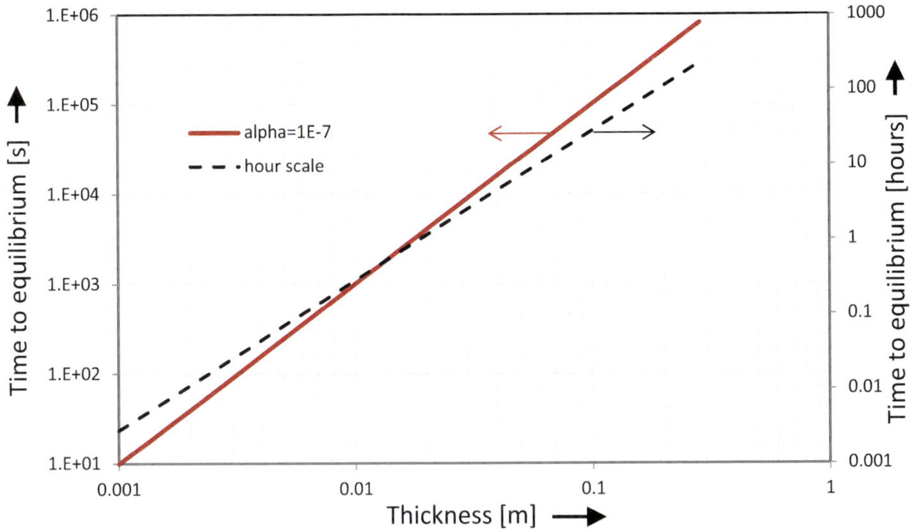

Figure 4.106 Time to thermal equilibrium versus thickness

A typical channel depth in the metering section is $0.1D$, where D is the diameter of the extruder barrel. This means that for a 50-mm extruder it will take about 250 seconds (4.17 minutes) to heat the polymer in a 5-mm deep screw channel. For a 150-mm extruder it will take 2250 seconds (37.5 minutes) to heat the polymer in a 15-mm deep channel. For a 90-mm extruder it will take 810 seconds (13.5 minutes) to heat the polymer in a 9-mm deep channel. If we add the barrel heating time to the polymer heating time, we get to a total heating time of about 55 minutes or just under 1 hour.

Preheating with Air Gap

If the polymer shrinks away from the barrel during cooling, there will be an air gap between the barrel and the polymer. This air gap will inhibit heat transfer from the barrel to the polymer because the thermal conductivity of air is much lower than that of a polymer. As a result, the preheat time will be increased significantly.

The heating rate of the polymer will be determined by the heat flux through the air gap. The heat flux is determined by the Fourier's law that states that the heat flux through the air (q_a) equals the thermal conductivity of air times the temperature gradient in the air gap.

$$q_a = \frac{k_a \left(T_o - T_i\right)}{\delta_a} \tag{4.4}$$

where k_a is the thermal conductivity of air.

The temperature gradient is determined by the temperature difference between the outside layer of air (T_o) and the inside layer of air (T_i) divided by the air gap thickness δ_a.

If we assume that the temperature difference is 200 °C and the air gap is 0.001 m, the temperature gradient is 200,000 °C/m. With the air thermal conductivity 0.024 J/ms°C, the heat flux becomes:

$$q_a = 0.024 \cdot 200{,}000 = 4800 \text{ W/m}^2$$

This corresponds to 4.8 kW/m². The surface of the polymer annulus (A) is the circumference (πD) times the length (L). If we take the diameter $D = 0.090$ m and the length $L = 0.01$ m, the surface area becomes:

$$A_t = \pi \cdot 0.090 \cdot 0.01 = 2.827\text{E}{-3} \text{ m}^2$$

The total amount of heat (Q_p) arriving at the polymer now becomes:

$$Q_p = q_a A_t = 4800 \cdot 2.827\text{E}{-3} = 13.57 \text{ W}$$

The temperature rise of the polymer (ΔT_p) will be determined by the specific heat of the polymer (C_p), the length of heating time (t_h), and the mass of polymer to be heated. The mass of polymer is determined by the volume (V_p) and the polymer density (ρ_p). The volume of the polymer annulus over a length of $L = 0.01$ m with outside diameter $D_o = 0.088$ m and inside diameter $D_i = 0.070$ m is:

$$V_p = 0.25\pi(D_o^2 - D_i^2)L = 2.2337\text{E}{-5} \text{ m}^3$$

The rate of temperature increase will slow down as the polymer heats up because the heat flux reduces as the polymer temperature goes up. With a reducing heat flux the time to heat up the polymer will increase.

For the polymer to achieve the desired process temperature, the screw has to heat up as well. Taking this into account, the polymer heating time can be expressed as:

$$t_h = \left(C_p \rho_p V_p + C_{ps} \rho_s V_s\right) \Delta T / Q_p \qquad (4.5)$$

with C_p = 1180 W·s/kg°C, ρ_p = 900 kg/m³, V_p = 2.2337E−5 m³, C_{ps} = 500 W·s/kg°C, ρ_s = 8000 kg/m³, V_s = 3.7E−5 m³, Q_p = 4.8 kW/m², the heating rate becomes 12.72 s/°C.

If the polymer temperature needs to be raised 200 °C, it will take 2544 seconds (42.4 minutes) at a constant heat flux of 4.8 kW/m². Obviously, the heat flux will not be constant; it will diminish as the polymer starts heating up. If we take the average heat flux to be 2.4 kW/m², the heating time will become 84.8 minutes – almost 1.5 hours.

If we add to this a barrel heating time of 45 minutes, we get to a total heating time of 130 minutes or 2.17 hours for the 90-mm extruder when an air gap forms between the barrel and the polymer. This compares to a total heating time of 55 minutes (0.917 hours) for preheating without an air gap. This indicates that formation of an air gap can more than double the time to heat up an extruder from room temperature to process temperature.

Air Gap Formation

Air gaps can form because the thermal contraction of the polymer is much greater than that of the steel of the barrel and screw. The linear and volumetric coefficient of thermal expansion for several materials is shown in Table 4.23.

Table 4.23 Values of Linear and Volumetric Coefficient of Thermal Expansion for Various Materials

Material	Linear coefficient at 20 °C	Volumetric coefficient at 20 °C
Units	10^{-6} °K^{-1}	10^{-6} °K^{-1}
PP	150	450
PVC	52	156
Steel	11–13	33–39
Stainless steel	10–17	30–52

Clearly, the thermal expansion coefficient for polymers is much greater than for steel – for semi-crystalline polymers the thermal expansion is about 10 times greater than steel. The thermal expansion of semi-crystalline polymers like PP is greater (about 3 times) than that of amorphous polymers like PVC. Figure 4.107 shows the specific volume versus temperature for PEEK KT820 (courtesy Solvay). The specific volume of PEEK at 390 °C is about 22% greater than the volume at room temperature.

During cooldown of an extruder filled with polymer it is likely that air gaps will form because the thermal contraction for semi-crystalline polymers is 5 to 20 times greater than that of steel. The volume change from melt to solid for semi-crystalline polymers ranges from 15–25% and from 5–10% for amorphous polymers. The typical temperature change from process temperature to room temperature for commodity polymers is about 200 °C. Over this temperature range the thermal contraction of steel is less than 1%. For engineering polymers and high temperature polymers the temperature change ranges from 200 to 400 °C.

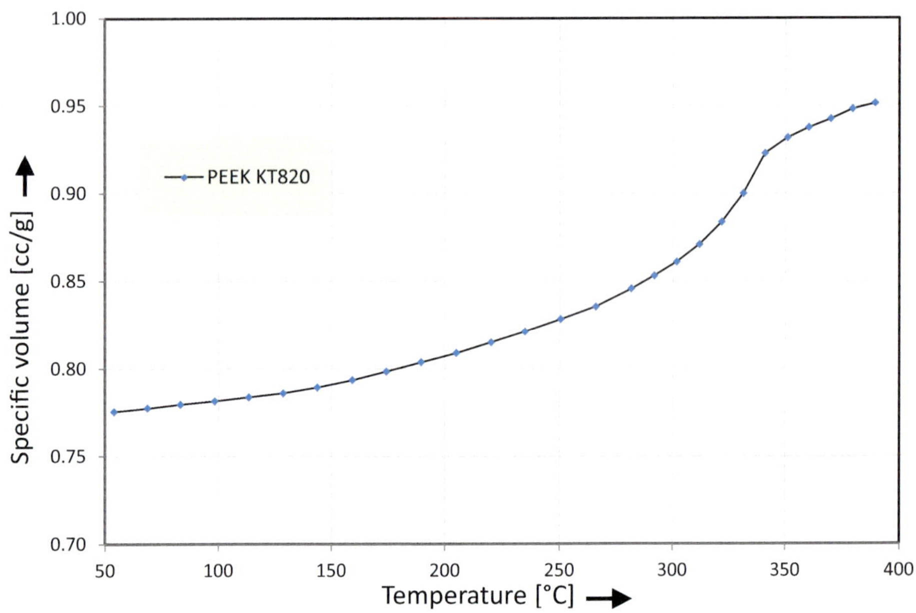

Figure 4.107 Specific volume versus temperature for PEEK KT820 (courtesy Solvay)

Air gaps can form either between the polymer and the screw or between the polymer and the barrel. In either case, the air gap will lengthen the preheating process because of the low thermal conductivity of air. Air gaps can form in the cooldown process but also during preheating. In the preheating process the barrel heats up before the polymer and the screw. The barrel expands as it heats up; this happens before the polymer heats up. As a result, an air gap may form during preheating. It is more likely that an air gap forms during cooldown than during preheating because the barrel expansion during preheating is much smaller than the polymer shrinkage during cooldown.

Because of the large difference in thermal contraction between the polymer and the steel of the barrel, the probability of an air gap forming is quite large. For that reason, to be on the safe side, the preheating procedure should be based on the assumption that an air gap is present. The thickness of the air gap is likely to be 10–20% of the channel depth of the extruder screw. Therefore, if the channel depth is 9 mm, the air gap is likely to be 1 mm or greater.

Instrumentation

Good instrumentation is essential in determining whether sufficient preheating has taken place. Unfortunately, most extruders do not have instrumentation that provides an accurate reading of the extent of preheating. The progress of preheating is typically judged by the barrel temperatures. Unfortunately, barrel temperatures do not provide a good indication of the extent of preheating. As discussed earlier, barrel temperatures increase before the polymer and the screw heat up. This is especially true when shallow-well thermocouples are used.

As a result, barrel temperatures can reach process temperatures long before the polymer heats up. If the extruder is started shortly after barrel temperatures reach process temperatures, it is likely that the polymer is not up to the required temperature and problems will occur. The temperature of the screw provides a better indication of the extent of preheating than the barrel temperature. Unfortunately, most extruders are not set up to measure screw temperature.

Inner machine temperatures are more useful in determining the extent of preheating than outer machine temperatures. This is also true for the extrusion die. For hollow products it is good practice to incorporate temperature sensors in the torpedo and spider supports. These sensors provide a good indication of melt temperatures during extrusion. These sensors also provide a good indication of the extent of preheating. As discussed earlier, the time to preheat the die can be substantially longer than the time to preheat the extruder. This will be discussed further in the discussion section.

Influence of Polymer Properties

The temperature rise directly affects the time required to heat up the extruder – the higher the temperature rise, the longer the preheating of the extruder will take. For commodity polymers this temperature rise is about 150–200 °C. For engineering polymers the temperature rise will be about 200–300 °C. For high temperature polymers, such as PEEK, the temperature rise will be 300–400 °C. Problems with insufficient preheating, therefore, are more likely to occur with high temperature polymers. The specific volume of PEEK versus temperature is shown in Figure 4.107.

For the same extruder a 400 °C temperature rise will take at least twice as long as an increase of 200 °C. In reality, a 400 °C temperature increase will take more than twice as long because the heat losses will increase at higher temperatures. This will depend on the thermal insulation that is available. If no insulation is used, heat losses will be substantial and heating times will be longer. With insulation heat losses can be reduced significantly and heating times will be shorter.

Figure 4.108 shows the heat flux in an extruder for three cases. Case 1 is no insulation, case 2 is 25 mm insulation, and case 3 is 38 mm insulation. The graph shows clearly that insulation reduces the heat flux dramatically – the reduction is by a factor of ten or higher! Barrel and die insulation usually represent one of the low hanging fruits when it comes to reducing power consumption and process variation. Die insulation is often more important than barrel insulation because die temperature variation tends to have a stronger effect on the extrusion process than barrel temperature variation.

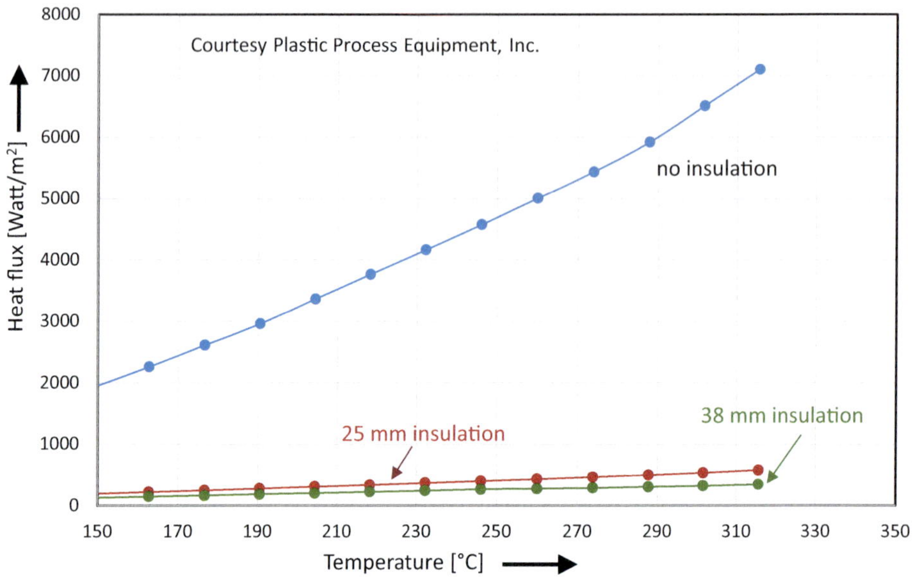

Figure 4.108 Heat flux in an extruder with and without barrel insulation (courtesy Plastic Process Equipment, Inc.)

Figure 4.109 shows the reduction in power consumption that can be achieved with barrel insulation. The reduction in power consumption in most cases is over 30% and can be as high as 50%.

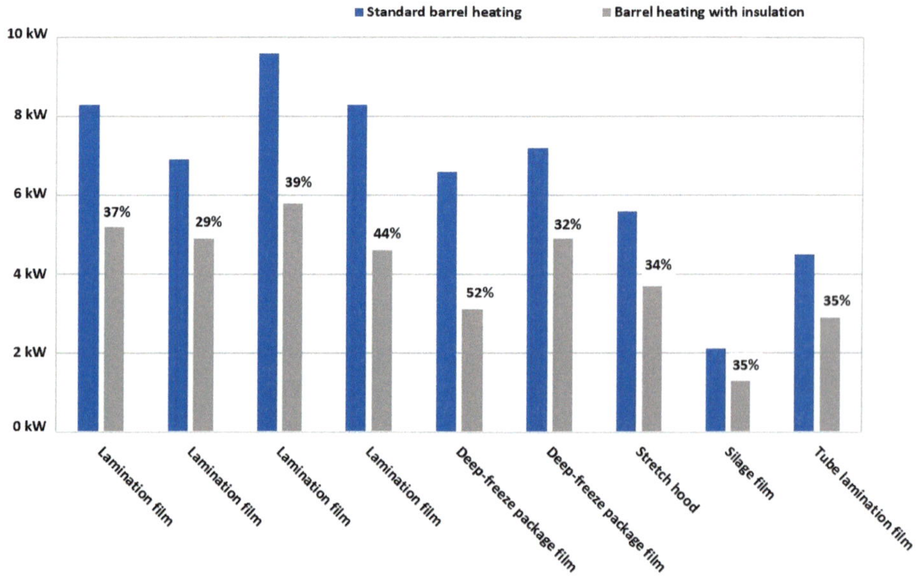

Figure 4.109 Power consumption with and without barrel insulation (courtesy Reifenhäuser-Kiefel)

Other important polymer properties that play a role in the heating and cooling process are the specific heat and the enthalpy. The specific heat is the energy needed to increase the temperature of a substance. The enthalpy is the total heat content of the substance. Figure 4.110 shows the specific heat and enthalpy of PEEK versus temperature (courtesy Solvay).

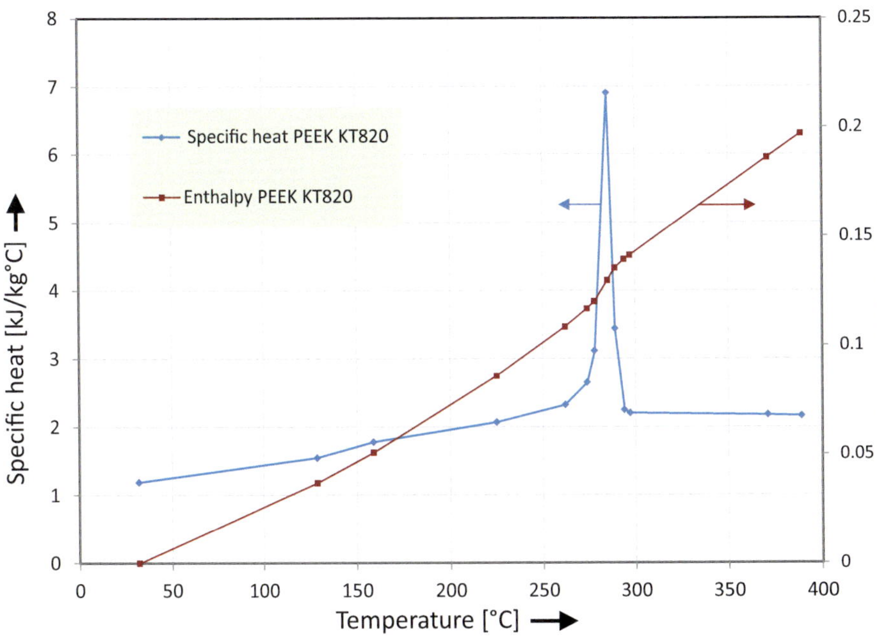

Figure 4.110 Specific heat and enthalpy of PEEK versus temperature (courtesy Solvay)

The data was determined at a cooling rate of 0.33 °C/s. As a result, the solidification peak has shifted to lower temperatures. Data obtained under heating would show a melting peak at higher temperatures.

The enthalpy is graphed relative to room temperature; in other words, the enthalpy at room temperature is taken as zero. The enthalpy rise of PEEK from room temperature to 390 °C is approximately 0.2 kWh/kg. From this enthalpy rise, the minimum specific energy requirement for PEEK becomes 0.2 kWh/kg. When an extruder processes PEEK at a rate of 100 kg/h, the minimum power requirement will be 20 kW (= $0.2 \cdot 100$). In extrusion essentially all energy introduced to the polymer is used to raise the temperature from feed temperature (usually room temperature) to process temperature. As a result, the enthalpy rise of the polymer is a measure of the minimum power requirement in extrusion.

Figure 4.111 shows the thermal conductivity and thermal diffusivity versus temperature for PEEK KT820.

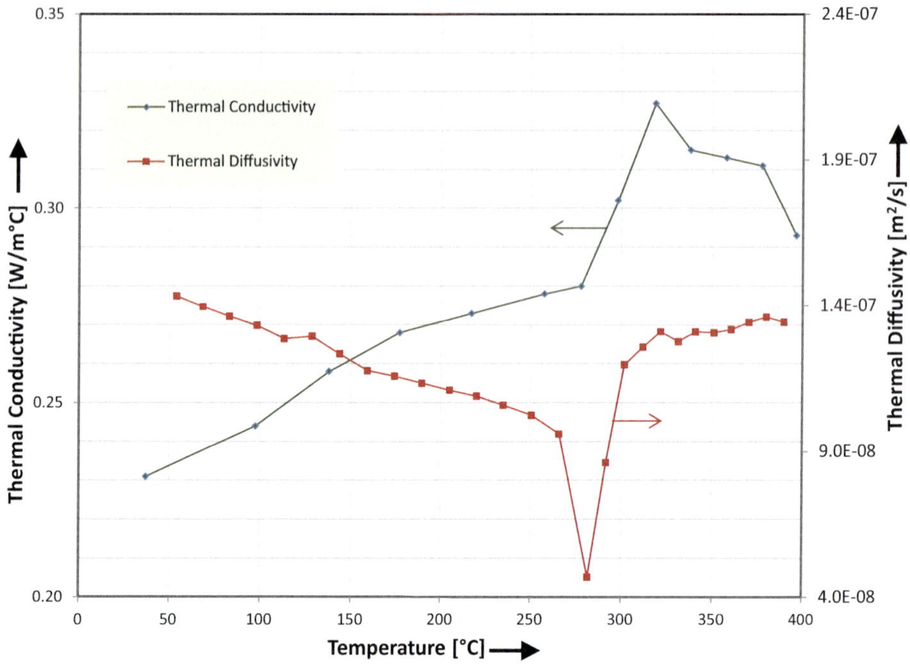

Figure 4.111 Thermal conductivity and diffusivity versus temperature for PEEK KT820 (courtesy Solvay)

The thermal conductivity increases from room temperature to the melting point. Beyond the melting point the thermal conductivity reduces with temperature. The thermal diffusivity reaches a minimum value at about 280 °C; this is the same temperature where the specific heat reaches a maximum, see Figure 4.110.

Computer Simulation

The heating process was simulated using commercial software chillWARE® developed by SHD plus GmbH in Germany. This software has a wire coating module that was used to simulate the heating of the polymer. In this case the conductor is actually the root of the screw. The thickness of the polymer was taken as 9 mm and the root diameter as 72 mm.

The initial polymer temperature was set at 20 °C with the polymer coming into contact with a heat transfer medium at 380 °C. Simulations were performed with and without an air gap between the barrel and the polymer. The polymer used in the simulation is PEEK KT820. This is one of the highest temperature polymers; preheating is more critical with this material than with lower temperature polymers.

Four layers were included in the analysis of heating with an air gap. The first layer is the steel barrel, the second layer is the air gap, the third layer is the polymer, and the fourth layer is the steel screw. Only three layers were used in the analysis of heating without an air gap.

Table 4.24 shows the dimensions of the layers and the material of each layer.

Table 4.24 Information on the Layers Used in the Simulation

Layer number	OD [mm]	ID [mm]	Material	Layer thickness [mm]
1	192	92	ST37 steel	50
2	92	88	air	2
3	88	70	PEEK KT820	9
4	70	0	ST37 steel	35

Figure 4.112 shows the temperatures at various radial positions versus time with a 2-mm air gap over a two hour period. Each unit on the horizontal scale represents 10 minutes. After 2 hours, the minimum polymer temperature is 306.3 °C and the maximum is 375.9 °C. The melting point of PEEK KT820 is 340 °C. The results indicate that the polymer has not fully melted after two hours.

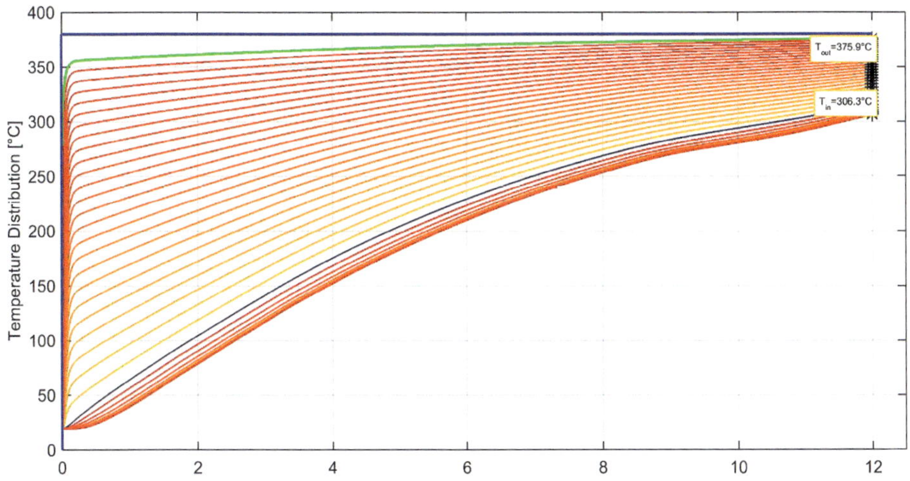

Figure 4.112 Temperatures at various radial positions versus time with air gap

Figure 4.113 shows cross-sectional color contour plots in 30 minute increments. The color contour plots provide a clear visualization of the progress of heating over the 2-hour heating time.

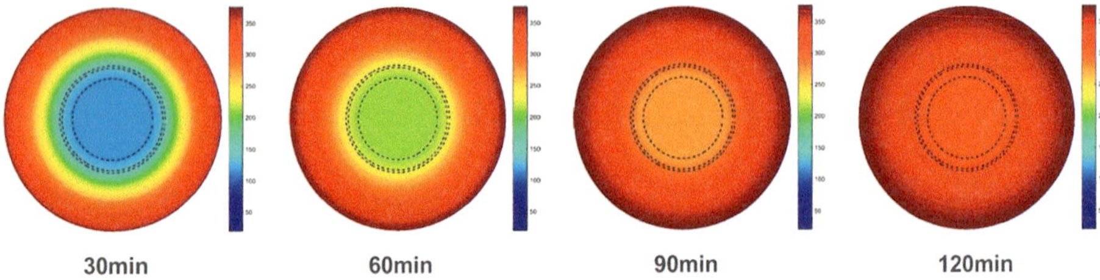

Figure 4.113 Cross-sectional color contour plots in 30 minute increments

Figure 4.114 shows the temperatures at various radial positions versus time without an air gap. After two hours, the minimum polymer temperature is 334.1 °C and the maximum is 377.5 °C. In this case, the polymer has almost completely melted. Within the same time frame the progress of heating without an air gap is substantially further than with an air gap.

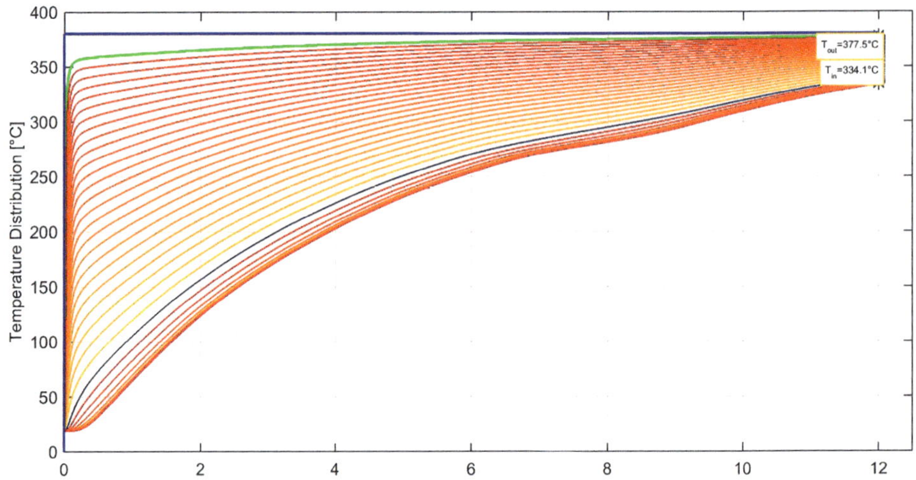

Figure 4.114 Temperatures at various radial positions versus time without air gap

Figure 4.115 shows the temperatures at various radial positions versus time with a 2-mm air gap over a two hour period with internal screw heating. The screw is maintained at 380 °C.

The polymer temperature at the screw increases to 377.4 °C after two hours, while the polymer temperature at the barrel increases to 366 °C. This figure indicates that internal screw heating can improve the heating process significantly.

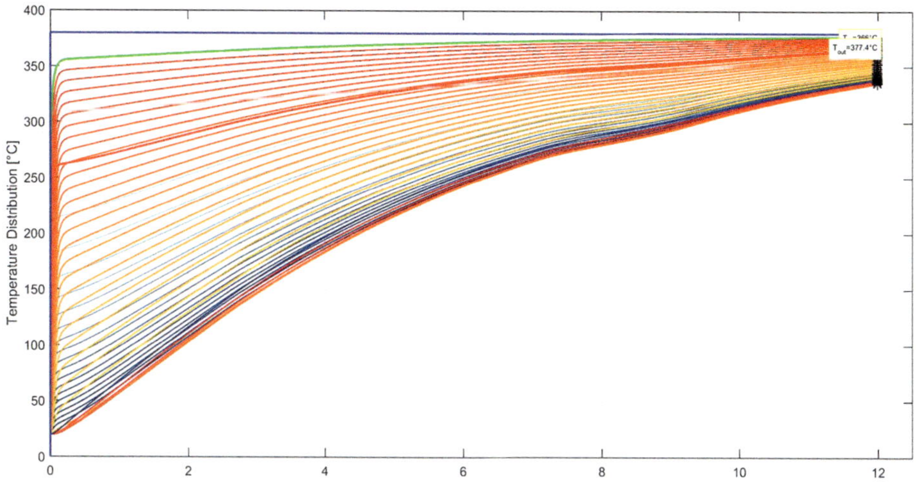

Figure 4.115 Temperatures at various radial positions vs. time with air gap and screw heating

Discussion

A complete extruder cooldown can be detrimental in several respects:

- It leads to long preheat times
- It leads to oxidative degradation of the polymer
- Increases chance of insufficient preheating; this can lead to:
  - Screw breakage
  - Die blowing off the extruder
  - Damage to screw hard-facing
  - Damage to screw coating

It may make more sense to keep the extruder at a temperature just above the polymer melting point. This will reduce preheat times and degradation. Cooling down below the melting point results in rapid volume shrinkage. In semi-crystalline polymers, the most rapid volume shrinkage occurs just below the melting point as shown in Figure 4.107; this shrinkage is about 5 times faster than at other temperatures.

Internal screw heating can reduce the preheat time significantly. Internal screw heating can be done with a cartridge heater with a built-in temperature sensor. This temperature reading of the screw actually provides a better indication of the extent of preheating than the barrel temperatures. During steady-state operation internal screw heating provides benefits as well; some of these are:

- Improved melt conveying when heating the melt conveying zone
- Reduced screw torque and motor load
- Increased control of the extrusion process

For preheating the extruder from room temperature, the following rule of thumb can be used: $t_h \sim 0.02D^2$, with D in mm and time in minutes. This would lead to a preheat time of 7.5 hours for a 150-mm extruder when an air gap has formed, and more than 13 hours for a 200-mm extruder. For a small 25-mm laboratory extruder the preheat time will be about 12.5 minutes. These numbers indicate that a complete cooldown for large machines may not make sense if the machine is down for only one or two days.

Figure 4.116 shows temperatures at startup for a 25-mm extruder [199].

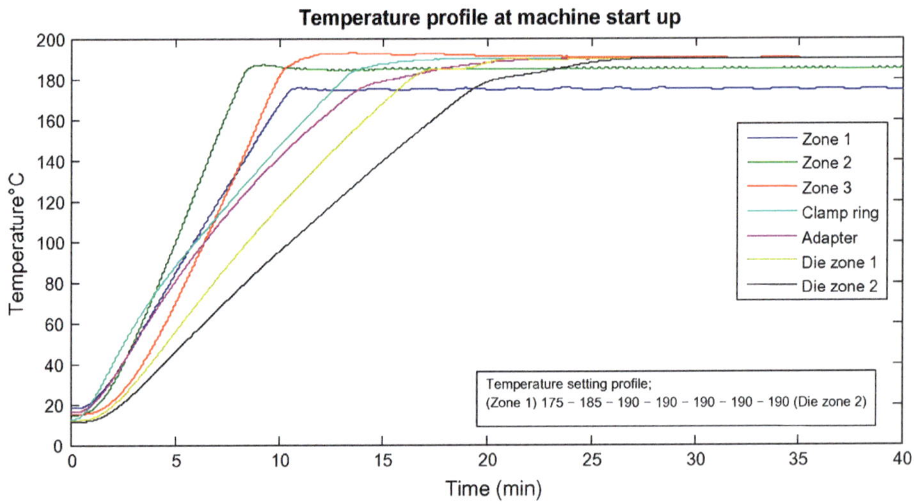

Figure 4.116 Temperatures versus time at startup for 25-mm extruder, courtesy Applied Energy Journal [199]

Figure 4.116 indicates that the three barrel temperatures reach setpoint after about 10 minutes. The die temperatures, however, take about 15–20 minutes. It is important to realize that die and adapter temperatures can take longer to heat, sometimes substantially longer, than the extruder barrel. Obviously, this depends on the die configuration and the heating capacity of the die heaters. Proper preheating procedures should take into account the heating characteristics of the extruder, the polymer, and the extrusion die and adapter.

If the tip of the extruder screw is not located close to the end of the barrel, this can create a large empty space. When this space is filled with polymer, it can take a very long time to heat this large volume of plastic. We can use the Fourier number to estimate the time needed to heat up this volume of plastic. For a 90-mm extruder and a polymer diffusivity of 10^{-7} m^2/s, the heating time will be about 20,000 seconds or 5.6 hours – see also Figure 4.106. This is much longer than the time needed to heat other parts of the extruder.

At the end of the extruder barrel the heating of the polymer will be faster when a breaker plate is used. The breaker plate is typically made of steel and steel has a thermal conductivity about 100 times greater than that of polymers. As a result, the breaker plate will improve the heating of the polymer by distributing the heat through the bulk of the polymer in close proximity to the breaker plate more effectively.

Conclusions

Preheating an extruder after shutdown can be problematic if the extruder filled with polymer is allowed to cool down to room temperature. For that reason, cooldown to room temperature is not always a good strategy. It may be better to cool down the extruder to a temperature just above the polymer melting point for semi-crystalline polymers, or just above the glass transition temperature for amorphous polymers. Internal screw heating can reduce preheating times and the chance of startup problems.

The required preheating time for a typical single-screw extruder when starting at room temperature can be estimated by the following rule of thumb: $t_h \sim 0.02D^2$. In this expression, the diameter D is expressed in mm and the preheating time in minutes. For very high temperature polymers the required preheating time can be significantly longer because much higher temperatures need to be reached.

Preheating times for large extruders can become very long; for instance more than seven hours for a 150-mm extruder. Obviously, if the extruder temperatures during shutdown are maintained just above the melting point, the preheating times will be significantly shorter, at least three times shorter.

Preheating times for the die and adapter may be longer than for the extruder. For that reason, the die and adapter have to be given equal importance with regard to establishing proper preheating procedures.

Barrel and die insulation reduce heat losses and power consumption. In preheating, barrel and die insulation will shorten preheat times and reduce process variation. Die insulation generally has a greater effect on process stability than barrel insulation. Considering the low cost of barrel and die insulation, the advantages greatly outweigh the disadvantages.

In this case study the problem with screws breaking upon startup was eliminated when proper preheating procedures were put in place.

A possible cause of high oil temperatures on water-to-oil cooled systems is poor heat exchange owing to scale deposits. Commercial scale treatments can significantly improve heat-exchange efficiency.

A2.2 Unusual Noises

Excessive noise from the input, intermediate, and thrust shaft bearings indicate a worn part that will only cause more damage if not replaced quickly. The same is true for abnormal sounds from the gears, oil pump, and drive belts.

It is important to know which bearing to replace in the case of bearing problems. If you hear a cyclical noise, time the number of cycles per turn of the thrust shaft. If the noise cycles about five times for each revolution of the thrust shaft, the problem is probably an intermediate pinion or a high speed gear or bearing on the intermediate shaft. If the noise cycles about twenty times for each revolution of the thrust shaft, this indicates a problem on the high speed input shaft. Vibration monitoring can provide a more detailed analysis of the drive system.

A2.3 Vibration Monitoring

Vibration monitoring is a technique that has proven itself very useful in analyzing machine problems. Vibration-monitoring systems analyze the sounds of rolling element bearings and meshing gear teeth to determine if unusual sounds are present, thus indicating impending failure of machine elements. These systems are often used on very large extruders involved in post-reactor finishing and large-volume compounding.

Vibration-monitoring systems are useful for at least two purposes. First, they can alert plant personnel to a cracked bearing or broken gear before a catastrophic failure occurs, which helps to minimize repair cost and downtime. It is possible to detect a gearbox problem a month before the problem would result in failure. Second, monitoring allows maintenance to be performed only when needed. No production time is lost disassembling a gearbox that does not need to have any parts replaced. Considering the high cost of downtime on large extruders in the range of 200 to 400 mm (8 to 16 in), it can be cost-effective to use a vibration-monitoring system to minimize machine downtime. Figure A2.1 shows some components of a vibration-monitoring system.

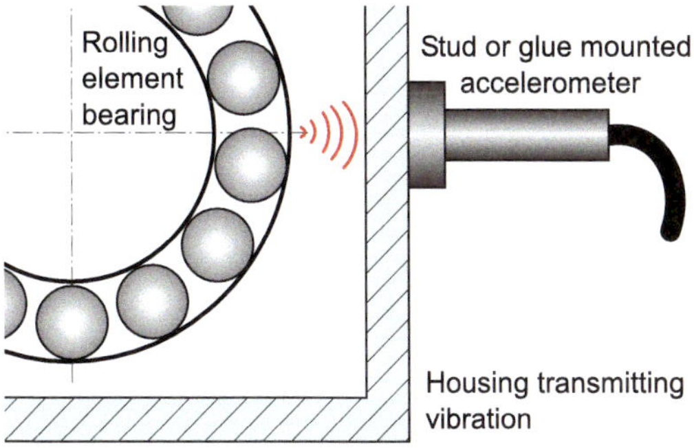

Figure A2.1
Components of vibration-monitoring system

The sensing element is an accelerometer; this can be stud-mounted or glued to the housing. The main element of the vibration-monitoring system is the analysis module. One supplier, MTC, has developed a system specifically for gear monitoring. This system can graphically display which gear tooth is pitted, chipped, or cracked, using a vibration analysis technique called high frequency envelope (HFE). Standard low-frequency spectrum analysis (0–10 kHz) is effective in detecting imbalance, misalignment, and looseness in rotating machinery. HFE filters out everything below 25 kHz, which makes it possible to detect small, high-frequency impact sounds caused by faulty bearings.

As the cost of vibration-monitoring systems comes down, this technique becomes more attractive for use on smaller extruders as well. A single 32-point extruder monitor system costs around $35,000 to $40,000. The system can be expanded relatively inexpensively to monitor up to 5000 points throughout a plant. Bearing temperature, product-flow rate, and motor load can be monitored as well.

Another good indicator of bearing problems is screw run-out. Regularly check the total runout (TIR) using a dial indicator between the gear-case and thrust housing under loaded conditions. TIR usually varies from 75 to 250 microns (0.003 to 0.010 in). A gradual deterioration of the TIR indicates bearing wear. A rapid increase in run-out indicates imminent bearing failure or a dirty screw shank, and corrective action should be taken as soon as possible.

A2.4 Drive Motors and Belts

Direct current brush motors require more maintenance than alternating current motors or brushless direct current motors. Primary symptoms of motor problems are excessive noise, brush sparking, discolored armature, high temperature, inadequate exhaust air flow, and vibration. Many motors can be fouled with dust. Be vigilant about checking for buildup in the motor housing, particularly when running flexible PVC. If the machine is not ventilated, a sticky film can foul the motor, reduce motor efficiency, and even cause arc-over.

A quick way to check belt tension is to deflect each belt in mid-span with your thumb. With correct tension, the deflection should be about 12 mm (0.5 in). Check all belts for fraying, cracking, or twisting, and replace damaged belts; make sure to replace belts as a full set! Newly installed belts should be adjusted during the first month of operation to compensate for normal stretching.

A2.5 Spare Parts

It is usually a good idea to have a spare screw available for each extruder. It is important to stock motor brushes (for direct current brush motors), drive belts (for belt drives), and drive fuses. Other useful parts to stock are extra oil and air filters, screens, die heaters, die- and barrel-temperature sensors, solenoids, shear pins or rupture disks, hoses, seals, and gaskets.

A2.6 Screw and Barrel

Worn screws and barrels often result in considerable loss in production. Screw and barrel wear usually results in increased melt temperature and pressure fluctuations at the extruder discharge. Such wear also reduces output. Screws usually wear more rapidly than barrels. It is a good idea to measure screws and barrels on an annual basis. When running abrasive compounds, the screw and barrel may have to be checked more frequently.

The butt end of the screw seats into a thrust pocket in the gearbox assembly. When the screw is pulled and the barrel is cleaned, remember to also clean out this thrust pocket. Contamination in this pocket causes a screw to run eccentrically and wear down the bushings, feed section, and barrel liner. Lubricate this area with anti-seize compound before installing the screw.

If the machine has been moved, or if the gearbox or barrel has been removed, it is good practice to bore-scope the machine to make sure that the barrel, feed section, and gearbox are properly aligned. If the radial screw wear is 0.5 mm (0.020 in), the screw probably should be replaced unless a noncritical product is being extruded. If the radial screw wear exceeds 1 mm (0.040 in), the screw should be replaced as quickly as possible. If the wear occurs within a short time period, the cause of wear should be eliminated before the screw and/or barrel are replaced.

A2.7 Extruder Maintenance Checklist

Checks performed during a normal run	Frequency
Gearbox:	
☐ Feel/record radial bearing cap temperatures	Weekly
☐ Track oil flow to all bearings	Weekly
☐ Observe/record oil temperatures and pressures	Weekly
☐ Feel/record thrust-housing temperature	Monthly
☐ Look for oil leaks at shaft bearings	Monthly
☐ Listen for unusual bearing and gear noises	Monthly
☐ Detect oil leaks at windows, joint lines, and elsewhere	Annually
☐ If instruments are available, record noise level of bearings	Quarterly
☐ Check/record screw run-out between gearbox and feed area	Quarterly
☐ Make sure that screw cooling rotary union seals are tight	Quarterly
Drive area:	
☐ Check air flow through direct current motor	Quarterly
☐ Look for sparking at motor brushes	Quarterly
☐ Record temperature at motor housing	Quarterly
☐ Check/record temperatures of motor bearing caps	Quarterly
☐ Observe condition of motor internal housing for foreign material	Weekly
☐ Check for excessive motor vibration	Quarterly
☐ Be sure that belt guards are securely in place	Weekly
Feed section:	
☐ Check for water flow	Quarterly
☐ Look for excessive powder leaks	Quarterly
☐ Check/record temperature of feed section casting	Weekly
Barrel, heating/cooling assembly:	
☐ Check amperage of barrel heaters	Quarterly
☐ Identify and isolate leaks in closed-loop systems	Annually
☐ Check pressure and temperature regulators	Annually
☐ Inspect pump seal for leaks	Annually
☐ Observe flow of tower/municipal water source	Annually
☐ Look for apparent solenoid problems	Quarterly
☐ Check water for possible scale problems	Monthly
☐ Feel/record temperatures of blower motors	Quarterly
Clamp area:	
☐ Check for leakage at seals	Monthly
☐ Make sure that pressure indicator is working properly	Daily
☐ Make sure that extruder output is normal at this screw speed	Daily

Checks performed during scheduled downtime	Frequency
Gearbox:	
☐ Visually inspect condition of gears	Quarterly
☐ Replace oil per manufacturer's instructions	Quarterly
☐ Clean oil sump, and examine for metallic inclusions	Quarterly
☐ Remove sludge from oil heat exchanger	Annually
☐ Change or clean oil filter	Annually
☐ Examine water side of oil heat exchanger; clean if needed	Annually
Drive area:	
☐ Check drive belts, and replace if necessary	Quarterly
☐ Be sure all belts have correct tension	Quarterly
☐ Clean filter on armature cooling blower intake	Monthly
☐ Observe color and surface imperfections on commutator	Quarterly
☐ Clean brush holder, commutator, and winding	Monthly
☐ Lubricate motor bearings as specified by manufacturer	Annually
☐ Make sure terminals inside motor conduit box are tight	Annually
☐ Check insulation tape in conduit box	Annually
☐ Make sure that guards are securely back in place	Annually
Feed section:	
☐ Check emergency stop switch and hopper barrier guards	Quarterly
☐ Look for gouges or excess wear in feed section casting	Annually
☐ Check condition of seal and bushing	Quarterly
Screw and barrel:	
☐ Clean barrel and look for cracks and other defects	Quarterly
☐ Measure/record inside diameter of barrel at one diameter intervals	Annually
☐ Measure/record screw diameter at one diameter intervals	Annually
☐ Visually check for abuse (cracks, chips, and such)	Annually
☐ Record screw and barrel clearance	Annually
☐ Clean screw-seat pocket and lubricate	Quarterly
Barrel, heating/cooling assembly:	
☐ Check corrosion pencil anodes (zincs)	Annually
☐ Clean out sump	Annually
☐ If treated water is used, check condition and level	Quarterly
☐ With municipal water, flush lines with scale removing compound	Annually
☐ With distilled water, check that chloride ion concentration < 10 ppm	Quarterly
☐ Check municipal side of water-to-water heat exchanger for scale	Annually
☐ With air cooling, clean filter screen and wheel; replace if necessary	Quarterly
☐ Lubricate blower motors per manufacturer's specifications	Annually
☐ Check seating of temperature sensors and heater terminal connections	Annually
☐ Measure resistance across heaters and to "ground"	Annually

Checks performed during scheduled downtime	Frequency
☐ Make sure that all heater clamping bolts or straps are tight	Annually
☐ Replace all required screws on barrel covers	Annually
Clamp area:	
☐ Check that over-pressure devices are functional	Monthly
☐ Inspect angles of clamp and flange	Annually
☐ Check breaker plate recess seal surface	Quarterly
☐ Verify pressure calibration of indicator/controller	Quarterly
Control panel:	
☐ Visually inspect power connections for heat buildup or arcing	Quarterly
☐ Check tightness of heat and drive power connections	Annually
☐ Verify continuity of all ground connections	Annually
☐ Check calibration of all temperature controllers and drive meters	Annually
☐ Calibrate pressure indicators/controllers	Quarterly
☐ Check function of door interlock, tightness of fuse clips	Quarterly
☐ Vacuum or blow out dust from panel, especially at contacts	Quarterly
☐ Clean air filter at ventilation input	Quarterly
☐ Check operation of ventilation fan	Annually
☐ Verify that drive potentiometers have steady increase in resistance	Annually
☐ Check drive acceleration; set 25 s minimum	Annually
☐ Make sure safety devices and overloads function properly	Quarterly
Plug plate:	
☐ See that all receptacle terminals are grounded	Annually
☐ Check condition of all head/die cords and plugs	Weekly
☐ Inspect the condition of die temperature sensors and wiring	Quarterly
☐ As a final safety check, make sure that all guards are back in place and fully secured, and that all safety signs are posted and legible	Weekly

Appendix 3: Extruder Barrel Temperatures

A3.1 Setting Extruder Barrel Temperatures

Feed section

Set barrel temperatures in the feed section to achieve high friction; this will result in:

- Increase in motor load
- Reduced pressure variation
- Increase in throughput (diehead pressure)

Feed housing (adjust water temperature and flow rate)

For polyolefins: water temperature as low as possible without condensation; these are relatively soft materials with $T_g < T_{ambient}$.

For PA and PET, higher temperatures work better; these are hard plastics with $T_g > T_{ambient}$.

Transition section (good melting)

Set temperatures in the transition section to achieve reduced melt temperature variation and improved stability.

- For olefins:

Feed	T-crystallization
Compression	in between
Metering	T_{mp} + 70 °C (T_{mp} is melting point temperature)

- For olefins with mixing section along the screw:

Feed	T-crystallization
Compression	in between
Metering	T_{mp} + 70 °C
Mixing	Heating turned off (adiabatic); this improves mixing (applies to dispersive misers)

T-crystallization for LDPE is about 90 °C, for HDPE about 100 °C.

- For olefins: circumferential speed about 0.5 m/s as first setting
- For PVC: circumferential speed about 0.2 m/s as first screw speed setting

Metering section

- For semi-crystalline plastics: T_{mp} + 70 °C
- For amorphous plastics: T_g + 100 °C

Design of experiments

The only truly correct method to determine optimum barrel temperature profile is to design experiments of use. This is necessary because often there are interaction effects between the different temperature zones. Unfortunately, this can be quite time-consuming and, therefore, it is not always practical.

A3.2 Extruder Barrel Temperature Profile Optimization

A3.2.1 Introduction

Determining the optimum barrel temperature profile is one of the most important tasks in setting up an extrusion process. Unfortunately, this is not a simple task; there a many myths and misconceptions about how extruder barrel temperatures should be set. In this section we will provide some guidelines and techniques to determine a good barrel temperature profile.

A3.2.2 Facts about Barrel Temperature Profile (BTP)

The listing below should be kept in mind when setting the BTP on an extruder.

1. A BTP that works well on one extruder may not work well on the "same" extruder right next to it; even if the extruders process the same plastic, are the same size, made by the same extruder manufacturer, use the same screw and die, and run at the same screw speed.
2. Extruders are like human beings; each extruder has its own unique characteristics. One extruder may run well with a reversed BTP, while the "same" extruder right next to it runs better with a flat BTP. The reasons for these differences are not always known; sometimes, very subtle differences can strongly affect process stability.
3. The BTP that works well is not fixed or static; it depends not only on the resin, but also on screw speed, screw design, diehead pressure, screw and barrel

wear, ambient conditions (humidity, temperature), resin inlet temperature, moisture level in the resin, etc. For instance, a BTP that works well at 50 rpm screw speed may not work well at 120 rpm.

4. When a new screw is installed in an extruder, particularly one with different flight geometry, it is often necessary to adjust the BTP to achieve optimum performance.
5. Some extruders are rather robust and run well with a number of different temperature profiles. Unfortunately, other extruders are quite sensitive to barrel temperature settings and require careful temperature adjustments.
6. In an extrusion plant with many extruders, trying to run all extruder at the same BTP will likely result in below optimum performance in certain extruders, see point 2!
7. In the metering section of the extruder, reducing the barrel temperature will improve the melt conveying capability of the screw. Unfortunately, no such general rule applies to the feed section of the extruder.
8. No person can accurately predict the correct BTP without having run the extruder under the appropriate process conditions. People claiming to be able to predict the correct BTP without such information should be regarded with caution.
9. Finding the correct BTP requires listening to the extruder. Only the extruder can tell you what temperatures will result in good performance. It is essential that the operator communicates with the machine and properly interprets the data from the extruder. Good man-machine communication is essential to effective BTP optimization.
10. One of the greatest detriments to finding the correct BTP is personnel with preconceived ideas about what temperature profile should work. Operating personnel trying to impose a preconceived BTP on the extruder will probably not achieve good extruder performance.

Listening to the extruder requires observing how the process reacts when changes are made. These changes can be intentional and obvious, such as a change in screw speed or barrel temperature setpoint. Sometimes these changes are unintentional and not necessarily obvious; for instance, when the extruder performance changes as a result of screw wear or as a result of buildup of contamination on the screen pack.

Listening requires not just using your ears; it requires visual observation but it may also involve your sense of smell or you sense of touch. For instance, fingertips can be used to feel the smoothness of a metal surface or the surface of the extruded product. Human fingertips are remarkably sensitive and can distinguish relatively small differences in surface quality. Fingertip sensing can also detect slight dimensional changes in the extrudate that may not be picked up by the measurement system.

The analysis of sound goes beyond an operator listening to an extruder. Today sophisticated tools are available to analyze airborne sound in determining impending failure of machine parts [175]. Fingertips can also be used to sense temperature. Of course, care has to be taken to avoid burns; for that reason it is better to use an infrared thermometer or camera. When the temperature is high, the fingers or hand should be kept some distance away from the hot surface to avoid burns. When the temperature is below 50 °C, the object can be touched to get a more accurate feel of the temperature.

A3.2.3 Typical Process Temperatures for Different Plastics

As a general rule, the melt temperature for semi-crystalline plastics is about 50–75 °C above the melting point of the plastic [176, 177]. For instance, HDPE (high density polyethylene) has a melting point of about 130 °C; it is typically processed at temperatures in the range of 180–205 °C. If the plastic is susceptible to degradation, it may be processed closer to the melting point to minimize degradation.

HDPE is a pretty forgiving material and it can be processed quite well at temperatures above 205 °C. High temperature semi-crystalline polymers such as PET and nylon tend to be processed closer to their melting point – about 20–30 °C above the melting point.

Amorphous plastics are usually processed about 100 °C above the glass transition temperature of the plastic [176, 177]. For instance, PS (polystyrene) has a glass transition temperature of around 100 °C; typical process temperatures for PS are around 200 °C. If an amorphous plastic is susceptible to degradation, it may be processed closer to the glass transition temperature to minimize degradation.

Some plastics are quite forgiving and can be processed at a wide range of temperatures – an example is low density polyethylene (LDPE). Other plastics are very temperature-sensitive and can only be successfuly processed in a narrow range of temperatures. An example is rigid PVC (polyvinylchloride); the acceptable melt temperatures for RPVC may range from 184 to 188 °C for a particular formulation.

The acceptable melt temperature range for PVC will depend on the formulation, especially the type and amount of stabilizers used. When the melt temperature is too low, the plastic will not be properly gelated; this will result in poor physical properties. When the melt temperature is too high degradation will occur; this will result in discoloration of the product and can also lead to poor physical properties. It is clear from these considerations that in the RPVC extrusion it is very important to have accurate melt temperature measurement.

A3.2.4 Guidelines and Considerations for Setting Barrel Temperatures

When considering the three major sections of the extruder (feed, transition, metering) the following guidelines can be used.

Feed section: set barrel temperatures to maximize motor load and minimize pressure variation at the discharge end of the extruder [177].

Transition section: set barrel temperatures to minimize melt temperature variation at the discharge end of the extruder [177].

Metering section: set barrel temperatures to the desired melt temperature for the resin, see Section A3.2.3. It should be noted that the discharge melt temperature can be significantly higher than the barrel temperatures in the metering section [177].

The feed section consists of zone 1 and perhaps zone 2 or part of zone 2. The metering section is usually made up of the last two barrel temperature zones. The temperature zones in-between make up the transition section. This is true for a non-vented (single-stage) extruder. For a vented (two-stage) extruder, the first stage (from the feed opening to the vent port) consists of the feed, transition, and metering section. The second stage (from the vent port to the end of the barrel) consists of the extraction section, the second compression, and the pump section.

In a vented extruder the extraction section should be maintained at high temperature because this facilitates the removal of volatiles. On the other hand, the second stage pump section should be maintained at relatively low temperature to improve the melt pumping and pressure development capability. Lower barrel temperatures in the pump section will reduce the chance of vent flow (molten plastic exiting from the vent port).

Short extruders (24–26 *L/D*) usually have 3–4 barrel temperature zones. Longer extruders (30–32 *L/D*) typically have 5–6 barrel temperature zones, while long extruders (34 *L/D* and longer) may have 6–10 barrel temperature zones.

It should be kept in mind that a certain BTP will often produce different results in different extruders. There are many possible reasons for this. One factor is the depth of the temperature sensors in the extruder barrel. If one extruder uses shallow-well thermocouples in the barrel and a second extruder deep well thermocouples, this will result in different temperatures along the internal wall of the barrel. The temperatures at the internal barrel wall are the ones that really affect the process.

Typical temperature profiles for several amorphous polymers are shown in Figure A3.1. The following figures use two temperature scales. The left vertical scale shows temperature in degrees centigrade; the right vertical scale shows temperature in degrees Fahrenheit. In the United States many companies still use Imperial units and there does not seem to be a rapid shift toward the use of the international system of units.

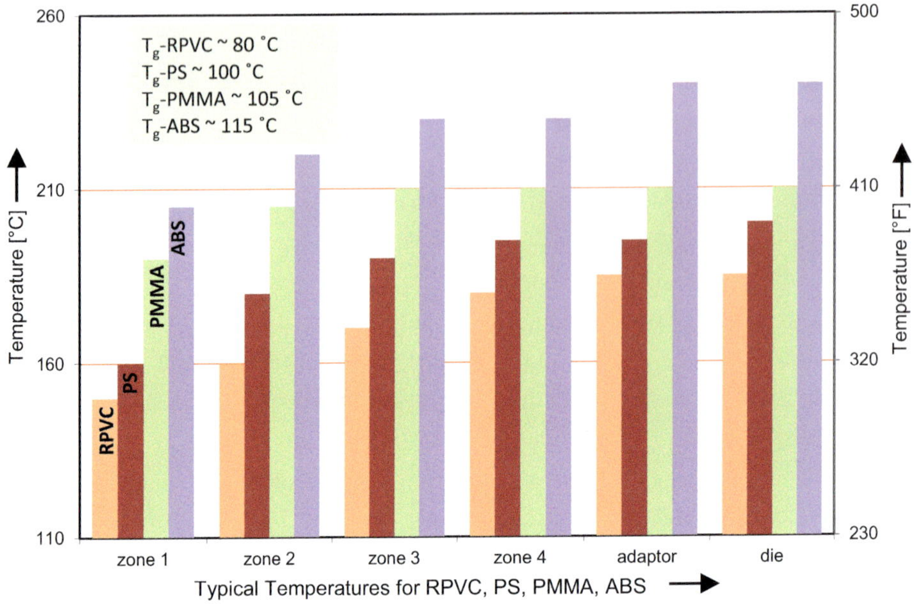

Figure A3.1 Typical temperature profiles for RPVC, PS, PMMA, and ABS

Rigid PVC is processed at relatively low temperatures because it is quite susceptible to degradation – the highest temperatures are 185 °C. On the other hand, ABS is processed at relatively high temperatures – as high as 240 °C. The right hand vertical scale shows the temperatures in degrees Fahrenheit. The U.S. extrusion industry is still using Imperial units on a large scale.

Figure A3.2 shows typical temperature profiles for LDPE blown film, LDPE cable, LDPE coating, and PP.

In LDPE blown film extrusion the barrel temperatures go up to about 185 °C. In LDPE cable extrusion the barrel temperatures tend to be slightly higher up to about 195 °C. Barrel temperatures for polypropylene range from 190 to 220 °C.

Figure A3.3 shows typical temperature profiles for HDPE for sheet, pipe, and film extrusion. For HDPE sheet and pipe most zones range from 200 to 220 °C. The barrel temperatures in HDPE pipe extrusion tend to be slightly higher (5–10 °C) than in sheet extrusion. For HDPE film the temperatures can go up to 240 °C. In film extrusion the diehead pressure can be a limiting factor. For that reason, it is advantageous to have a high melt temperature. A high melt temperature will reduce the melt viscosity and this will lower the diehead pressure.

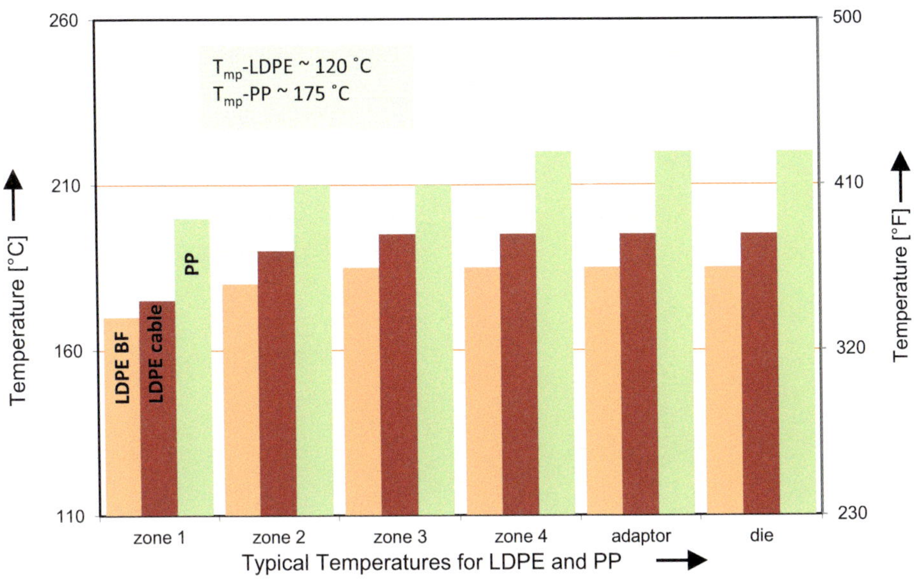

Figure A3.2 Typical temperature profiles for LDPE and PP

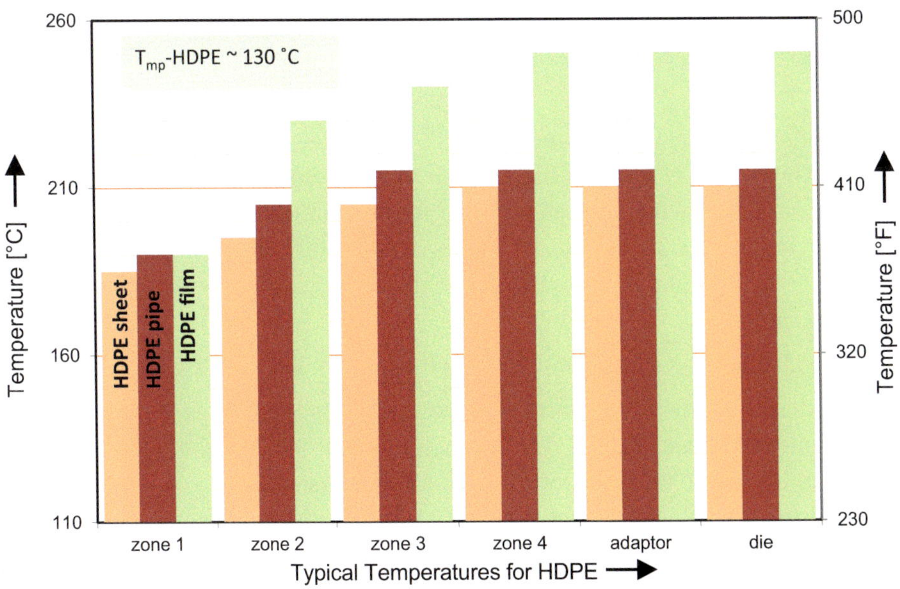

Figure A3.3 Typical temperature profiles for HDPE for sheet, pipe, and film extrusion

A typical temperature profile for LLDPE blown film is shown in Figure A3.4.

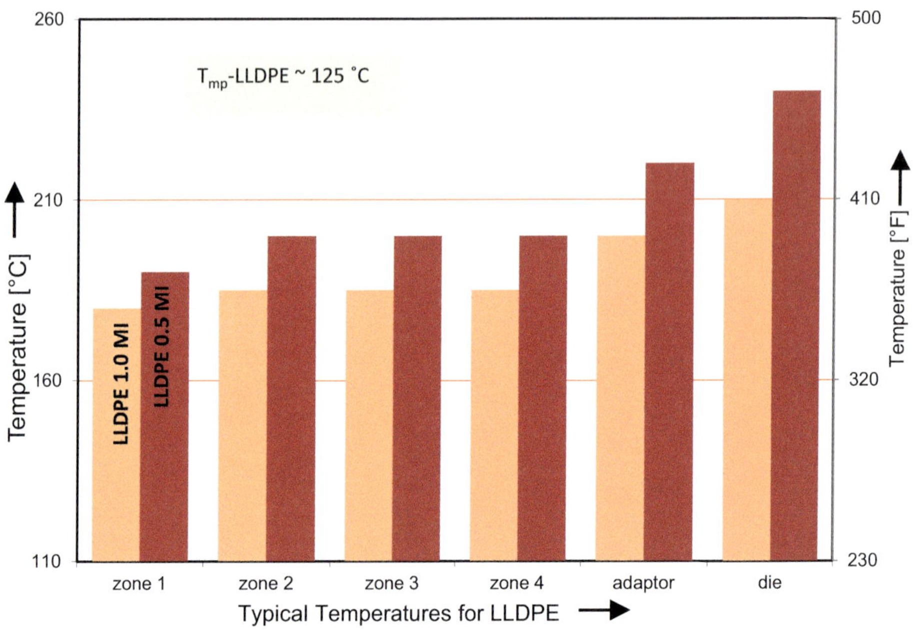

Figure A3.4 Typical temperature profile for LLDPE

Temperatures for the 0.5 melt index LLDPE are higher than for the 1.0 melt index LLDPE. The 0.5 MI LLDPE has higher melt viscosity than the 1.0 MI LLDPE – for that reason higher temperatures are used to reduce the melt viscosity. Generally, low melt index plastics are run at higher temperatures than higher melt index plastics to achieve lower melt viscosity. This will result in lower motor load and discharge pressure.

Figure A3.5 shows a typical temperature profile for nylon and PET.

The PET barrel temperature starts at about 260 °C and increases to about 280 °C. Nylon 6 melts at about 220 °C. The barrel temperatures for nylon 6 start at about 220 °C and increase to about 250 °C. Nylon 6-6 melts at about 269 °C. The barrel temperatures for nylon 6 start at about 275 °C and increase to about 285 °C.

Figure A3.6 shows typical temperatures for three fluoropolymers, PVDF, FEP, and PFA. These are some of the highest temperature polymers. Process temperatures for PVDF are relatively low – about the same as those for PMMA and ABS. Process temperatures for FEP and PFA are much higher – in the range of 370 to 400 °C. These two polymers are also very corrosive for the extrusion equipment. For that reason the extruder equipment should be made out of highly corrosion-resistant materials such as Hastelloy, Monel, and Inconel.

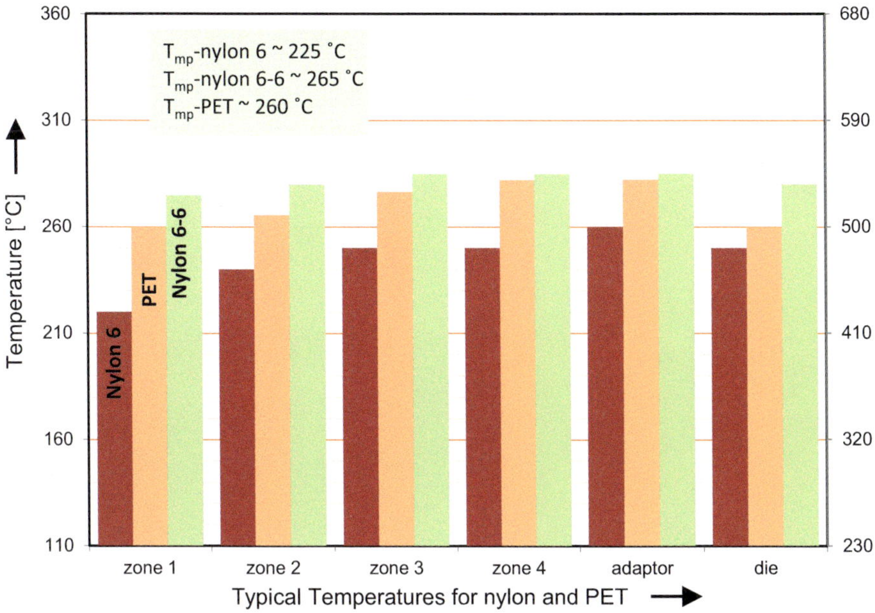

Figure A3.5 Typical temperature profile for nylon and PET

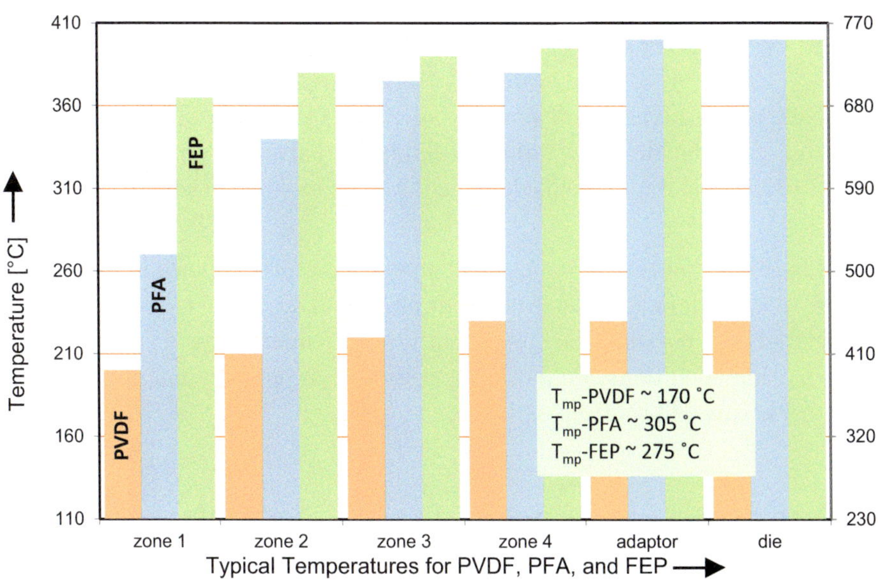

Figure A3.6 Typical temperatures for PVDF, FEP, and PFA

Figure A3.7 shows typical temperatures for polycarbonate, LDPE for coating, and polyether ether ketone (PEEK).

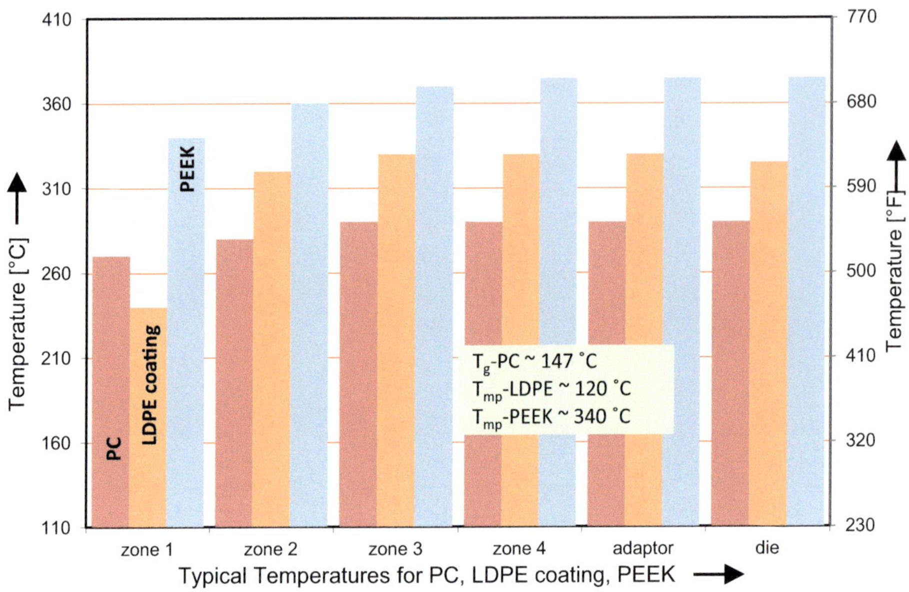

Figure A3.7 Typical temperatures for polycarbonate, LDPE for coating, and PEEK

PEEK is one of the highest temperature polymers. LDPE normally is not a high temperature polymer. However, in LDPE coating the barrel temperatures are very high, up to 320 °C, to intentionally degrade the LDPE to improve adhesion to a substrate.

It should be noted that the typical barrel temperature profiles shown in the figures above may not result in acceptable extruder performance. These temperatures can be considered as a possible starting point. If these temperatures do not result in stable extrusion conditions, we have to go through an optimization of the barrel temperatures. This will be discussed next.

A3.2.5 BTP Optimization by Design of Experiments (DOE)

The only technically correct method of finding the best BTP is to perform design of experiments (DOE) incorporating all barrel temperature zones. This should be done with a full factorial design [177] because it is possible that interaction effects exist between the various barrel zones.

A full factorial DOE is reasonable when there are only three or four barrel zones. When a two-level factorial design is performed with four factors (barrel temperature zones), this will require 16 (2^4) experiments. If each experiment takes 30 minutes, this will take a total of 8 hours - a full day!

When there are five or more zones, a full factorial DOE becomes impractical in most situations, especially when we are dealing with a large size production extruder. For instance, a two-level factorial design with six factors requires 64 experiments (2^6). If each experiment takes 30 minutes, the total factorial design will take 32 hours - almost a whole week. As a result, the full factorial DOE approach to BTP optimization is rarely used in industrial extrusion operations - it is too time-consuming and too expensive.

A3.2.6 BTP Optimization by One-at-a-Time Experiments (OTE)

The OTE method is the one most commonly used in industrial BTP optimization [176, 177]. In the OTE method the temperature steps are usually small, e.g., 5 °C. This method cannot uncover interaction effects. However, it works well in many industrial extrusion operations. The problem with this approach is that it is time-consuming and, therefore, expensive.

Each time a temperature change is made one has to wait until the extruder stabilizes after the barrel zone reaches setpoint. Reaching the setpoint can take 5-10 minutes on a small extruder (20-40 mm). However, it may take 30-60 minutes or longer on a large extruder (over 100 mm). The time the extruder takes to stabilize with the new BTP can take considerable time as well. Again, on a small extruder it can take 5-10 minutes but on a large extruder it can take 30-60 minutes or longer.

As a result, making five or six changes can take an entire day or longer when working with a large extruder. This can become prohibitively expensive in a large production line.

A3.2.7 Dynamic BTP Optimization

The DOE and OTE methods both suffer from the problem that they are time-consuming, in some cases, making these methods unacceptable. Dynamic optimization involves making large changes (20-40 °C or more) and tracking the dynamic response of the extruder. This method was developed by Rauwendaal Extrusion Engineering as a fast but robust method of process optimization that works even for very large extruders operating in a production environment [220].

It should be noted that when the setpoint of a barrel temperature zone is changed by a large amount, the temperature control system may not be able to achieve the setpoint temperature. For instance, if the setpoint for zone 3 is changed from 220 °C down to 160 °C, the actual barrel temperature may only go down to 184 °C.

If the cooling system is on full blast at this condition, the barrel temperature cannot be reduced further even if the setpoint is set at a much lower setting. The only way to achieve further temperature reduction would be to increase the cooling capacity or to change certain process conditions, e.g., reduce the screw speed.

As an example of dynamic optimization, suppose zone 1 barrel temperature is changed from 200 °C to 150 °C in a 100-mm extruder. It may take the extruder 15–20 minutes to bring down the actual barrel temperature to 150 °C. In dynamic optimization the actual barrel temperature is recorded every 15–30 seconds, whatever time interval allows an accurate determination of the transient behavior of the extruder. When a data acquisition system (DAS) is available, the data is recorded automatically. With a DAS it is important to make sure that the data-sampling frequency is high enough so that the transient behavior can be determined accurately.

It is very important to closely monitor the extruder when a large setpoint change is made. During the transient response it is possible that the motor load increases to a level that may cause the extruder to shut down. In that case, an intervention by the extruder operator is required to keep the motor load from reaching a critical level. In the example above this intervention would increase the setpoint from 150 °C to a level that will not result in excessive motor load.

At each barrel temperature the corresponding melt pressure variation is recorded. This allows construction of the graph showing pressure variation versus barrel temperature, see for example Figure A3.8.

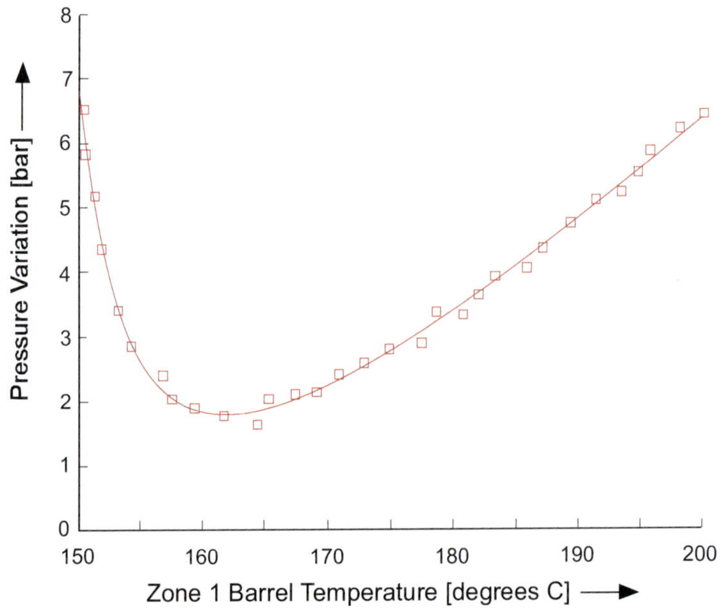

Figure A3.8 Example of pressure variation versus barrel temperature

Figure A3.8 shows that the minimum pressure variation is reached between 160 and 165 °C. The pressure variation increases rapidly at temperatures below 155 °C. Therefore, it would be prudent to set zone 1 temperature at 165 °C because we want to avoid the steep part of the curve between 150 and 155 °C. Obviously, when a data acquisition system with trending capability is available it is not necessary to construct a graph as shown in Figure A3.8. The graphical display will show changes in the width of the pressure band.

It is possible that under steady-state conditions the pressure variation is not the same as under transient conditions. If the steady-state pressure variation at 165 °C is much higher than the transient pressure variation, it may be necessary to do a few OTE runs around the 165 °C. In most cases, however, this will not be necessary and the extruder will run well at the setpoint determined by dynamic optimization.

The zone 1 barrel temperature in many cases has the strongest effect on extrusion process stability. Therefore, it is often not necessary to do further experiments with the BTP. In a 100-mm extruder, therefore, it may be possible to determine the optimum BTP in less than one hour; perhaps even less than 30 minutes.

A3.2.8 Other Studies on Optimization of Extruder Barrel Temperature Profiles

Abeykoon et al. [230] developed a model-based approach for prediction and optimization of thermal homogeneity in single-screw extrusion. Abeykoon also published [231] a comprehensive review of control of single-screw extrusion with more than one hundred references. A significant portion of this review deals with fuzzy logic control and model-based extrusion with fuzzy logic.

Crabtree et al. [217] published a study on optimization of barrel temperatures using a 63.5-mm single-screw extruder with three barrel temperature zones. The screw that was used in this study was a simple conveying screw with three zones without mixing elements. The screw geometry is shown in Table A3.1. The reported drag flow rate is 0.71 kg/h · rpm; this corresponds to 35.5 kg/h at 50 rpm.

Table A3.1 Screw Geometry

screw section	feed depth (mm)	feed length	flight lead
feed section	8.89 (0.140*D*)	6*D*	1*D*
compression section	↓	8*D*	1*D*
metering section	3.18 (0.050*D*)	7*D*	1*D*

One polypropylene (PP) resin and one high density polyethylene (HDPE) resin were used for the study; the resins were manufactured by The Dow Chemical Company. The PP resin was a homopolymer with a solid density of 0.900 g/cm^3 and a melt flow rate of 8.8 g/10 min (230 °C, 2.16 kg). The HDPE resin had a solid density of 0.958 g/cm^3 and a melt flow index of 5.9 g/10 min (190 °C, 21.6 kg). The melting peak temperature was 142 °C for the PP and 133 °C for the HDPE. It should be noted that the melt index for HDPE is a high load melt index measured at 21.6 kg load – this is ten times the standard load!

A melt index of 5.9 g/10 min at 21.6 kg load indicates a polymer with a very high melt viscosity. The results of the study show that the high melt viscosity of HDPE result in very high melt temperatures – much higher than the melt temperatures for the PP. Table A3.2 shows the initial conditions for the extruder. These conditions were based on common industry practice. The screw was kept at 50 rpm through all the trials.

Table A3.2 Initial Conditions Used in the Trials

Setpoint [°C]	PP	HDPE
Barrel zone 1	200	170
Barrel zone 2	210	195
Barrel zone 3	220	220
Die zone	220	220
Screw speed, rpm	50	50
Feed housing cooling	on	on

The solids conveying zone temperature, zone 1, was increased (or decreased) in 10 °C increments to determine the optimum temperature. In these experiments the temperature of barrel zone 3 was kept at 220 °C; barrel zone 2 was kept at the midpoint of zones 1 and 3. In this study the temperature was optimized with respect to output. In most extrusion processes, process stability is more critical than output. For that reason, it usually makes more sense to optimize with respect to process stability as discussed in the previous section. The process stability can be assessed by die pressure variation, dimensional variation of the extrudate, motor load variation, or another measure of process stability.

Figure A3.9 shows the effect of barrel zone 1 on output. The output increases with barrel temperature initially. The output reaches a maximum at about 240–250 °C; beyond this temperature there is a slight reduction in output. The optimum zone 1 temperature for PP (blue line) is about the same as it is for HDPE (red line). Figure A3.9 also shows a quadratic curve fit – there is a reasonable fit with the actual data. It was found that the motor current reduced with zone 1 temperature for PP but increased for HDPE.

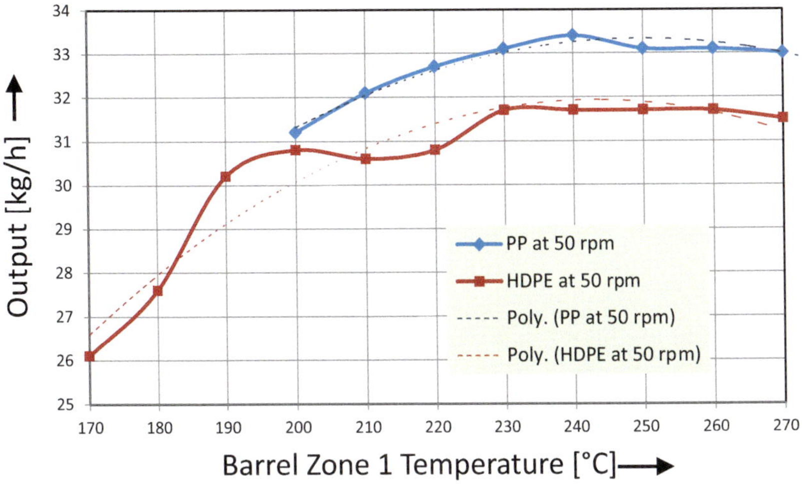

Figure A3.9 Output versus barrel zone 1 temperature

The discharge melt temperature was measured with a hand-held thermocouple into the extrudate. Figure A3.10 shows the melt temperatures at different barrel zone 1 temperatures. The melt temperature trends up with barrel zone 1 temperature. The increase in melt temperature for PP is larger than for HDPE. This figure also shows that the HDPE melt temperature is substantially above the temperature of barrel zone 3 (220 °C). This indicates that there is an excessive amount of viscous heating with the HDPE. This is caused by the high melt viscosity of the HDPE resin.

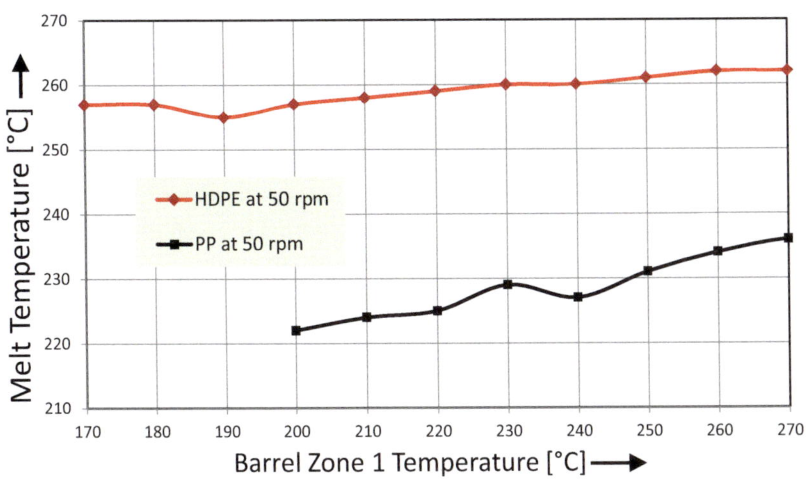

Figure A3.10 Melt temperature versus barrel zone 1 temperature

The very high melt temperatures of HDPE indicate that the extruder screw is not appropriate for this high viscosity polymer. Figure A3.11 shows how the melt temperature changes with the temperature of barrel zone 3.

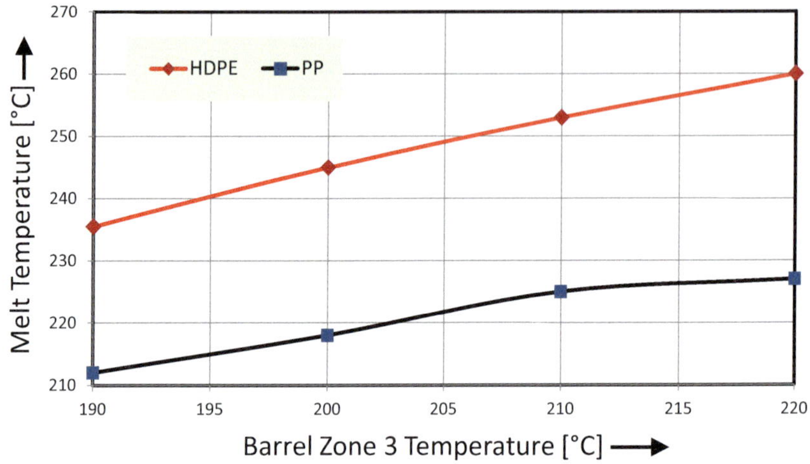

Figure A3.11 Melt temperature versus barrel zone 3 temperature

The melt temperature of both polymers increases with barrel zone 3 temperature. The increase in melt temperature for HDPE (~25 °C) is greater than for PP (~15 °C). The HDPE melt temperature is about 40 °C greater than the temperature of barrel zone 3. For PP the melt temperature is about 10–20 °C higher than zone 3 temperature. This difference is largely caused by the high viscosity of HDPE resulting in high viscous heat generation.

Figure A3.12 shows the output versus barrel zone 3 temperature.

Figure A3.12 shows that the output tends to go up with barrel zone 3 temperature, although the trend changes with PP beyond 210 °C. For HDPE the increase in output accelerates beyond 210 °C. The increase in output with barrel zone 3 temperature is related to the melt temperature. When the melt temperature increases with zone 3 temperature, the melt viscosity reduces with a corresponding reduction in discharge pressure. A lower discharge pressure will result in higher output. No data was included in the paper [217] on the effect of zone 3 temperature on discharge pressure.

The pressures along the length of the barrel were measured with pressure transducers at 12 locations. Figure A3.13 shows the barrel pressures for PP at two values of barrel zone 1 temperature.

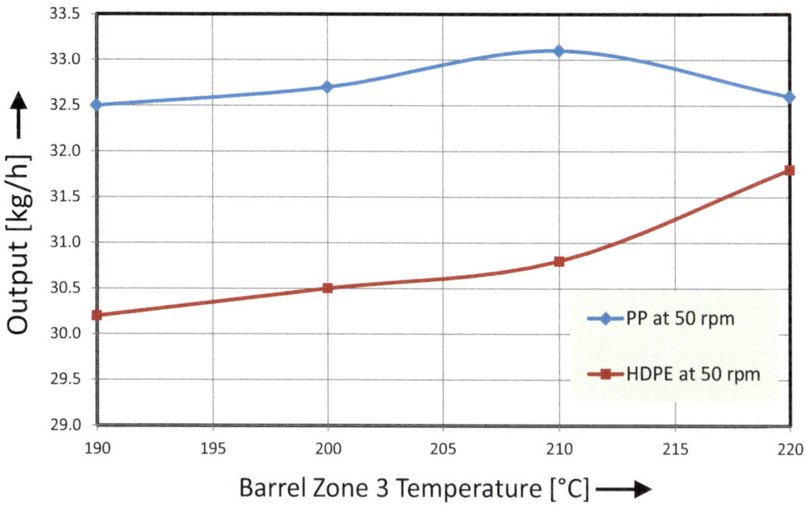

Figure A3.12 Output versus barrel zone 3 temperature

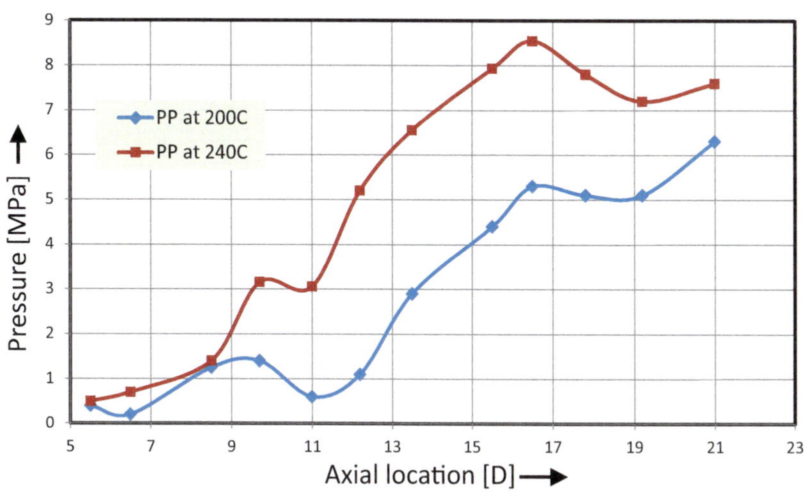

Figure A3.13 Barrel pressures for PP at 50 rpm at two zone 1 temperatures

The barrel pressures at 240 °C are substantially higher than at 200 °C. This is to be expected since the output at 240 °C was higher than at 200 °C. Higher output will go along with higher pressures. Figure A3.14 shows the barrel pressures for HDPE at two values of barrel zone 1 temperature.

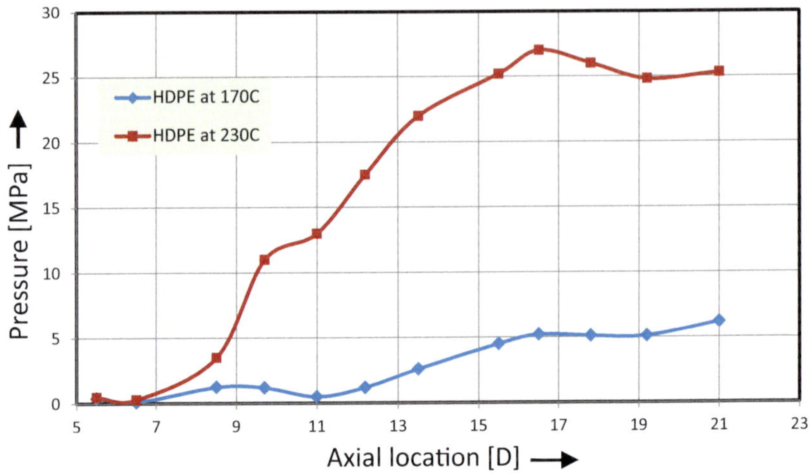

Figure A3.14 Barrel pressures for HDPE at 50 rpm at two zone 1 temperatures

The barrel pressures for HDPE follow the same trend as PP with the difference that the pressures at 230 °C are significantly higher than at 170 °C. The higher pressures with HDPE are to be expected considering that the HDPE melt viscosity is much greater than the PP melt viscosity. At the same output a polymer with higher melt viscosity will require higher discharge pressure than a polymer with lower melt viscosity.

The pressure profiles show that there is little pressure rise in the feed section of the screw (0–6*D*). There is significant pressure rise in the compression section of the screw (6D–14*D*) but little pressure rise in the metering section of the screw (14*D*–21*D*). The large pressure rise in the compression section is caused by the gradual reduction in channel depth. The large increase in pressure in the compression section of the screw is common in non-grooved feed single-screw extruders.

A substantial amount of research on the effect of extruder barrel temperatures has been done recently at the Kunststofftechnik Paderborn [221–227] by V. Resonnek and V. Schöppner. This research represents a significant extension and advancement of the earlier work by Rauwendaal [220] on dynamic optimization of extruder barrel temperatures. The effect of barrel temperatures on extruder performance was determined on a 45-mm 32*D* single-screw extruder [221, 222]. The experimental setup is shown in Figure A3.15. Two extruder screws were used in the study. One screw was a simple conveying screw shown in Figure A3.15; the other, a barrier type extruder screw.

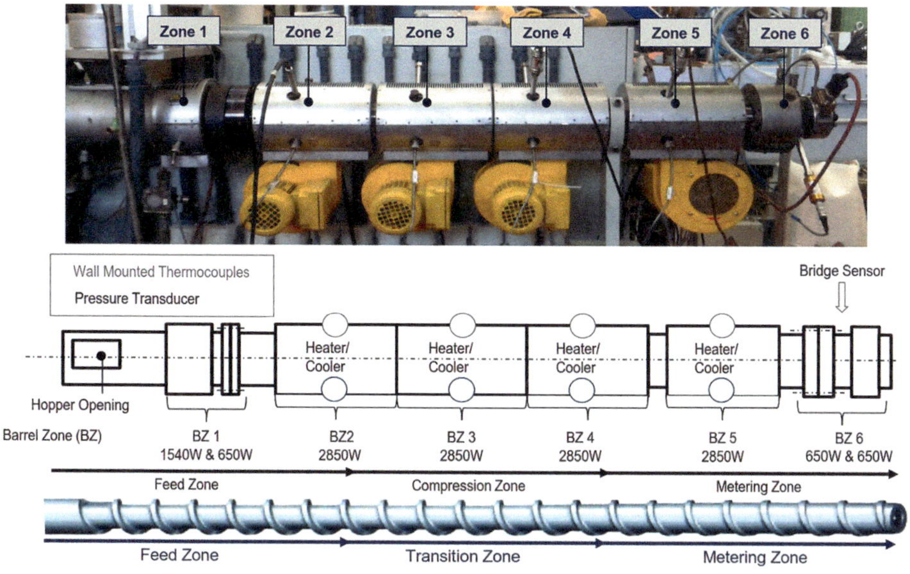

Figure A3.15 Experimental setup of 45-mm extruder

Several barrel temperature profiles (BTP) were tested: a flat BTP, an increasing BTP, a hump BPT, and a decreasing BTP. For each type of BTP four tests were performed at different temperature levels, low, medium-low, medium high, and high, see Figure A3.16.

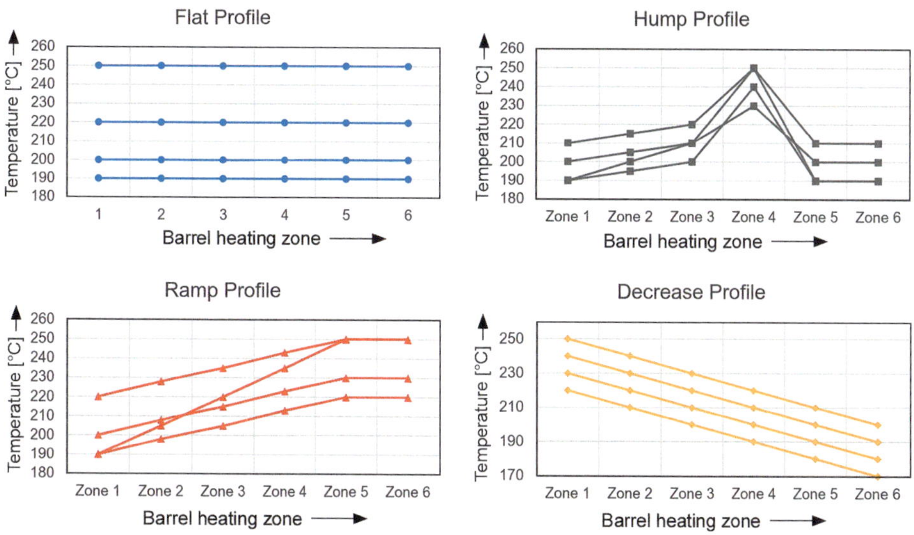

Figure A3.16 Four types of BTP with four temperature levels

Results with barrier screw at screw speed of 75 rpm:

- Output increase as high as 27%
- Reduced performance with hump BTP
- Reduced pressure variation with lower feed temperatures
- Melt temperature increases with high metering zone temperatures

Results of the effect of feed barrel temperatures on process variation with polypropylene using a single stage screw are shown in Figure A3.17 and Figure A3.18. At low screw speed (25 rpm) the effect of feed barrel temperatures on temperature and pressure variation is relatively weak, see Figure A3.17. At high screw speed (100 rpm) the effect is stronger. These figures also show that the melt temperature and pressure variation are strongly dependent on screw speed. At 25 rpm the temperature variation is less than 0.2% and the pressure variation less than 1%. At 100 rpm the temperature variation ranges from 0 to 6% and the pressure variation ranges from 12 to 23%.

The process stability at 100 rpm is rather poor with the simple conveying screw. Pressure variation over 10% is generally considered unacceptable. However, it is well known that simple conveying screws generally do not achieve good process stability especially at high screw speed. Barrier type extruder screws tend to result in better process stability. This is an important reason for the widespread use of barrier screw in the plastics extrusion industry.

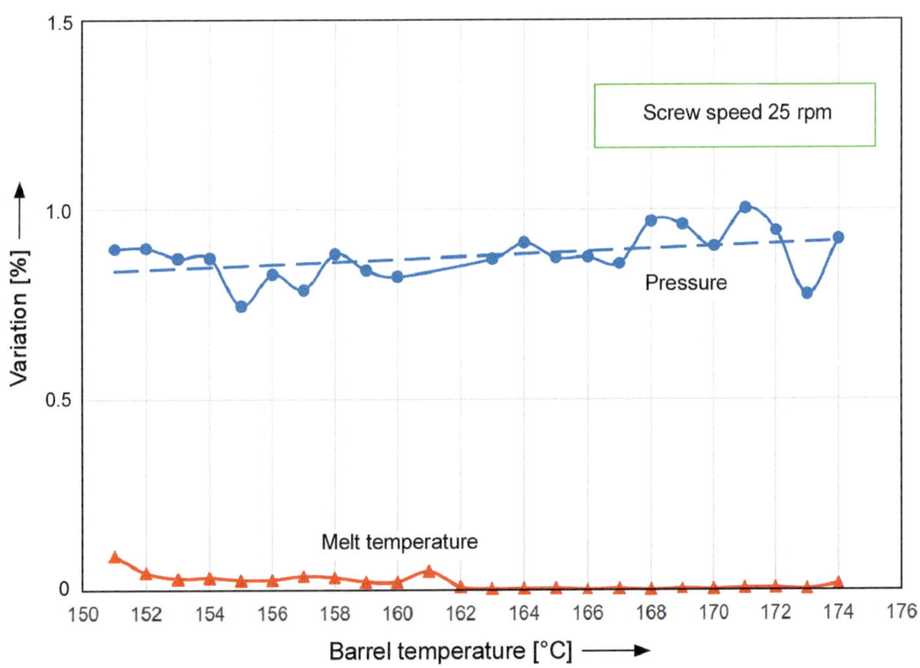

Figure A3.17 Pressure and melt temperature variation versus feed barrel temperature, 25 rpm

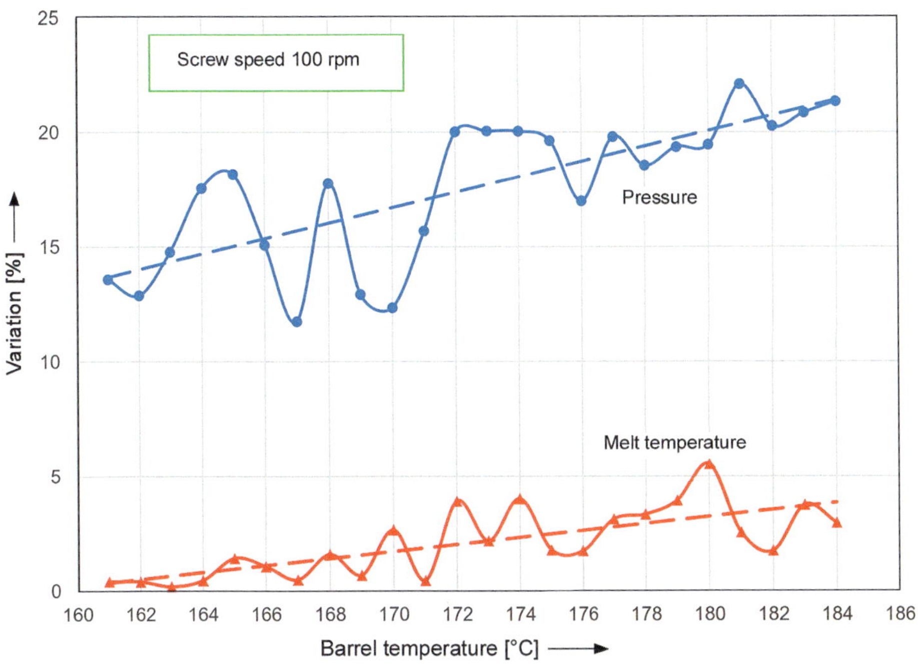

Figure A3.18 Pressure and melt temperature variation versus feed barrel temperature, 100 rpm

These results show the following trends:

- The effect of feed barrel temperature increases with screw speed
- Lower feed barrel temperature reduces pressure and temperature variation, especially at higher screw speed
- The level of pressure and temperature variation increases strongly with screw speed
- The change in variation is not monotonic; there are distinct peaks and valleys in the variation-temperature curves
- Small changes in barrel temperature can produce large changes in pressure and melt temperature stability

It should be noted that the distinct peaks and valleys in Figure A3.17 and Figure A3.18 would not be seen if the zone temperature were increased in steps of 10 °C, see for instance the study by Crabtree et al. [217] discussed earlier. This is a distinct drawback of the stepwise approach to barrel temperature optimization because significant information can be lost with the stepwise approach. The greater the step size, the more information is lost.

Extruder output tends to increase with transition barrel temperatures as shown in Figure A3.19. This data is for polypropylene (LyondellBasell Moplen 420M).

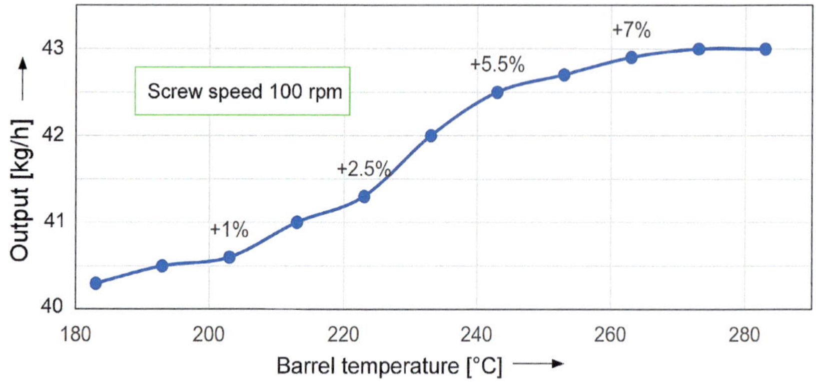

Figure A3.19 Output versus transition barrel temperatures at 100 rpm

These results show that higher temperatures in the transition zone increase extruder output. Higher temperatures in the transition zone also reduce pressure variation but increase melt temperature variation.

Figure A3.20 shows the effect of metering zone barrel temperature on pressure (blue) and melt temperature (red) variation. Pressure variation reduces with metering temperature but melt temperature variation increases with metering temperature. Higher metering zone temperatures result in increased melt temperature and lower discharge pressure.

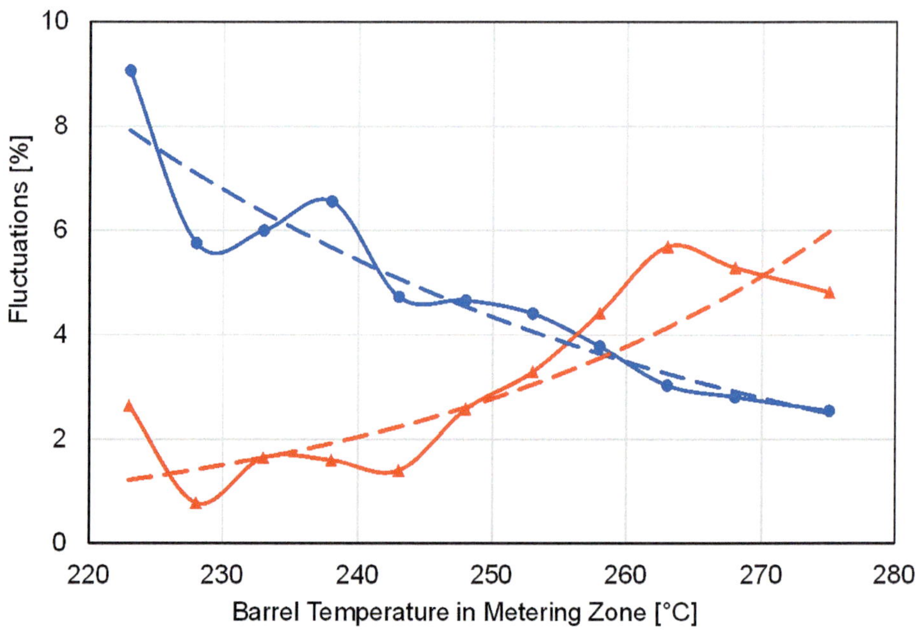

Figure A3.20 Pressure (blue) and melt temperature (red) variation versus metering barrel temperature

A fuzzy control scheme was developed to achieve a self-optimizing control system [223–227]. Figure A3.21 shows the basic elements of a fuzzy controller.

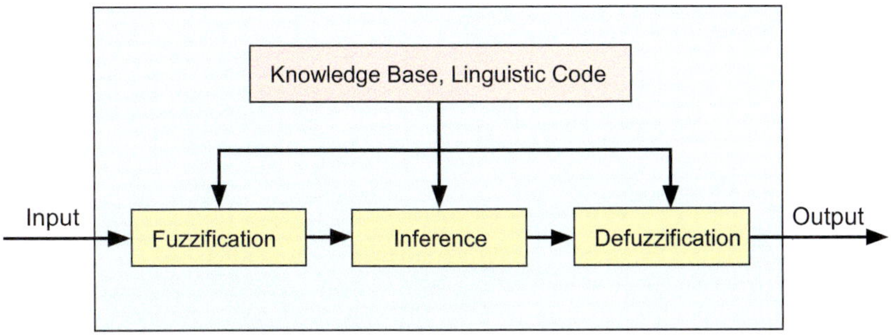

Figure A3.21 Structure of a fuzzy controller

The fuzzy controller consists of three function blocks: fuzzification, inference, and defuzzification. Figure A3.22 shows the transmission to a barrel temperature fuzzy controller. In the fuzzification step precise input values are matched to fuzzy linguistic terms. The temperature and pressure variation values are concatenated to fuzzy domains as in this example to "very low" (sn) and "low" (n), "medium" (m), "high" (h), and "very high" (sh) fluctuations using trapezoidal shapes. Figure A3.22 shows an overview of input and output values for the extruder.

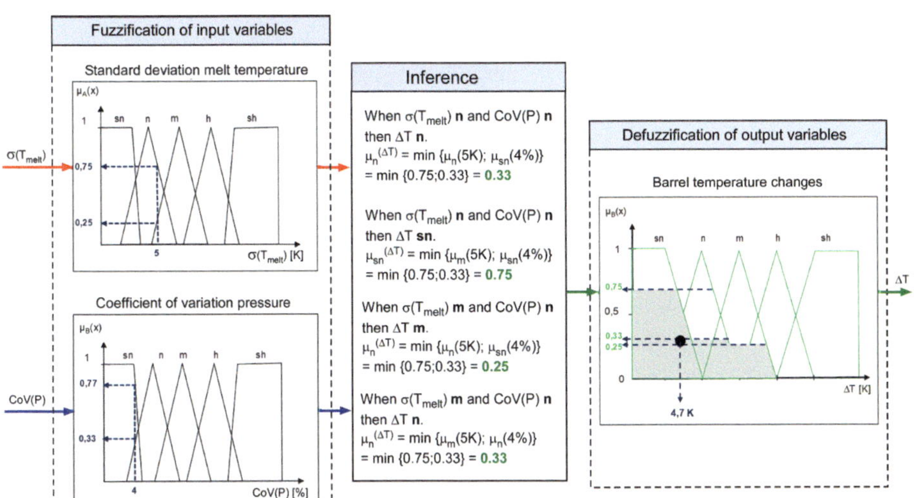

Figure A3.22 Schematic of fuzzy controller to determine temperature changes

The database has two aspects. On one side, the fuzzy controller needs to accomplish the fuzzification step. On the other side, the actual optimization can start with a basic temperature recommendation. With this approach fewer optimization loops need to be performed. Figure A3.23 shows the structure of the circuit diagram, showing interfaces and processing of measurement signals.

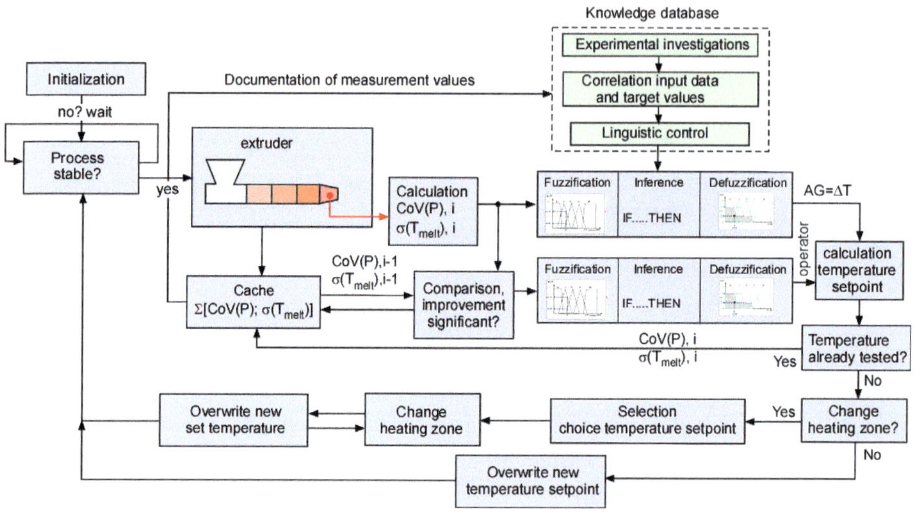

Figure A3.23 Schematic of barrel temperature control

The first step is to set a basic barrel temperature profile. After a certain period of time, the pressure and temperature fluctuations are determined. The crisp values are forwarded to the fuzzy controller. The controller is responsible for adjusting the barrel temperature. Here, the fluctuation values are processed further. For example, the request for zone 1 could be "change the barrel temperature by 10 °C". The second fuzzy controller is responsible for deciding whether to increase or decrease the temperature ("mathematical operator").

This signal is transmitted to the heating/cooling control system that adjusts the barrel temperature in zone 1 accordingly. After a defined period of time, the fluctuations are determined once again to evaluate the optimization step. If the barrel temperature adjustment in zone 1 leads to a decrease of the fluctuations and the fluctuations are now in an area of "regular", the next barrel zone will be optimized. On the other hand, if an increase in fluctuation has occurred, the barrel temperature change will be canceled and a different barrel temperature will be checked.

The decision of the final barrel temperature in the respective zone is chosen by weighting the pressure and temperature fluctuations (50–50). After every trial a key number is calculated and the optimal barrel temperature is chosen by taking the best value. The same procedure is applied for the remaining barrel zones. The optimization is finished when the last zone is reached. Figure A3.23 demonstrates a synopsis of the fuzzy system and the described operational sequence.

Table A3.3 shows three different BTPs with resulting temperature and pressure fluctuation. The first BTP is the starting profile. The second BTP was the result of experiments. The third BTP is the result of auto-optimizing using fuzzy control.

Table A3.3 Three Barrel Temperature Profiles and Resulting Temperature and Pressure Fluctuation

	BZ1 [°C]	BZ2 [°C]	BZ3 [°C]	BZ4 [°C]	BZ5 [°C]	ΔT [°C]	ΔP [%]
starting BTP	180	190	220	190	180	4.3	7.5
prior testing	150	140	210	250	250	6.5	2.3
auto-optimizing	149	170	220	245	190	5.6	3.4

Table A3.3 shows that the self-optimizing control system is capable of producing conditions with low pressure variation, less than half of the pressure variation with the starting BTP.

Figure A3.24 shows the representation of the user interface in the LabVIEW14 software. The process consists of several phases. The individual phases are visualized as tabs in LabVIEW. To mark the current phase, a green light is used. Overall, five tabs are listed in the user interface.

The first tab is used to configure the boundary conditions. At this point, general default values are defined, such as the maximum standard deviation of the actual barrel temperature, the barrel temperature ranges of the individual heating zones in which it is to be varied, or the number of variation passes per heating zone, see Figure A3.24.

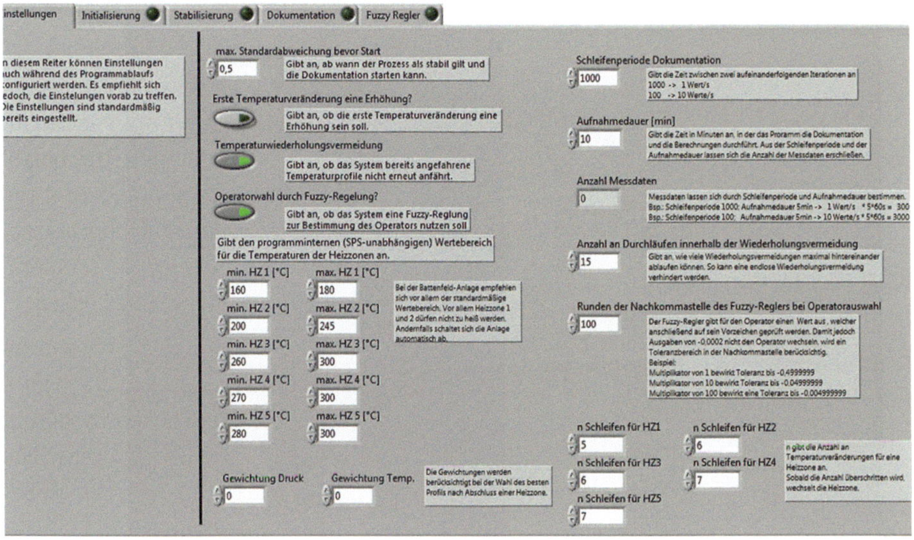

Figure A3.24 User interface for setting boundary conditions in LabVIEW14

In the next tab, "Fuzzy controller", of the user interface, the most important data such as the heating zone, the corresponding heating zone temperature, and the pressure and temperature fluctuations are clearly displayed (see Figure A3.25). In addition to these, the decisions of the fuzzy controllers are made visible in the same line. The decisions can be modeled on the "temperature change" and the mathematical "operator". Once the predefined boundary conditions are met (such as a predefined number of iterations), the change of the heating zone takes place.

Before the setting of the next heating zone is raised, the selection of the final temperature setting for the corresponding heating zone takes place. For this purpose, the pressure and temperature fluctuations are taken from the buffer and weighted in equal proportions. Then they are added together. Smaller fluctuations indicate better the barrel temperature setting. Therefore, the setting is adopted, which shows the lowest sum. In general, all temperature settings are stored as an xls file and in the LabVIEW UI. This way, measured values can be subsequently reviewed and reconstructed. In addition, the barrel temperature controller detects whether the temperature setting has already been started and can thus avoid repeated measurement of the same temperature setting.

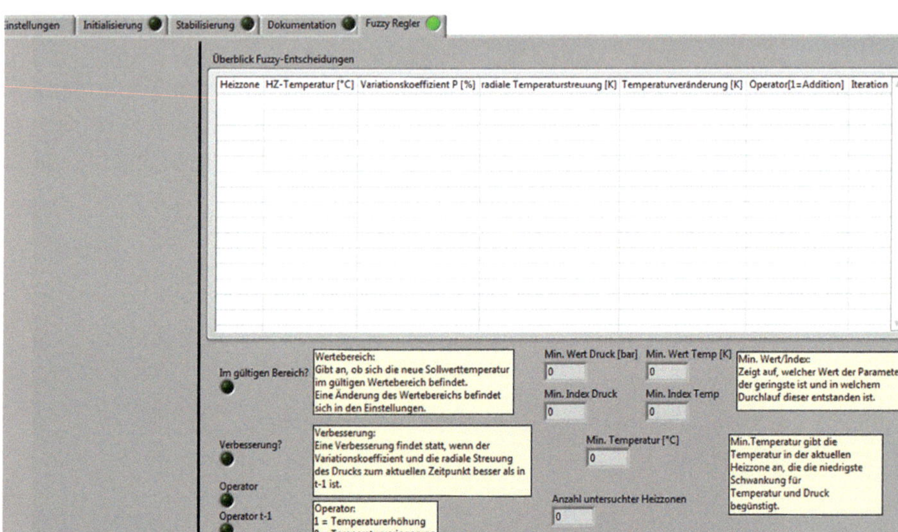

Figure A3.25 User interface in LabVIEW14 with further documentation

For a more detailed description of this work, the readers are referred to the Ph.D. thesis of V. Resonnek [227].

Outlook

With the development of self-optimizing controllers for the extruder barrel temperature profile, it can be expected that artificial intelligence in process control for extrusion will become a reality and be incorporated in commercial extruders. It can also be expected that the artificial intelligence will be used in extrusion troubleshooting.

A3.2.9 Conclusions

Setting correct barrel temperatures is a critical aspect of running a successful extrusion operation. If standard or typical barrel temperature profiles do not result in acceptable performance, the temperature settings will have to be optimized. DOE and OTE optimization methods are time-consuming and, therefore, unattractive and often impractical. This is particularly true for large production extruders where it may take several hours before an extruder stabilizes after a change in barrel temperature setpoint is made.

The dynamic optimization method was developed by Rauwendaal [220] to allow fast and accurate optimization of the barrel temperatures. This method has been used for many years and has proven its effectiveness in a wide variety of extrusion operations. In working with a large number of extrusion companies we have learned that the dynamic optimization method is not widely known.

V. Resonnek and V. Schöppner [221–227] have recently extended and advanced the optimization of barrel temperatures by developing a control scheme capable of automatic optimization of barrel temperatures. This is an important development because it removes the optimization burden from the operator or the process engineer. In practice, there are relatively few people that have the ability to effectively determine optimum barrel temperatures. As a result, many production extruders operate in a sub-optimum fashion.

Appendix 4: Process Signal Analysis Using Fast Fourier Transform

When we analyze a process signal in the form of a sine wave, see Figure A4.1, it is easy to see that this signal has a repeating pattern. The signal could be melt pressure, melt temperature, motor load, product dimension, or any other signal.

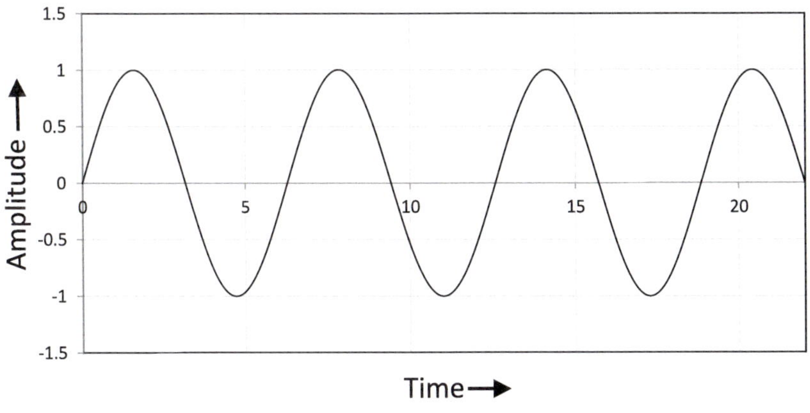

Figure A4.1 Graph of sine wave, $y = \sin(t)$, amplitude versus time

In extrusion process signals are generally not as simple as the one shown in Figure A4.1. There are multiple causes of variation in the extrusion process, somewhere between 20 and 40. At any time, 3 to 6 or more can have a measurable effect on the process. When different causes of variation affect the extrusion process the process signal will become more complicated, see for example Figure A4.2.

The signal shown in Figure A4.2 does not show a clear repeating pattern and it would be tempting to conclude that this is a random variation. The signal shown in Figure A4.2 is actually the sum of three sine waves, each with a different frequency and amplitude. The function plotted is $y = 4 + \sin(t) + 1.6\sin(0.3t + 0.25) + 0.7\sin(1.8t + 0.4)$. Examination of Figure A4.2 would not easily reveal that the signal is a combination of three sine waves with different amplitude and frequency.

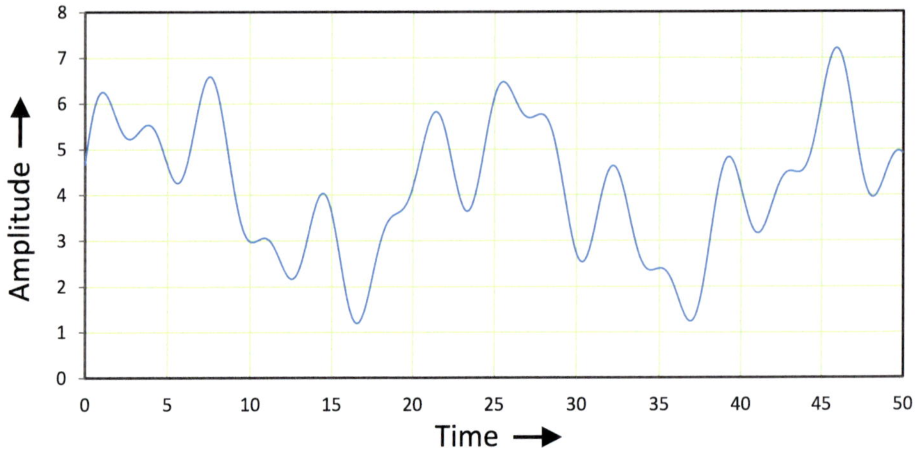

Figure A4.2 Graph of sum of three sine waves

One technique that is very useful in the analysis of signals is the Fast Fourier Transform or FFT. This is an algorithm that computes the discrete Fourier transform (DFT) of a sequence. With Fourier analysis a signal in the time or space domain is converted to a representation in the frequency domain. In the example above, FFT analysis makes it possible to find the base frequencies that make up the complex signal.

A relatively easy method to use FFT is to take advantage of this capability in the Microsoft spreadsheet program Excel. The FFT is available through an add-in called ToolPak. The FFT can be found under "Data Analysis" when the "DATA" tab is open, see Figure A4.3.

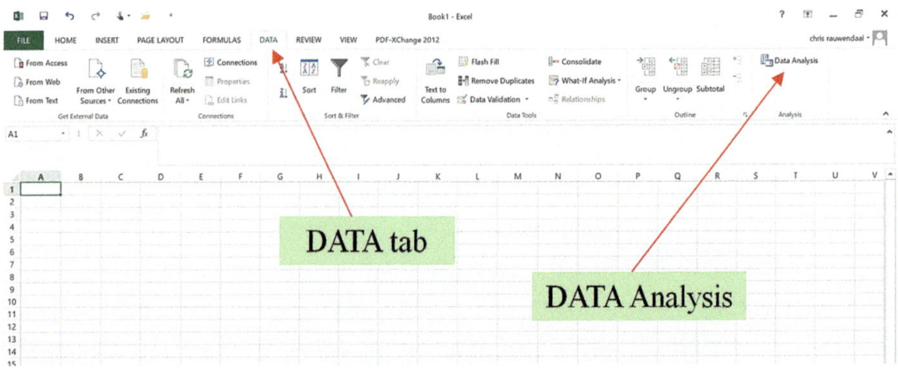

Figure A4.3 Microsoft Excel showing DATA tab with Data Analysis feature

If the data analysis feature is not shown, this indicates that the ToolPak add-in is not installed. The ToolPak feature can be installed by going to Excel Options by clicking on the Office button in Excel, see Figure A4.4.

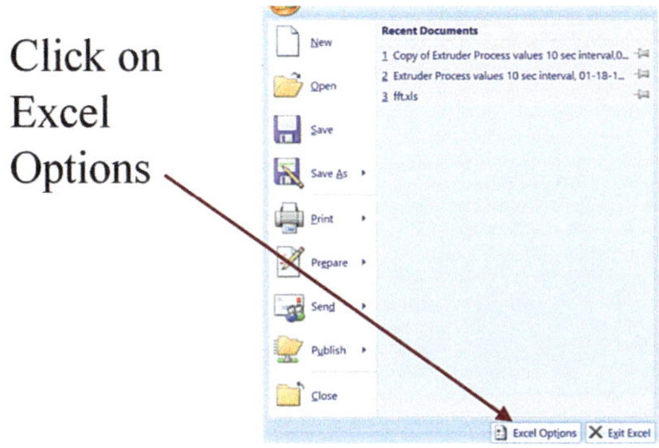

Figure A4.4 Excel Options

After clicking on Excel Options the following window will open, see Figure A4.5.

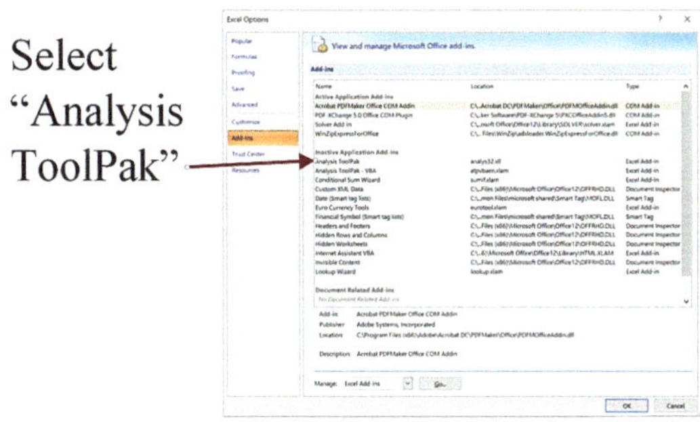

Figure A4.5 The Excel Options window showing Analysis ToolPak

When the Add-Ins window opens, select "Analysis ToolPak" and click OK.

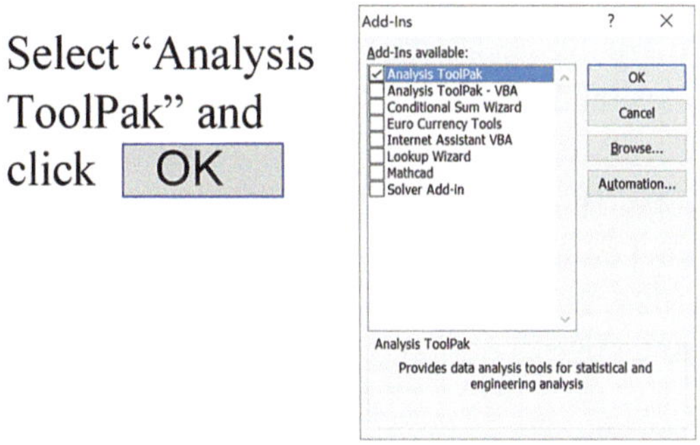

Figure A4.6 The Excel Add-Ins window

We will take the function $y = 4 + \sin(t) + 1.6\sin(0.3t + 0.25) + 0.7\sin(1.8t + 0.4)$ as an example of how to use the Excel FFT tool. A graph of this function is shown in Figure A4.2. We will use column A for time with a time increment of 0.01 seconds, see Figure A4.7, with A2 = 0, A3 = 0.01, A4 = 0.02, etc. until A4097 = 40.95.

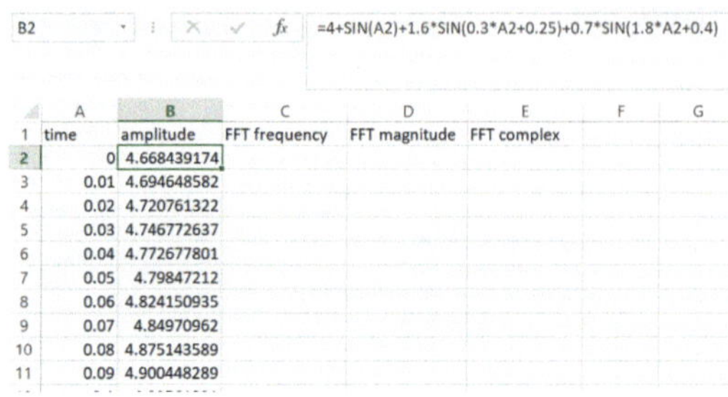

Figure A4.7 Column A, time, and column B, amplitude

Column B will be the amplitude. For cell B2 we will set the value of the cell to B2 = $4 + \sin(t) + 1.6\ \sin(0.3t + 0.25) + 0.7\ \sin(1.8t + 0.4)$. We will do this for all B cells from B2 through B4097. There are 4096 data in column A and B. When we plot the data in columns A and B we produce the graph shown in Figure A4.2.

When the Analysis ToolPak is installed, click on Data Analysis, Figure A4.3. This will open the Data Analysis window, see Figure A4.8.

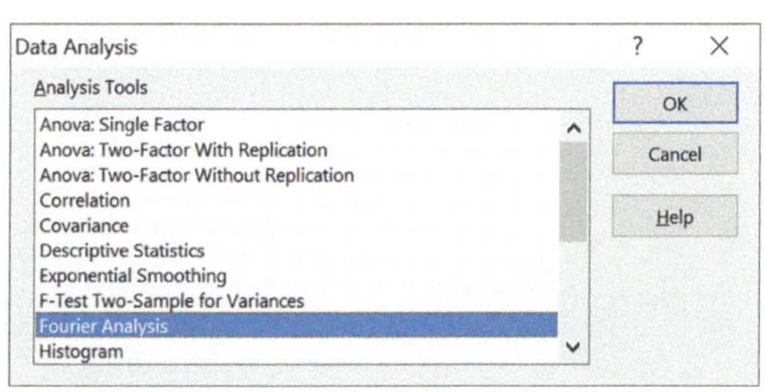

Figure A4.8 Data Analysis window

In the Data Analysis window, select Fourier Analysis and click OK. This will open the Fourier Analysis window, see Figure A4.9.

We have to select the input range and the output range. The input range will be the amplitude of the signal. The range should contain the amplitude data. The number of data points operated on must be a number that is a power of two (2, 4, 8, 16, 32, 64, 128, 256, 512, 1024, 2048, or 4096). The maximum number of data that Excel FFT can handle is 4096. In this example we will use the maximum number of data, 4096.

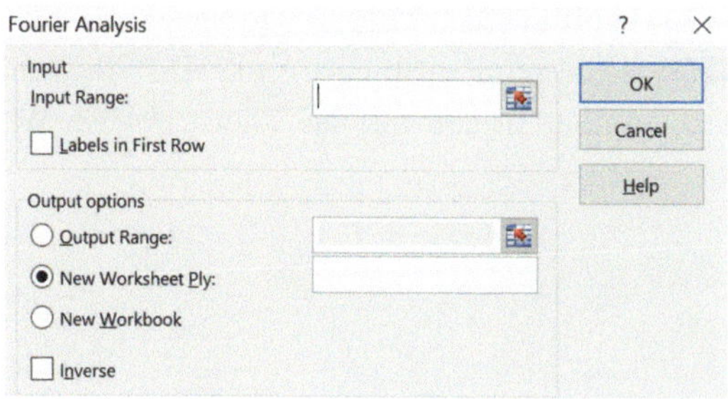

Figure A4.9 Fourier Analysis window

The output range will be the FFT complex. We will put this data in column E from E2 → E4097, see Figure A4.10.

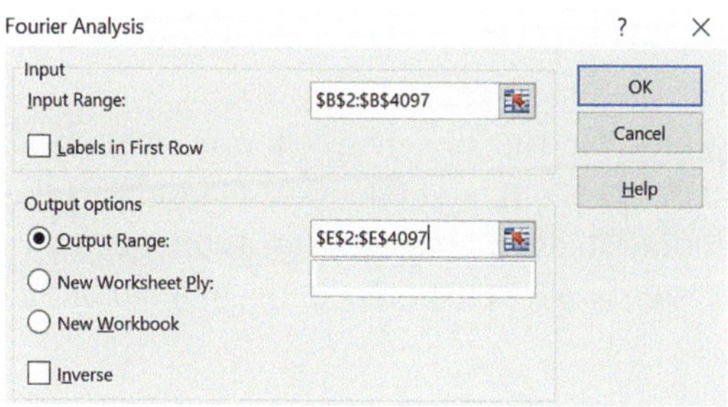

Figure A4.10 Fourier Analysis window with input and output range filled in

When we click OK the following data will show up in column E, see Figure A4.11. This column is called FFT complex and it contains complex numbers resulting from the Fast Fourier Transform.

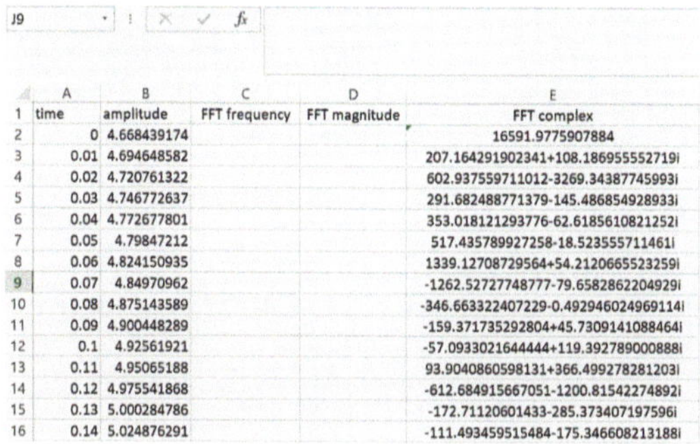

J9

	A	B	C	D	E
1	time	amplitude	FFT frequency	FFT magnitude	FFT complex
2	0	4.668439174			16591.9775907884
3	0.01	4.694648582			207.164291902341+108.186955552719i
4	0.02	4.720761322			602.937559711012-3269.34387745993i
5	0.03	4.746772637			291.682488771379-145.486854928933i
6	0.04	4.772677801			353.018121293776-62.6185610821252i
7	0.05	4.79847212			517.435789927258-18.523555711461i
8	0.06	4.824150935			1339.12708729564+54.2120665523259i
9	0.07	4.84970962			-1262.52727748777-79.6582862204929i
10	0.08	4.875143589			-346.663322407229-0.492946024969114i
11	0.09	4.900448289			-159.371735292804+45.7309141088464i
12	0.1	4.92561921			-57.0933021644444+119.392789000888i
13	0.11	4.95065188			93.9040860598131+366.499278281203i
14	0.12	4.975541868			-612.684915667051-1200.81542274892i
15	0.13	5.000284786			-172.71120601433-285.373407197596i
16	0.14	5.024876291			-111.493459515484-175.346608213188i

Figure A4.11 FFT complex data in column E

We now enter the FFT magnitude in column D. We set cell D2 = 2/4096*IMABS(E2). The denominator 4096 is the number of data and IMABS(E2) returns the absolute value of the complex number in cell E2, see Figure A4.12. We do this for D2 → D4097.

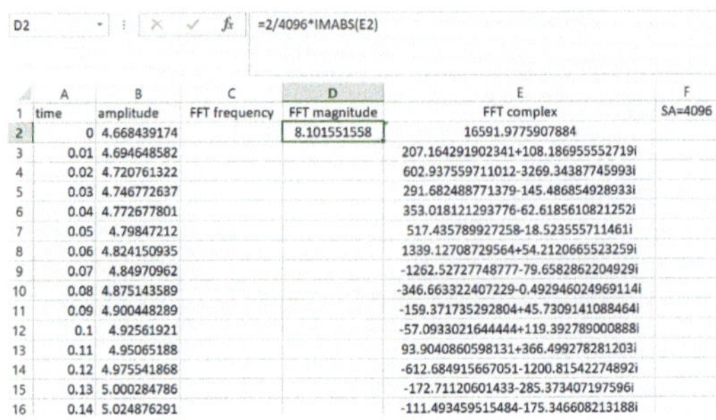

D2 =2/4096*IMABS(E2)

	A	B	C	D	E	F
1	time	amplitude	FFT frequency	FFT magnitude	FFT complex	SA=4096
2	0	4.668439174		8.101551558	16591.9775907884	
3	0.01	4.694648582			207.164291902341+108.186955552719i	
4	0.02	4.720761322			602.937559711012-3269.34387745993i	
5	0.03	4.746772637			291.682488771379-145.486854928933i	
6	0.04	4.772677801			353.018121293776-62.6185610821252i	
7	0.05	4.79847212			517.435789927258-18.523555711461i	
8	0.06	4.824150935			1339.12708729564+54.2120665523259i	
9	0.07	4.84970962			-1262.52727748777-79.6582862204929i	
10	0.08	4.875143589			-346.663322407229-0.492946024969114i	
11	0.09	4.900448289			-159.371735292804+45.7309141088464i	
12	0.1	4.92561921			-57.0933021644444+119.392789000888i	
13	0.11	4.95065188			93.9040860598131+366.499278281203i	
14	0.12	4.975541868			-612.684915667051-1200.81542274892i	
15	0.13	5.000284786			-172.7112060143 3-285.373407197596i	
16	0.14	5.024876291			-111.493459515484-175.346608213188i	

Figure A4.12 The value of cell D2 is set to D2 = 2/4096*IMABS(E2)

We now set the values for the FFT frequency in column C. We set C2 = 0, C3 = 100/4096 (= 0.024414063), and C4 = 0.024414063 + C3. The numerator for C3 is the number of time steps per second (Nt = 100); it is the reciprocal time step (ΔT). When the time step is ΔT = 0.01 second, the number of time steps per second Nt = 100.

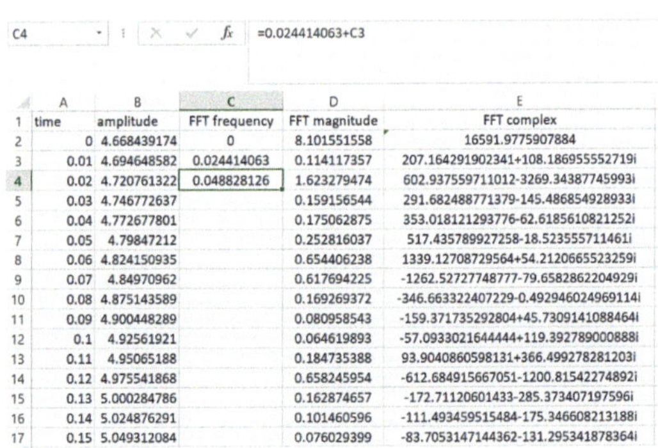

C4 =0.024414063+C3

	A	B	C	D	E
1	time	amplitude	FFT frequency	FFT magnitude	FFT complex
2	0	4.668439174	0	8.101551558	16591.9775907884
3	0.01	4.694648582	0.024414063	0.114117357	207.164291902341+108.186955552719i
4	0.02	4.720761322	0.048828126	1.623279474	602.937559711012-3269.34387745993i
5	0.03	4.746772637		0.159156544	291.682488771379-145.486854928933i
6	0.04	4.772677801		0.175062875	353.018121293776-62.6185610821252i
7	0.05	4.79847212		0.252816037	517.435789927258-18.523555711461i
8	0.06	4.824150935		0.654406238	1339.12708729564+54.2120665523259i
9	0.07	4.84970962		0.617694225	-1262.52727748777-79.6582862204929i
10	0.08	4.875143589		0.169269372	-346.663322407229-0.492946024969114i
11	0.09	4.900448289		0.080958543	-159.371735292804+45.7309141088464i
12	0.1	4.92561921		0.064619893	-57.0933021644444+119.392789000888i
13	0.11	4.95065188		0.184735388	93.9040860598131+366.499278281203i
14	0.12	4.975541868		0.658245954	-612.684915667051-1200.81542274892i
15	0.13	5.000284786		0.162874657	-172.71120601433-285.373407197596i
16	0.14	5.024876291		0.101460596	-111.493459515484-175.346608213188i
17	0.15	5.049312084		0.076029399	-83.7053147144362-131.295341878364i

Figure A4.13 The value of cell C4 is set to C4 = 0.024414063 + C3

We now calculate the FFT frequency for C4→C4097, see Figure A4.14.

H4

	A	B	C	D	E
1	time	amplitude	FFT frequency	FFT magnitude	FFT complex
2	0	4.668439174	0	8.101551558	16591.9775907884
3	0.01	4.694648582	0.024414063	0.114117357	207.164291902341+108.186955552719i
4	0.02	4.720761322	0.048828126	1.623279474	602.937559711012-3269.34387745993i
5	0.03	4.746772637	0.073242189	0.159156544	291.682488771379-145.486854928933i
6	0.04	4.772677801	0.097656252	0.175062875	353.018121293776-62.6185610821252i
7	0.05	4.79847212	0.122070315	0.252816037	517.435789927258-18.523555711461i
8	0.06	4.824150935	0.146484378	0.654406238	1339.12708729564+54.2120665523259i
9	0.07	4.84970962	0.170898441	0.617694225	-1262.52727748777-79.6582862204929i
10	0.08	4.875143589	0.195312504	0.169269372	-346.663322407229-0.492946024969114i
11	0.09	4.900448289	0.219726567	0.080958543	-159.371735292804+45.7309141088464i
12	0.1	4.92561921	0.24414063	0.064619893	-57.0933021644444+119.392789000888i
13	0.11	4.95065188	0.268554693	0.184735388	93.9040860598131+366.499278281203i
14	0.12	4.975541868	0.292968756	0.658245954	-612.684915667051-1200.81542274892i
15	0.13	5.000284786	0.317382819	0.162874657	-172.71120601433-285.373407197596i

Figure A4.14 FFT frequency and magnitude calculations completed

When we plot the FFT magnitude (column D) versus frequency (column C), we obtain the frequency domain, see Figure A4.15.

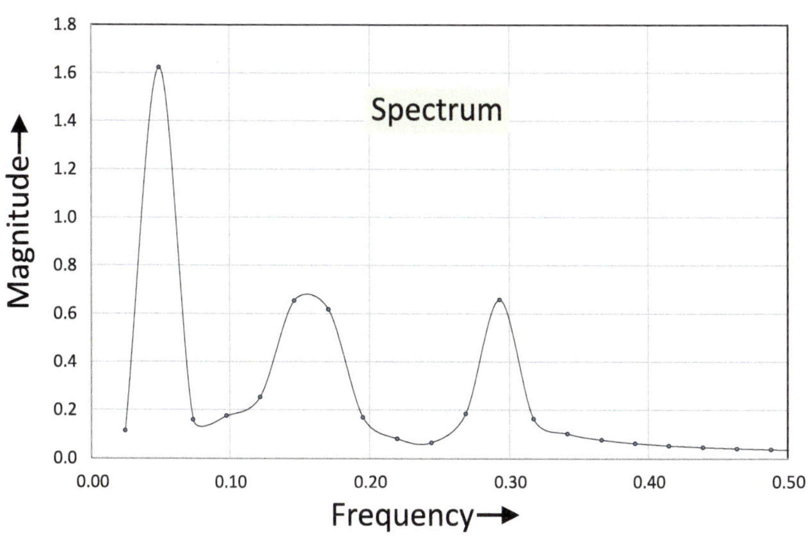

Figure A4.15 FFT magnitude versus frequency (step size 0.01, Ns = 4096)

The spectrum shows three distinct maxima. The first one is at frequency ~ 0.05, the second one at frequency ~ 0.16, and the third one at frequency ~ 0.29. The three frequencies correspond to the three sine functions. The function y can be written as follows:

$$y(t) = 4 + \sin(2\pi f_1 t) + 1.6 \sin(2\pi f_2 t + 0.25) + 0.7 \sin(2\pi f_3 t + 0.4) \quad (4.1)$$

where $f_1 = 0.159$, $f_2 = 0.0477$, and $f_3 = 0.286$ [rad/s].

The frequency-magnitude plot in Figure A4.15 was determined with a time step size of $\Delta T = 0.01$ seconds and a sample size of Ns = 4096. We can repeat the same procedure using a time step size of 1.0 and a sample size of 256. The resulting spectrum is shown in Figure A4.16. The peaks in this spectrum are significantly sharper, providing a better definition of the three frequencies making up the signal.

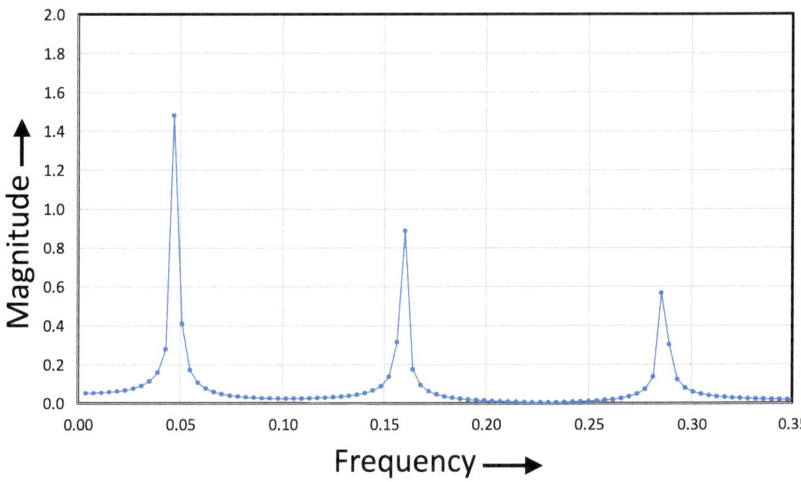

Figure A4.16 FFT magnitude versus frequency (step size $\Delta T = 1.0$, Ns = 256)

The first peak is the $1.6\sin(0.3t + 0.25)$ term, the second peak is the $\sin(t)$ term, and the third peak is the $0.7\sin(1.8t + 0.4)$ term. The height of the peak reflects the constant before the sine term. The ratio of peak height one to peak height two to peak height three equals 1.6:1.0:0.7. This example shows how the three frequencies that make up the signal shown in Figure A4.2 can be determined through FFT analysis. Also, the height of the peak reflects the amplification factor of the sine wave.

When we plot the frequency spectrum we should not plot more than half the rows of the sample. For instance, in Figure A4.16 we plotted 128 rows of the 256 rows. If we plot more rows, the FFT representation starts to duplicate itself. This is shown in the frequency spectrum in Figure A4.17, where we plot all 256 rows. Of course, we can avoid this issue by formatting the horizontal axis and selecting the maximum value to be 0.3 rather than 1.0.

Figure A4.18 shows function $y(t)$ for two time step sizes. The blue line is for a time step of 0.01 seconds, the red line is for a time step of 10 seconds. There is a significant and obvious difference in the shape of the two curves.

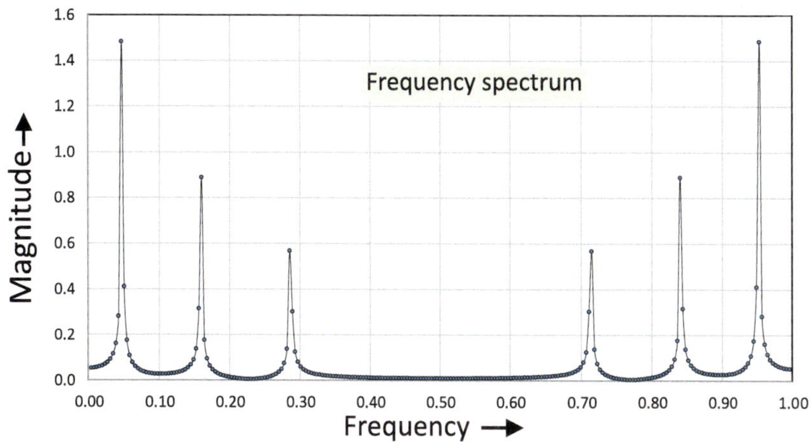

Figure A4.17 FFT magnitude versus frequency with full sample size plotted

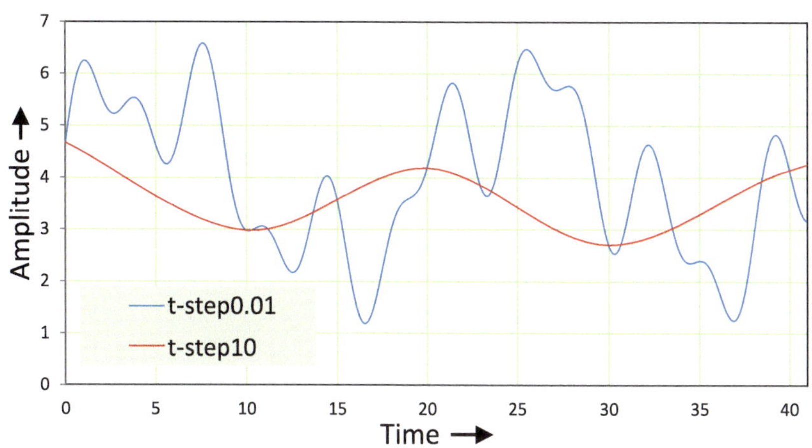

Figure A4.18 Function $y = 4 + \sin(t) + 1.6\sin(0.3t + 0.25) + 0.7\sin(1.8t + 0.4)$ plotted with two time steps; blue line: time step 0.01 seconds, red line: time step 10 seconds

The signal is modified significantly when the time step is increased from 0.01 to 10 seconds. The short-term variation can no longer be captured, only long-term variation can be observed and significant information is lost. In signal processing this problem is called aliasing. The problem is not only that information is lost; we are actually getting wrong information about the signal.

This example has practical implications in actual extrusion operations. Modern extrusion lines should have data acquisition systems (DAS). Unfortunately, the number of extrusion lines with good DAS capability is small. Most of the DAS used today have low speed data acquisition, perhaps capturing data once per minute or once per 10 seconds.

Figure A4.18 illustrates quite graphically that with low speed data acquisition a large portion of the pertinent process data is lost and the information we get is actually wrong! Every extrusion process experiences short-term variations occurring at a time scale of less than one second. Examples of process variables with short-term variation are melt pressure, melt temperature, and motor load; these are the vital signs of extrusion process.

A good DAS should have high speed data acquisition capability. This means that the DAS should have the ability to capture certain critical variables at least 10 times per second – a capture rate of one data point every 0.1 seconds. Unfortunately, many DAS have a data capture rate of 10 seconds or longer. With such a slow data capture rate a significant part of the process variation is not accessible and the graphical information presented to the user is actually wrong!

The frequency spectrum obtained with a time step of 10 seconds is shown in Figure A4.19 as the red curve. Figure A4.19 also shows the frequency spectrum with a time step of 1 second, the blue curve.

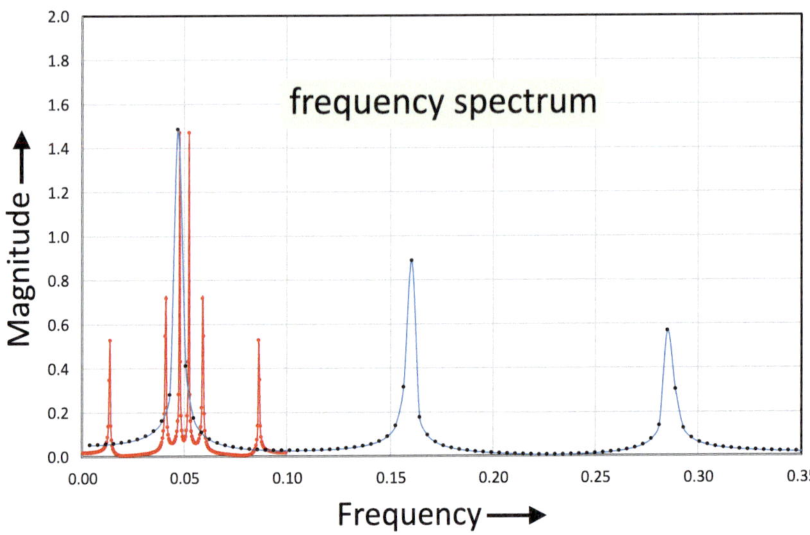

Figure A4.19 FFT frequency spectrum; blue line: time step 1 second, red line: time step 10 seconds

The red curve resulting from a time step of 10 seconds shows six peaks between 0.0 and 0.10. These peaks do not reflect the true frequency spectrum reflected by the blue line; these peaks result from a time step that is too large to properly capture the actual variation. This example illustrates how important it is to use the proper time steps when analyzing process variation.

There is an important rule when it comes to setting the time steps; this is the Nyquist sampling theorem. Olshausen [210] discusses this theorem in his paper on aliasing. Aliasing occurs when a signal is discretely sampled at a rate insufficient to capture the changes in the signal. The Nyquist sampling theorem can be stated as follows:

"*The sampling frequency should be at least twice the highest frequency contained in the signal.*"

If f_C is the highest frequency contained in the signal, the sampling frequency f_S should be greater or equal to $2f_C$. In mathematical terms:

$$f_S \geq 2f_C \tag{4.2}$$

This theorem can be understood by considering a sine wave at a frequency of 1 Hz, see Figure A4.20.

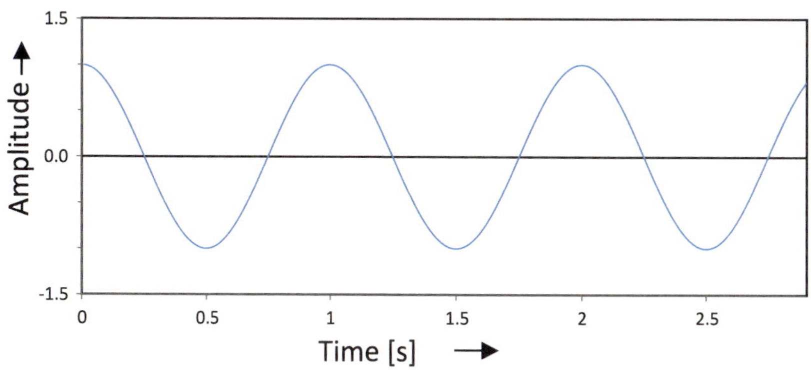

Figure A4.20 Sine wave at frequency 1 Hz

If the waveform is sampled at 2 Hz, see Figure A4.21, this is sufficient to capture each peak and trough of the signal.

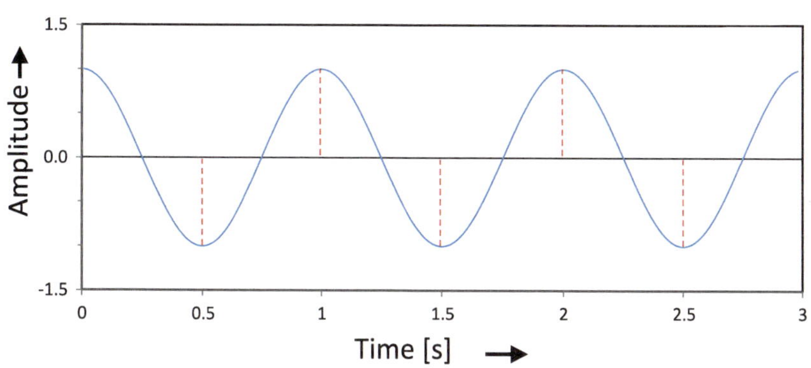

Figure A4.21 Sine wave at frequency 1 Hz sampled at 2Hz

If we sample at a frequency lower than 2Hz, then there are not enough samples to capture all the peaks and troughs in the signal. Figure A4.22 shows a sine wave sampled at 1.5 Hz.

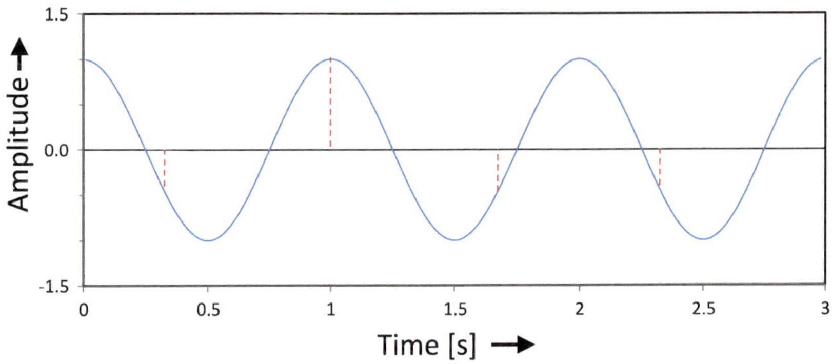

Figure A4.22 Sine wave at frequency 1 Hz sampled at 1.5 Hz

At a sampling frequency of 1.5 Hz there are not enough samples to capture all the peaks and troughs in the signal. In this situation we are not only losing information, we are actually getting incorrect information about the signal. If we have a DAS that captures process data every 10 seconds, we will only be able to analyze a variation occurring over a time period longer than 20 seconds. In other words, we can only see variations at frequencies less than 0.05 Hz. All variations occurring at higher frequencies will not be captured by the DAS. That means that a variation caused by screw rotation, such as screw beat, cannot be seen unless the screw speed is less than 3 rpm. Typical screw speeds are one or two orders of magnitude higher.

Conclusions

Fast Fourier Transform is a powerful tool to analyze process signals. It allows determination of the frequencies that make up a complex signal. This is critical in troubleshooting extrusion problems because when we can determine the frequency, the number of possible causes can be narrowed down significantly. For instance, when we see that the discharge pressure varies at a frequency of 1 Hz and the screw speed is 60 rpm (1 Hz), then the pressure variation is most likely caused by the screw rotation.

The Microsoft spreadsheet program Excel has Fast Fourier Transform capability built-in as part of the ToolPak add-in. When DAS data is imported into Excel, it allows the user to perform FFT without the need for other tools. In order to use FFT properly the sampling frequency should be at least twice the highest frequency contained in the signal. If this condition is not fulfilled, the analysis will produce incorrect results.

References

1. C. Rauwendaal, "Statistical Process Control in Injection Molding and Extrusion", Hanser Publishers, Munich (2000)
2. C. Rauwendaal and K. Cantor, "Computer-Based Interactive Training in Extrusion", 56th SPE ANTEC, Atlanta, GA, 340–344 (1998), also published in *Plastics Engineering*, June issue, pp. 41–43 (1998)
3. J. Ogando, "Portable Analyzers Find Out What Ails Your Process", *Plastics Technology*, February, pp. 54–63 (1995)
4. G. Mennig, "Measuring of Local Mass Temperature Inside of Extruders by Temperature Indicating Materials", SPE ANTEC, 529–531 (1978)
5. C. Rauwendaal, Unpublished study (1978)
6. W. W. Wendlandt, "Thermal Methods of Analysis", Wiley, NY (1979)
7. T. Daniels, "Thermal Analysis", Anchor Press, London (1973)
8. E. A. Turi (Ed.), "Thermal Characterization of Polymeric Materials", Academic Press, NY (1981)
9. G. W. Miller and R. S. Porter, "Analytical Calorimetry", Vol. 2, Plenum Press, NY, p. 407 (1970)
10. V. B. F. Mathot (Ed.), "Calorimetry and Thermal Analysis of Polymers", Hanser Publishers, Munich (1994)
11. W. Mehdorn, *Kunststoffe*, 34, pp. 133–136 (1944)
12. G. Mennig, "Wear in Plastics Processing", Hanser Publishers, Munich (1995)
13. E. Wagner, Ph. D. thesis, Technische Hochschule Darmstadt, Germany (1978)
14. M. M. Kruschchov, *Ind. Lab* (USSR), 20, pp. 372–376 (1962)
15. H. Schuele and H. G. Fritz, *Kunststoffe*, 73, pp. 603–605 (1983)
16. G. Mennig and P. Volz, *Kunststoffe*, 70, pp. 385–390 (1980)
17. W. Bauer, K. Eichler, and W. John, *Kunststoffe*, 57, pp. 53–55 (1967)
18. K. Eichler and W. Frank, *Ind. Anzeiger*, 95, pp. 2033–2035 (1973)

19. W.D. Mahler, Ph.D. thesis, Technische Hochschule Darmstadt, Germany (1975)
20. P. Volz, W.D. Mahler, and G. Mennig, *Kunststoffe*, 66, pp. 428–434 (1976)
21. W. Knappe and W.D. Mahler, *Kunstoff-Rundschau*, 19, pp. 45–51 (1972)
22. R.S. Plumb and W.A. Glaeser, *Wear*, 46, pp. 219–229 (1978)
23. E.L. Moon and R.A. Hunter, "An Abrasion Study of Surface Treated Calcium Carbonate Fillers in Rigid PVC", Technical Bulletin, Georgia Marble Corp., Atlanta, GA
24. H.G. Fritz, *Kunststoffe*, 65, 176–182 and 258–274 (1975)
25. P. Volz, *Kunststoffe*, 69, pp. 758–771 (1979)
26. P. Volz, *Kunststoffe*, 67, pp. 279–283 (1976)
27. G.P. Calloway, E.D. Morrison, and R.F. Williams, Jr., SPE ANTEC, pp. 354–360 (1972)
28. D. Braun and G. Maelhammar, *Angew, Makromol. Chem.*, 69, pp. 157–167 (1978)
29. H.G. Moslé, H.F. Schmidt, and J. Schroeder, *Kunststoffe*, 67, pp. 220–223 (1977)
30. G. Maelhammar, Ph.D. thesis, Technische Hochschule Darmstadt, Germany (1978)
31. P. Volz, *Kunststoffe*, 69, pp. 259–262 (1979)
32. E. Broszeit, Ph.D. thesis, Technische Hochschule Darmstadt, Germany (1972)
33. G.A. Saltzman and J.H. Olson, SPE ANTEC, pp. 173–175 (1974)
34. V. Murer and G.A. Saltzman, *Kunststoffe*, 66, pp. 219–220 (1976)
35. W.W. McCandles and W.D. Maddy, *Plastics Technology*, February, pp. 89–93 (1981)
36. S.H. Collins, *Plastics Compounding*, May/June, pp. 113–124 (1982)
37. W.D. Mahler, *Kunststoffe*, 67, pp. 224–226 (1977)
38. B.A. Olmsted, *SPE Journal*, 26, pp. 42–43 (1970)
39. P. Luelsdorf, VDI-Lehrgang: "Grundlagen der Extrudertechnik" (1975)
40. R.A. Lai Fook and R.A. Worth, SPE ANTEC, Washington, DC, pp. 450–252 (1978)
41. H.H. Winter, SPE ANTEC, New Orleans, pp. 170–175 (1979)
42. K. O'Brien, *Plast. Technology*, February, pp. 73–74 (1982)
43. G. Thursfield, *Modern Plastics*, October, pp. 94–96 (1975)
44. M. Hoffmann, *Plastics Technology*, April, pp. 67–72 (1982)
45. K. Luker, Paper presented at the TAPPI Paper Synthetics Conference in Lake Buena Vista, Florida (1983)

46. S. Levy, Plast. *Mach. Equip.*, March (1984)
47. S. Trompler, *Kunststoffe*, 73, 596 (1983)
48. W. Lucius, *Kunststoffe*, 63, pp. 433-435 (1973)
49. "Plasticating Components 2000", a publication from Spirex Corporation, Youngstown, Ohio (2000)
50. C. Panzera and G.A. Saltzman, Proceedings of the 2nd International Conference on Wear of Materials, pp. 441-448 (1979)
51. R. Bonetti, *Metal Progress*, June (1981)
52. V. Anand and K. Das, "The Selection of Screw Base and Hard Facing Materials," publication CEC SSB 493.01 from Canterbury Engineering Company, Inc., Champlee, Georgia
53. C. Rauwendaal and P. Gramann, "Why Corrosion Resistant Screws Can Bind in the Extruder Barrel", 56th SPE ANTEC, Atlanta, GA, pp. 102-106 (1998)
54. L.B.P.M. Janssen, G.H. Noomen, and J.M. Smith, "The Temperature Distribution across a Single-Screw Extruder Channel", *Plastics & Polymers*, August, pp. 135-140 (1975)
55. C. Rauwendaal and J. Anderson, "FEA of Flow in Single Screw Extruders," 52nd SPE ANTEC, pp. 298-305 (1994)
56. D.I. Marshall, I. Klein, and R.H. Uhl, "Measurement of Screw and Plastic Temperature Profiles in Extruders", *SPE Journal*, 20, No. 4, April, 329 (1964)
57. S.A. Klein, W.A. Beckman, and G.E. Myers, FEHT – Finite Element Analysis, v6.98, copyright 1996-97, University of Wisconsin-Madison (1997)
58. A. Casale and R.S. Porter, "Polymer Stress Reactions", Vol. 1 and 2, Academic Press, New York (1978)
59. Y.I. Frenkel, *Acta Physicochim (USSR)*, 19, 51 (1944)
60. W.J. Kauzmann and H. Eyring, *J. Am. Chem. Soc.*, 62, 3113 (1940)
61. F. Bueche, "Physical Properties of Polymers", Wiley, New York (1962)
62. A. Holmstrom, A. Andersszon, and E.M. Sorvik, *Polym. Eng. Sci.*, 17, pp. 728-273 (1977)
63. P.W. Springer, R.S. Bradley, and R.E. Lynn, *Polym. Eng. Sci.*, 15, pp. 583-587 (1975)
64. V.L. Folt, *Rubber Chem. Technol.*, 42, 1294 (1969)
65. R.W. Ford, R.A. Scott, and R.J.B. Wilson, *J. Appl. Polym. Sci.*, 12, 547 (1968)
66. K. Arisawa and R.S. Porter, *J. Appl. Polym. Sci.*, 14, 879 (1970)
67. R.S. Porter and A. Casale, *Polym. Eng. Sci.*, 25, pp. 129-156 (1985)
68. J.A. Odell, A. Keller, and M.J. Miles, *Polymer Comm.*, 24, pp. 7-10 (1983)

69. G. Pinto and Z. Tadmor, *Polym. Eng. Sci.*, 10, pp. 279–288 (1970)

70. D. Bigg and S. Middleman, *Ind. Eng. Chem. Fundam.*, 13, pp. 66–71 (1974)

71. G. Lidor and Z. Tadmor, *Polym. Eng. Sci.*, 16, pp. 450–462 (1976)

72. D. Wolf and D. H. White, *AIChE J.*, 22, pp. 122–131 (1976)

73. N. R. Schott and D. V. Saleh, SPE, ANTEC, Washington, DC, pp. 536–539 (1978)

74. C. Rauwendaal, *Polym. Eng. Sci.*, 21, pp. 1982–1100 (1981)

75. J. C. Golba, SPE ANTEC, New York, pp. 83–87 (1980)

76. Z. Kemblowski and J. Sek, *Polym. Eng. Sci.*, 21, pp. 1194–1202 (1981)

77. D. B. Todd, *Polym. Eng. Sci.*, 15, pp. 437–443 (1975)

78. L. P. B. M. Janssen, "Twin Screw Extrusion", Elsevier, New York (1978)

79. L. P. B. M. Janssen, R. W. Hollander, M. W. Spoor, and J. M. Smith, *AIChE J.*, 25, pp. 345–351 (1979)

80. C. J. Walk, SPE ANTEC, San Francisco, pp. 423–426 (1982)

81. R. J. Nichols, J. C. Golba, and P. K. Shete, paper presented at the AIChE Annual Meeting, Paper No. 59F (1983)

82. O. Levenspiel, "Chemical Reaction Engineering", Wiley, New York (1965)

83. D. M. Himmelblau and K. A. Bischoff, "Chemical Process Analysis", Wiley, New York (1966)

84. P. V. Danckwerts, *Chem. Eng. Sci.*, 2, 1, (1953)

85. L. P. B. M. Janssen, G. H. Noomen, and J. M. Smith, *Plastics & Polymers*, August, pp. 135–140 (1975)

86. S. Anders, D. Brunner, and F. Panhaus, *Plaste und Kautschuk*, 23, pp. 593–598 (1976)

87. T. Osswald, P. Gramann, B. Davis, M. del P. Noriega, and O. A. Estrada, "Experimental Study of a New Dispersive Mixer", 57th SPE ANTEC, New York, pp. 167–176 (1999), also the annual meeting of the Polymer Processing Society in Den Bosch, the Netherlands (1999)

88. H. E. H. Meijer, J. F. Ingen-Housz, and W. C. M. Gorissen, *Polym. Eng. Sci.*, 18, pp. 288–292 (1978)

89. C. Rauwendaal, 47th SPE ANTEC, New York, NY, pp. 108–110 (1989)

90. J. F. T. Pittman, *Dev. Plast. Technol.*, 3, pp. 203–273 (1986)

91. C. Rauwendaal and J. F. Ingen-Housz, *Intern. Polym. Proc.*, 3, pp. 123–133 (1988)

92. C. Rauwendaal, "Leakage Flow in Extruders", Doctoral Thesis, Twente University, the Netherlands (1988)

93. N. C. Wheeler, Annual Conv. Wire Ass., Baltimore, MD, October 22–25 (1962)

94. R.T. Fenner, A.P D. Cox, and D. P Isherwood, *Polymer*, 20, pp. 733–736 (1979)
95. F.N. Cogswell, *J. Non-Newtonian Fluid Mech.*, 2, pp. 373–47 (1977)
96. M.T. Dennison, *Trans. J. Plastics Inst.*, 35, pp. 803–808 (1967)
97. J.J. Benbow and E.R. Howells, *Trans. J. Plastics Inst.*, 30, pp. 240–254 (1960)
98. British Patent 32559/72
99. J.P. Tordella, in "Rheology", Vol. 4, Chapter 3, F.R. Eirich (Ed.), Academic Press, New York (1969)
100. J.L. White, *Appl. Polym. Symp.*, 20, 155 (1973)
101. J.M. Lupton and J.W. Regester, *Polym. Eng. Sci.*, 5, 235 (1965)
102. L.L. Blyler and A.C. Hart, *Polym. Eng. Sci.*, 10, 193 (1970)
103. S.M. Barnett, *Polym. Eng. Sci.*, 7, 168 (1967)
104. E. Boudreaux and J.A. Cuculo, *J. Macromol. Sci. – Rev. Macromol Chem.*, C16, pp. 39–77 (1977–1978)
105. C.D. Han and R. Lamonte, *Polym. Eng. Sci.*, 11, 385 (1971)
106. J.L. den Otter, *Rheol. Acta*, 10, pp. 200–207 (1971)
107. T.W. Huseby, *Trans. Soc. Rheol.*, 10, pp. 181–190 (1966)
108. A.P. Metzger and C.W. Hamilton, *SPE Trans.*, 4, pp. 107–112 (1964)
109. G.V. Vinogradov et al., *Polym. Eng. Sci.*, 12, pp. 323–334 (1972)
110. J.J. Benbow and P. Lamb, *SPE Trans*, 3, pp. 7–17 (1963)
111. W. Phillippoff and F.H. Gaskins, *Trans. Soc. Rheol.*, **I**, pp. 263–284 (1958)
112. C.J.S. Petrie and M.M. Denn, *AIChE J.*, 22, pp. 109–236 (1976)
113. W. Gleissle, *Rheol. Acta*, 21, 484 (1982)
114. J.J. Benbow, R.V. Charley, and P. Lamb, *Nature*, 192, 223 (1961)
115. L.A. Utracki and R. Gendron, *J. Rheol.*, 5, 28, pp. 601–623 (1984)
116. U.S. Patent 2,991,508 by R.T. Fields and C.F.W. Wolf to E.I. du Pont de Nemours and Company, issued July 11 (1961)
117. A.M. Kraynik and W.R. Schowalter, *J. Rheol.*, 25, pp. 95–114 (1981)
118. R.A. Worth, J. Parnaby, and H.A.A. Helmy, *Polym. Eng. Sci.*, 17, 257 (1977)
119. C.J.S. Petrie and M.M. Denn, *AIChE J.*, 22, pp. 236–246 (1976)
120. J.C. Miller, *SPE Trans.*, 3, 134 (1963)
121. S. Kase, *J. Appl Polym. Sci.*, 18, 3279 (1974)
122. G.F. Cruz-Saenz, G.J. Donnelly, and C.B. Weinberger, *AIChE J.*, 22, 441 (1976)
123. J.R.A. Pearson and Y.T. Shah, *Trans. Soc. Rheol.*, 16, 519 (1972)

124. J.R.A. Pearson and Y.T. Shah, *Ind. Eng. Chem. Fundam.*, 13, 134 (1979)
125. R.J. Fischer and M.M. Denn, *Chem. Eng. Sci.*, 30, 1129 (1975)
126. Y. Ide and J.L. White, *J. Appl. Polym. Sci.*, 20, pp. 2511–2531 (1976)
127. Y. Ide and J.L. White, *J. Non-Newt. Fluid Mech.*, 2, pp. 281–298 (1977)
128. Y. Ide and J.L. White, *J. Appl. Polym. Sci.*, 22, pp. 1061–1079 (1978)
129. J.L. White and Y. Ide, *J. Appl. Polym. Sci.*, 22, pp. 3058–3074 (1978)
130. R.S. Lenk, *J. Appl. Polym. Sci.*, 22, pp. 1781–1785 (1970)
131. N.C. Wheeler, Annual Convention Wire Association, Baltimore, MD, October 22–25 (1962)
132. U.S. Patent 3,006,029 (1961)
133. P. Gramann, T. Osswald, B. Davis, and C. Rauwendaal, 56th SPE ANTEC, Atlanta, GA, pp. 277–283 (1998)
134. C. Rauwendaal, Annual Meeting of the Polymer Processing Society, Yokohama, Japan, June 8–12 (1998)
135. C. Rauwendaal, *Plastics Additives and Compounding*, September issue (1999)
136. C. Rauwendaal, Film Conference 99, Somerset, NJ, 141–148, December 7–9 (1999)
137. C. Rauwendaal, Compounding Conference, Cleveland, OH, 219–228, November 14–15 (2000)
138. U.S. Patent 5,932,159, C. Rauwendaal (1999)
139. U.S. Patent 6,136,246, C. Rauwendaal, P. Gramann, B. Davis, T. Osswald (2000)
140. Z. Tadmor and I. Klein, "Engineering Principles of Plasticating Extrusion", Van Nostrand Reinhold, New York (1970), p. 413
141. I.R. Edmondson and R.T. Fenner, *Polymer*, 16, pp. 49–56 (1975)
142. B.H. Maddock, *SPE J.*, 20, 1277 (1964)
143. H.W. Gitschner and J. Lutterbeck, *Kunststoffe*, 74, pp. 12–14 (1984)
144. J. Becker, P. Bengtsson, C. Klason, J. Kubat, and P. Saha, *Intern. Polymer Processing*, VI, 4, pp. 318–325 (1991)
145. J. Becker, C. Klason, J. Kubat, and P. Saha, *Intern. Polymer Processing*, VI, 4, pp. 326–331 (1991)
146. R.H. Reinhard, A.R. Crixell, C.L. Loney, and P.W. Tidwell, 32nd SPE ANTEC, pp. 40–42 (1974)
147. G.M. Gale, 49th SPE ANTEC, 95 (1991)
148. G.M. Gale, *Adv. Polym. Techn.*, 16(4), 251 (1997)
149. M.H. Mack, *Plastics Engineering*, 47/12, 39 (1991)

150. M.R. Thompson, G. Donoian, and J.C. Christiano, 57th SPE ANTEC, 145 (1999)
151. P. Elemans and J.M. van Wunnik, 58th SPE ANTEC, pp. 265–267 (2000)
152. P. Elemans, 58th SPE ANTEC, pp. 2582–2586 (2000)
153. K. Luker, paper presented at the Continuous Compounding Conference, Beachwood, Ohio, November 14–15 (2000)
154. K. Luker, paper presented at the Wood Composite Conference, December 5–6 (2000)
155. C. Rauwendaal, "Polymer Mixing, A Self-Study Guide", Hanser Publishers, Munich (1998)
156. P. Limbach, in "Extrusion Solutions", publication by the SPE Extrusion Division, p. 6 (1999)
157. S.E. Amos, G.M. Giacoletto, J.H. Horns, C. Lavallée, and S.S. Woods, "Polymer Processing Aids" in "Plastics Additives Handbook", 5th Edition, H. Zweifel, Hanser Publishers, Munich (2000)
158. T.J. Blong, K. Focquet, C. Lavallée, SPE ANTEC Technical Papers, p. 3011 (1997)
159. C.M. Chan, *Intern. Pol. Proc.*, V, 10 (1995)
160. D.E. Hauenstein, D.J. Cimbalik, and P.G. Pape, SPE ANTEC Techn. Papers, pp. 3002–3010 (1997)
161. D.E. Priester, G.R. Chapman, "Reducing Die Build Up in Extrusion Applications", Technical Information, DuPont Dow Elastomers
162. S.E. Amos, SPE RETEC Techn. Papers, pp. 133–143 (1997)
163. M. Kanekiyo, M. Kobayashi, et al., *Macromolecules*, 33 (21), pp. 7971–7976 (2000)
164. Eval Americas – Kuraray, "Processing EVAL resins", Technical Bulletin No. 120, 07 (2000)
165. S. Derezinski, *J. Plastic Film and Sheeting*, 24, pp. 125–136 (2008)
166. S. Jikan, M.S.F. Samsudin, Z.M. Ariff, et al., *J. Reinforced Plastics and Composites*, 28, 2577 (2009)
167. C. Rauwendaal, ANTEC 2000, Extrusion Division, Vol. 1, Orlando, USA (2000)
168. T. Osswald, L.S. Turng, et al., "Injection Molding Handbook, 2nd ed.", Hanser Publishers, Munich (2007)
169. Guojun Xu, "Study of Thin-Wall Injection Molding", Ph.D. Dissertation, The Ohio State University, USA, (2004)
170. J.F. Agassant, P.D. Anderson, et al., in: I. Manas-Zloczower (Ed.) "Mixing and Compounding of Polymers: Theory and Practice, 2nd ed.", Hanser Publishers, Munich (2009)

171. E. Grünschloss, Polymer Single Screw Extrusion: Modeling", Encyclopedia of Materials: Science and Technology, Elsevier Science Ltd., pp. 7491–7504 (2001)
172. W. Michaeli, "Extrusion Dies for Plastics and Rubber, 3rd ed.", Hanser Publishers, Munich (2003)
173. C. Rauwendaal "How to fix vent-flow problems", Troubleshooter, *Plastics Technology* (2002)
174. Maag Pump Systems, "Gear pumps for thermoplastic applications", *http://www.maag.ch*, Zurich, (2009)
175. J. Liebetrau and S. Grollmisch, "Predictive Maintenance with Airborne Sound Analysis", *Process Control & Automation*, September 11 (2017)
176. C. Rauwendaal, "Polymer Extrusion", 5th edition, Hanser Publishers, Munich (2014)
177. C. Rauwendaal, "Understanding Extrusion", 3rd edition, Hanser Publishers, Munich (2019)
178. J. Perdikoulias, "Dimensional Analysis of Extruded Profiles Using a Mobile Phone Camera", *Extrusion* 2017, October 18–20 (2017)
179. K. Chaloupková and M. Zatloukal, "Effect of Die Design on Die Drool Phenomenon for Metallocene based LLDPE: Theoretical and Experimental Investigation", *Journal of Applied Polymer Science* (2009)
180. K. Chaloupková and M. Zatloukal, "Theoretical and Experimental Analysis of the Die Drool Phenomenon for Metallocene LLDPE", *Polymer Engineering & Science* (2007)
181. M. Zatloukal and J. Musil, "Die Design Effect on Internal Die Drool Phenomenon", https://www.4spe.org/files/Chapters/D22/Technical Articles/2013-08-POTM.pdf (2013)
182. J. Musil and M. Zatloukal, "Historical Review of Die Drool Phenomenon in Plastics Extrusion", *Polymer Reviews*, 54, pp. 139–184 (2014)
183. H. Overeijnder, "PVC processing with counter-rotating twin screw extruders", Extrusion seminar Twente University, April 15–17 (2019)
184. C. Rauwendaal, "Leakage flow in extruders", Ph. D. thesis, Twente University (the Netherlands) (1988)
185. A. Kelly et al., "Melt temperature field measurement in single screw extrusion: influence of melt pressure and die geometry", SPE ANTEC, pp. 291–295 (2005)
186. J. Vera-Sorroche et al., "Thermal optimization of polymer extrusion using in-process monitoring techniques", *Applied Thermal Engineering*, Volume 53, Issue 2, May 2, pp. 405–413 (2013)

187. J. Vera-Sorroche et al., “Infrared melt temperature measurement of single screw extrusion”, *Polymer Engineering & Science*, May (2015)

188. G. Menges, Haberstroh, and W. Michaeli: Werkstoffkunde Kunststoffe, Hanser Publishers, Munich (2011)

189. G. Hiesgen, K. Saul, W. Spitz, and P. Weiss, “Pipe cooling simulation for energy savings and enhanced product quality”, Proceedings of the Polymer Processing Society 28th Annual Meeting, PPS-28, December 11–15, Pattaya, Thailand (2012)

190. G. Hiesgen, K. Saul, W. Spitz, P. Weiss, and J. Wortberg, “Application of the finite difference method for the cooling process simulation of multi-layer pipes and cables”, Proceedings of the Polymer Processing Society Annual Meeting PPS-29, July 15–19, Nuremberg, Germany (2013)

191. P. Weiss, G. Hiesgen, K. Saul, and W. Spitz, “Cooling simulation for the prediction of quality properties and production costs of semi-finished extruded products like pipes”, Plastics Engineering, June 2014 & SPE ANTEC, April, Las Vegas, Nevada, USA (2014)

192. C. Rauwendaal and J. Anderson, “FEA of flow in single screw extruders”, 52nd SPE ANTEC, pp. 298–305, San Francisco, CA (1994)

193. H. Potente and P. Fischer, “Modellgesetze für die Auslegung von Plastifiziereinschneckenextrudern”, *Kunststoffe* 67, pp. 242–247 (1977)

194. H. Potente, “Auslegen von Schneckenmaschinen – Baureihen – Modellgesetze und ihre Anwendung”, Hanser Publishers, Munich (1981)

195. Paakinaho et al., “Melt Spinning of Polylactide 96/4: Effects of MW and melt processing on hydrolytic degradation”, Poly Degradation and Stability, 94, pp. 438–442 (2009)

196. C. Rauwendaal, “How to Collect and Interpret Extrusion Process Data – Part 3”, *Plastics Technology*, pp. 40–43, July (2017)

197. C. Rauwendaal, “Extrusion without Discolored Specks”, *Plastics Technology*, April (2013)

198. Optical Control Systems website, https://www.ocsgmbh.com/all-products/pellet-powder-analysers/ps25c.html

199. J. Deng, K. Li, E. Harkin-Jones, M. Price, N. Karnachi, A. Kelly, J. Vera-Sorroche, P. Coates, E. Brown, and M. Fei, “Energy monitoring and quality control of a single screw extruder”, *Applied Energy* 113, pp. 1775–1785 (2014)

200. C. Capone, L. Di Landro, F. Inzoli, M. Penco, and L. Sartore, “Thermal and Mechanical Degradation during Polymer Extrusion Processing”, *Polymer Engineering and Science*, Vol. 47, Issue 11, pp. 1813–1819 (2007)

201. A. Farahanchi, R. Malloy, and M.J. Sobkowicz, "Effects of Ultra-high Speed Twin Screw Extrusion on the Thermal and Mechanical Degradation of Polystyrene", *Polymer Engineering & Science*, 56, 7, pp. 743–751, (2016)
202. A. Farahanchi and M.J. Sobkowicz, "Kinetic and Process Modeling of Thermal and Mechanical Degradation in Ultra-high Speed Twin Screw Extrusion", *Polymer Degradation and Stability*, 138, pp. 40–46 (2017)
203. C. Rauwendaal and G. Hiesgen, "Analysis of Preheating Extruders", Extrusion 2018 Conference, Cleveland, OH, September 18–20 (2018)
204. C. Rauwendaal, "Analysis of Temperature Induced Voids in Extrusion", Annual Meeting of Polymer Processing Society, Cancun, Mexico, December 10–13 (2017)
205. C. Rauwendaal, "Analysis of Shrink Voids in Extrusion", Extrusion 2017 Conference, Charlotte, NC, October 18–20 (2017)
206. C. Rauwendaal, "How to Collect and Interpret Process Data in Extrusion, Part II", *Plastics Technology*, pp. 40–44, March issue (2017)
207. C. Rauwendaal, "How to Collect and Interpret Process Data in Extrusion, Part I", *Plastics Technology*, pp. 38–43, January issue (2017)
208. C. Rauwendaal, "Dimensional Variation by Melt Temperature Variation in Pipe Extrusion", Extrusion 2015 Conference, Charlotte, NC, November 2–3 (2015)
209. K. Saul, G. Hiesgen, and C. Rauwendaal, "Temperature Induced Dimensional Variation in Extrusion", meeting of Polymer Processing Society in Graz, Austria, September 21–25 (2015)
210. B.A. Olshausen, "Aliasing", PSC 129 – Sensory Processes, October (2000) [http://www.rctn.org/bruno/npb261/aliasing.pdf]
211. W. Sundblad, "Intelligent Industrial Automation", Extrusion 2018 conference, Cleveland, Ohio, September (2018)
212. W. Sundblad, "Industry 4.0 in Extrusion", Extrusion 2017 conference, Charlotte, NC, October (2017)
213. C. Rauwendaal, "Extrusion Data and Industry Best Practices", Oden Extrusion Data Conference, September 13 (2017)
214. J. Christiano, "Instrumenting your Extruder for the IIoT with a Focus on Predictive and Preventative Maintenance", Extrusion 2018 conference, Cleveland, Ohio, September (2018)
215. J. Zinski, "Securely Monitor Equipment, Add Analytics and Software Apps – No Cloud Expertise Required with Microsoft Azure", Extrusion 2018 conference, Cleveland, Ohio, September (2018)
216. B. Pal, "Insights into Predictive Maintenance and Power Issues", Extrusion 2015 conference, Charlotte, NC, November (2015)

217. S. Crabtree, M.A. Spalding, and C.L. Pavlicek: "Single-Screw Extruder Zone Temperature Selection for Optimized Performance", SPE ANTEC, 1410 (2008)

218. J. Taur, C. Tao, and C. Tsai, "Temperature Control of a Plastic Extrusion Barrel Using PID Fuzzy Controllers", Proceedings of 1995 International IEEE/IAS conference on Industrial Automation and Control: Emerging Technologies, Taipei, Taiwan, pp. 370–375 (1995)

219. T. Womer, "Optimize Barrel Temperatures for Barrier Screws", *Plastics Technology*, pp. 75–78, May (2008)

220. C. Rauwendaal, "Time to Learn about Dynamic Optimization of Extruder Barrel Temperatures", *Plastics Technology*, pp. 72–75, May (2008)

221. V. Schöppner and V. Resonnek, "Einfluss des Zylindertemperaturprofils auf das Prozessverhalten und Strategien zur Ermittlung einer optimalen Temperaturführung", Extrusionstechnik 2017: Anlagentechnik und Prozessführung 4.0, VDI-Jahrestagung, Köln (2017)

222. V. Schöppner and V. Resonnek, "Investigation of the Barrel Temperature Profile on the Process Behavior of Single Screw Extruders and Strategies to Determine the Optimal Temperature Control", Polymer Processing Society Annual Meeting, Cancun, Mexico, December (2017)

223. V. Resonnek and V. Schöppner, "Self-optimizing Barrel Temperature Setting Control of Single Screw Extruders for Improving the Melt Quality", AIP Conference Proceedings 2065, 030010 (2019); https://doi.org/10.1063/1.5088268, published online: February 6 (2019)

224. V. Schöppner and V. Resonnek, "Extruder optimier dich! Schmelzequalitität mit maschinellen Lernverfahren verbessern", *Kunststoffe* 1, pp. 56–61 (2019)

225. V. Schöppner, V. Resonnek, and J. Trippe, "Self-Optimizing Barrel Temperature Setting Control of Single Screw Extruders for Improving Melt Quality", Polymer Processing Society Annual Meeting, Taipei, Taiwan, May (2018)

226. V. Schöppner and V. Resonnek, "Determination of the Barrel Temperature Setting of Single Screw Extruders Using Fuzzy Logic", SPE ANTEC, Detroit, March (2019)

227. V. Resonnek, Ph.D. thesis, "Entwicklung einer selbstregelnden Zylindertemperatureinstellung auf Basis von Fuzzy-Logik zur Erhöhung der Schmelzehomogenität von Einschnecken-extrudern", University of Paderborn (2019)

228. C. Rauwendaal, "How to Determine Viscosity Data Using a Slit-Die Viscometer – part 1", March issue of *Plastics Technology*, pp. 46–47 (2019)

229. C. Rauwendaal, "How to Determine Viscosity Data Using a Slit-Die Viscometer – part 2", May issue of *Plastics Technology*, pp. 42–48 (2019)

230. C. Abeykoon, K. Li, M. McAfee, P.J. Martin, Q. Niu, A.L. Kelly, and J. Deng, "A new model based approach for the prediction and optimization of thermal homogeneity in single screw extrusion", *Control Engineering Practice 19* (8), pp. 862–874, July (2011)

231. C. Abeykoon, "Single Screw Extrusion Control: A Comprehensive Review and Directions for Improvements", *Control Engineering Practice*, volume 51, pp. 69–80, June (2016)

232. E.C. Brown, A.L. Kelly, and P.D. Coates, "Melt temperature field measurement in single screw extrusion using thermocouple meshes", *Review of Scientific Instruments* 75, 4742 (2004), https://doi.org/10.1063/1.1808895

Index

A

abrasion and corrosion 72
adjustable melt temperature probe 17
agglomerates 196
agglomerates and grammage variation in a PP sheet, case study 200
air entrapment 151

B

backpressure 202, 203, 209, 213, 224
barrel temperature profile (BTP) 296
black speck *see* black speck, case study
black speck, case study 247
breaker plate 12
BTP optimization
- DOE 304
- dynamic 305
- OTE 305

C

case studies *see* under each case study
- agglomerates and grammage variation in a PP sheet 201
- black speck 247
- color variation in polypropylene carpet fiber 172
- deficient solids conveying and dispersion 213
- degradation in a long adapter 255
- dispersion problem in a high-density polyethylene bottle 181
- film coextrusion, degradation in the middle layer 165
- film coextrusion with interfacial problems 169
- gear pump speed variation in sheet extrusion 238
- gel formation in a coextruded film 199
- heat-sealing problems in a coextruded film 187
- high melt temperature and insufficient output in a coextrusion 208
- high melt temperature in sheet extrusion 234
- improper preheating of extruder 268
- instability of formation at die 216
- insufficient melting and mixing in a plasticating unit 204
- lines in the extruded film 170
- masterbatch selection 194
- mechanical degradation 252
- melt fracture or sharkskin in m-PE 227
- melt temperature variation 244
- multilayer film, several appearance problems 179
- non-homogeneous melt in blow molding 232
- output problem in a blown film line 190
- pipe extrusion problem 196
- plastic film with poor transparency 175

- polymer degradation 184
- scale-up of LLDPE single-screw extruder 229
- shrink voids 258
- unstable pumping in a vented extruder 222
- wear problem in film extrusion 178
cloud-based
- server 28
- system 28
color contour plot 19
colorimetry 33, 34, 180
color variation in polypropylene carpet fiber, case study 172
contiguous solids melting (CSM) 16
cooling process 140
cooling simulation 143
CSM 16

D

data acquisition
- system (DAS) 9
data acquisition systems (DASs) 28
- capabilities of 30
- comparison among 32
- features of 31
- fixed-station data acquisition systems 30
- portable data collectors (PDCs) 28
- portable machine analyzers (PMAs) 30
data analysis
- window 324
data collection rate 9, 24
DC motor 52
- troubleshooting guide 52
deficient solids conveying and dispersion, case study 213
degradation in a long adapter, case study 255
degradation *see* polymer degradation devolatilization 126
diaphragm 11
die-flow problems 156
- mixing, improvement in 160
- polymer degradation 184
die inlet pressure 10
die-lip buildup *see* die-flow problems differential scanning calorimetry (DSC) 35
differential scanning calorimetry (DSC)
- application field 36
differential thermal analysis (DTA) 35
dispersion problem in a high-density polyethylene bottle, case study 181
dispersive mixers *see* mixers 119
distributive mixers *see* mixers 119
draw resonance 119

E

energy efficiency 10
enthalpy 10
equipment
- condition of 1
extruder barrel temperature
- other studies on optimization 307
- profile optimization 296
- setting 295
extruder performance 54
extruder temperatures, setting of 34
extrusion instabilities
- causes 115
- devolatilization 126
- draw resonance 119
- fluctuations, random 123
- fluctuations, slow 123
- frequency of 116
- functional instabilities 124
- low-frequency instabilities 122
- melt-conveying 126
- melt fracture 117
- mixing related 127
- plasticating 125
- screw frequency instabilities 120
- shark skin 117
- solids-conveying 125
- solving of 149

extrusion process 1
extrusion process, understanding of 1
- time line, construction of 4

F

Fast Fourier Transform 24, 323
feed stock
- and performance problems 6
- importance of 6
feed system, and trouble with feed hopper 53
FFT 24
film coextrusion, degradation in the middle layer, case study 165
film coextrusion with interfacial problems, case study 169
fishbone diagram 8, 152, 167
Fourier analysis 324
Fourier transform infrared spectroscopy 38
- application field 39
- equipment, characteristics of 38
functional zone of the extruder 55

G

gear pump speed *see* gear pump speed variation in sheet extrusion, case study
gear pump speed variation in sheet extrusion, case study 238
gel formation in a coextruded film, case study 199
gel problems 153, 155
- avoid creation of 154
- cause of 160
- degradation, reduction of 160
- die design, changing of 158
- die-lip buildup, reduction of 158
- for voids 152
- gel formation 153
- incompatible component, removal of 158
- measuring gels 153
- melt fracture 156
- patterns, V-and W- 159
- process, changing of 158
- reduction of 156
- removing gels 155
- shear stress, reduction of 156
- specks and discoloration 160
grammage variation 200
grooved feed extruder 213

H

hard-facing materials
- application of 84
hard-facing techniques 84
- comparison among 84
- laser hard-facing 87
- metal inert-gas welding 87
- oxyacetylene welding 85
- plasma transfer arc (PTA) welding 86
heating and cooling system 53
- checklist 54
- for heating and cooling systems 54
heat-sealing problems in a coextruded film, case study 187
helical solid ribbon 16
high melt temperature *see* high melt temperature in sheet extrusion, case study
high melt temperature and insufficient output in a coextrusion, case study 208
high melt temperature in sheet extrusion, case study 234
hot spot 20

I

ID waviness 24
immersion melt temperature probe 16
improper preheating *see* improper preheating of extruder, case study
improper preheating of extruder, case study 268
infrared radiation 17

infrared thermography 47
infrared thermometer 17, 27
inhomogeneities 206, 216, 222, 225
instability of formation at die 124
instability of formation at die, case study 216
instrumentation 1
insufficient melting and mixing 115, 204
- case study 204
intermittent pumping in extruder, case study 222
IR temperature measurement 19

L

laser hard-facing 87
light microscope 33
light microscopy 33
lines in the extruded film, case study 170
lines in the extruded product 158
low-frequency instabilities 122

M

machine-related problems 51
- feed system 53
- heating and cooling system 53
- mechanisms 66
- solutions to problems 77
- testing for 67
- troubleshooting guide 52
- wear 54
machine troubleshooting and maintenance 287
- drive motors and belts 289
- extruder maintenance checklist 291
- noises, unusual 288
- oil, checking of 287
- screw and barrel 290
- spare parts 290
- vibrations, monitoring of 288
masterbatch selection, case study 193
mechanical degradation *see* mechanical degradation, case study
mechanical degradation, case study 252
melt-conveying functional instabilities 126
melt fracture or sharkskin in m-PE, case study 227
melt pool 16
melt pressure 9, 10
melt temperature 9, 16
- distribution 19, 140
- in extruder 142
- non-uniform 141
- variation 19, 140
melt temperature variation *see* melt temperature variation, case study
melt temperature variation, case study 244
mercury 12
metal-to-metal wear 74
- and flight geometry 75
mixers 127
- blister ring 132
- cavity 127
- Chris Rauwendaal Dispersive (CRD) mixer 136
- comparison among dispersive 138
- comparison among distributive 132
- dispersive 131
- distributive 127
- elongational mixer 138
- fluted; CRD-Z; dispersive; blockhead 63
- pin 129
- planetary gear 135
- slotted flight 129
- variable depth 130
mixing problems, solving of 139
- and functional instabilities 127
Mohs scale 72
motor
- load 9
multilayer film, several appearance problems, case study 179

N

non-homogeneous melt in blow molding, case study 232
non-isothermal analysis 19
non-Newtonian
- polymer 19

O

optical and appearance properties 163
output problem in a blown film line, case study 190
overpressure shutdown 13
oxyacetylene welding 85

P

pipe extrusion problem, case study 196
plasma transfer arc (PTA) welding 86
plasticating 125
plastic film with poor transparency 175
polymer degradation
- chemical 93
- in extrusion 97
- mechanical 93
- residence time distribution (RTD) 97
- temperature distribution, numerical calculations of 107
- temperature distribution, simple calculation of 101
- thermal 93
- types of 93
polymer degradation, case study 184
portable data collectors (PDCs) 29
portable machine analyzers (PMAs) 29
power measurements 49
pressure
- drop 133, 217
- feedback control 12
- transducer 11
problem tree 8
process signal 323
process signal analysis 323
process temperature 298
pushing flight flank 20
pyrometer 27

R

residence time 23
rheometer
- capillary 44
- Couette 43
- rotational 43
rheometry 35
- high pressure capillary rheometry 41
- torque rheometry 40
rod extrusion 258
rupture disk 13

S

sapphire window 18
scale-up *see* scale-up of LLDPE single-screw extruder, case study
scale-up of LLDPE single-screw extruder, case study 229
screen
- pack 12
screw speed 23
screw binding 87
- and compressive load 92
- and temperature 88
- materials, thermal conductivity of 88
- mechanics of 88
- temperature distribution, analysis of 91
screw design 54
- consideration 58
- features 61
- variables 59
screw flight materials 79
screw frequency instabilities 120, 121
screws and barrels
- hard-facing materials, and properties of 80
- rebuilding of 83

SEC 10
setting barrel temperature 299
shark skin 117
shear rate 41, 102, 221
shear stress 41, 118, 156, 221
short-term melt temperature variations 23
short-term pressure variations 13
shrink void *see* shrink voids, case study
shrink voids, case study 258
smartphone 48
solid bed 16
solid bed width profile (SBP) 202
solids-conveying functional instabilities 125
specific energy
- consumption (SEC) 10
strain gage capillary pressure 12

T

teams for repair, building of 5
temperature 17
- measurement devices 2
thermal analysis 35
- application field 33, 36, 38, 39
- differential scanning calorimetry (DSC) 35
- differential thermal analysis (DTA) 35
- equipment, characteristics of 38
- thermal optical analysis (TOA) 46
- thermogravimetric analysis 37
- thermomechanical analysis (TMA) 39
- tools, miscellaneous 46
- torque rheometer 40
- torsion braid analysis (TBA) 46
thermally sensitive materials 17
thermal optical analysis (TOA) 46
thermochromic materials, and temperature-related problems 34
thermocouple
- mesh 20
thermographic camera 47
ToolPak add-in 325
tools, miscellaneous 46
torque rheometer 40
- application field 40
TPE extrusion 252
trailing flank 20
training 25
transparent window 18
troubleshooting guides and checklists
- for extrusion instabilities 151
- for wear problems 78
- tungsten inert-gas welding 85
tubing extrusion 244, 247
tungsten inert-gas welding 85

V

vented extruder 222
vital signs 9

W

wall thickness variation 141
waviness 23
wear 65
- abrasion and corrosion 72
- abrasive wear 66
- adhesive wear 66
- Bauer and coworkers 69
- Broszeit 71
- Calloway and coworkers 71
- causes of 72
- corrosive 66
- Deutsches Kunststoff Institut 69
- factors affecting 72
- fillers, surface treatment of 73
- laminar wear 66
- Maelhammar 71
- mechanisms 66
- Mehdorn 68
- Mennig and Volz 68
- metal-to-metal wear 74
- Mohs scale 72
- Moslé 71
- Plumb and Glaeser 70

- screw binding 87
- screw flight materials 78
- solutions 77
- surface-fatigue 67
- testing, types of 68
- troubleshooting flow chart 78
- universal disk tribometer 70
wear problem in film extrusion 178